Ruchi Kohli

Hydrogen bonding abilities of hydroxamic acid and its isosteres

Anchor Academic
Publishing

Kohli, Ruchi: Hydrogen bonding abilities of hydroxamic acid and its isosteres, Hamburg, Anchor Academic Publishing 2016

Buch-ISBN: 978-3-96067-004-9
PDF-eBook-ISBN: 978-3-96067-504-4
Druck/Herstellung: Anchor Academic Publishing, Hamburg, 2016

Bibliografische Information der Deutschen Nationalbibliothek:
Die Deutsche Nationalbibliothek verzeichnet diese Publikation in der Deutschen Nationalbibliografie; detaillierte bibliografische Daten sind im Internet über http://dnb.d-nb.de abrufbar.

Bibliographical Information of the German National Library:
The German National Library lists this publication in the German National Bibliography. Detailed bibliographic data can be found at: http://dnb.d-nb.de

All rights reserved. This publication may not be reproduced, stored in a retrieval system or transmitted, in any form or by any means, electronic, mechanical, photocopying, recording or otherwise, without the prior permission of the publishers.

Das Werk einschließlich aller seiner Teile ist urheberrechtlich geschützt. Jede Verwertung außerhalb der Grenzen des Urheberrechtsgesetzes ist ohne Zustimmung des Verlages unzulässig und strafbar. Dies gilt insbesondere für Vervielfältigungen, Übersetzungen, Mikroverfilmungen und die Einspeicherung und Bearbeitung in elektronischen Systemen.

Die Wiedergabe von Gebrauchsnamen, Handelsnamen, Warenbezeichnungen usw. in diesem Werk berechtigt auch ohne besondere Kennzeichnung nicht zu der Annahme, dass solche Namen im Sinne der Warenzeichen- und Markenschutz-Gesetzgebung als frei zu betrachten wären und daher von jedermann benutzt werden dürften.

Die Informationen in diesem Werk wurden mit Sorgfalt erarbeitet. Dennoch können Fehler nicht vollständig ausgeschlossen werden und die Diplomica Verlag GmbH, die Autoren oder Übersetzer übernehmen keine juristische Verantwortung oder irgendeine Haftung für evtl. verbliebene fehlerhafte Angaben und deren Folgen.

Alle Rechte vorbehalten

© Anchor Academic Publishing, Imprint der Diplomica Verlag GmbH
Hermannstal 119k, 22119 Hamburg
http://www.diplomica-verlag.de, Hamburg 2016
Printed in Germany

ABSTRACT

Hydrogen bonding (H-bonding) being the strongest non covalent interaction has established its importance in several areas of science. This study deals with the identification of active H-bonding sites and evaluation of intra and intermolecular H-bonding parameters and strengths for H-bonding of formohydroxamic acid (FHA), its isosteres thioformohydroxamic acid (TFHA) and formylphosphinous acid (FPA) with some simple molecules. The hydroxamic acids (HAs) are regarded molecules of prime importance because of their similarity to amide group of proteins and numerous applications in diverse areas. The HAs are known for their role as inhibitors of a range of enzymes including peroxidases, hydrolases, ureases, lipoxygenases, cyclooxygenases and matrix metalloproteinases etc. Their enzyme inhibitory action is suggested to involve H-bonding. Thus drug designing with an aim to inhibit specific enzyme can utilize the H-bonding sites present on the enzyme as well as the inhibitor. The evaluated H-bond parameters can prove useful while designing artificial inhibitors and drugs. Hence exploring the H-bonding abilities of HAs becomes important. Many drugs based on HAs are useful for treatment of cancer, metal poisoning, iron overload, malaria, allergic diseases and tuberculosis. HAs are chelating agents for many metal ions, hence used in their detection and quantitative estimation of metals in analytical chemistry. Solvent is known to affect the tautomeric equilibrium and the population ratio of various isomers of HAs by competitive H-bonding. There is competition between solute-solute and solute-solvent intermolecular H-bonding which is further complicated by intramolecular H-bonding, so effects of medium are studied both explicitly and implicitly. The protonation and deprotonation studies is carried out which is important for understanding the reactivity and chelating properties of HAs respectively.

This book has been divided into seven chapters. In the introductory chapter i.e. chapter 1, various non covalent interactions are briefly listed with discussion on H-bonding in context to its geometric requirements, properties, molecular orbital treatment and biological applications including theoretical and experimental techniques to study H-bonding. In the chapter, an overview of chemistry and biological functions of HAs is also presented. This chapter provides a review of experimental and theoretical studies done by various research groups on FHA and its isosteres till recently. Brief description of the principles of quantum mechanics underlying the computational methods employed to explore the chemistry of HAs is also given in the chapter. The next chapter i.e. chapter 2 describes the theoretical results obtained for intra and intermolecular H-bonding interactions between FHA and single water molecule and the dimerization among the various isomeric forms. Chapter 3 involves the comparative study on intra- and intermolecular H-bonding ability of TFHA relative to that of FHA. The deprotonation enthalpies of different sites of FHA and TFHA are explored in chapter 4. The possibilities of interconversions among various anions in gas phase are described. Effect of aqueous medium on deprotonation by using explicit and implicit solvation methods is also described. The intermolecular H-bonding in anions of FHA and TFHA with water is also analyzed. Further insight into H-bonded aggregates and dimers of HAs is gained through the analysis of stabilization energies associated with these and their comparison to similar H-bonded functionalities. The reasons behind the H-bond cooperativity in the aggregates and dimers are explored in chapter 5. Chapter 6 deals with the study of properties like conformational analysis, isomerism, protonation, deprotonation and H-bonding of FPA isostere of FHA and a comparative study is carried out. In chapter 7, the aggregation of the most stable keto and enol conformer of FHA and TFHA with five amino acid side chain groups that are commonly occurring functionalities present at active sites of enzyme is studied. The factors responsible for the observed stabilization energies are explored through analysis of atomic charges and NBO analysis.

CONTENTS

ABSTRACT 5

CONTENTS 7

IMPORTANT LIST OF ABBREVIATIONS 11

1 INTRODUCTION AND QUANTUM MECHANICAL BACKGROUND ...13
   1.1 INTRODUCTION 13
      1.1.1 HYDROGEN BONDING..14
      1.1.2 HYDROXAMIC ACIDS 18
      1.1.3 THIOHYDROXAMIC ACIDS 21
      1.1.4 FORMYLPHOSPHINOUS ACID 22
   1.2 REVIEW OF LITERATURE 23
   1.3 QUANTUM MECHANICAL BACKGROUND 32
      1.3.1 THE SCHRODINGER EQUATION 32
      1.3.2 THE BORN OPPENHEIMER (BO) APPROXIMATION............33
      1.3.3 THE VARIATIONAL PRINCIPLE 34
      1.3.4 THE HARTREE FOCK (HF) SELF CONSISTENT FIELD APPROACH 34
      1.3.5 BASIS FUNCTIONS AND BASIS SETS: 37
      1.3.6 ELECTRON CORRELATION...39
      1.3.7 DENSITY FUNCTIONAL THEORY (DFT) 41
   1.4 COMPUTATIONAL DETAILS 43
   1.5 METHODOLOGY TO CALCULATE VARIOUS PROPERTIES 44
      1.5.1 CALCULATION OF STABILIZATION ENERGIES 44
      1.5.2 CORRECTION FOR BASIS SET INCOMPLETENESS 44
      1.5.3 ATOMIC CHARGES AND ORBITAL INTERACTIONS 45
      1.5.4 INTERACTIONS WITH MEDIUM BASED ON DIELECTRIC CONTINUUM MODEL 46
      1.5.5 PROTON AFFINITIES 48
      1.5.6 GAS PHASE ACIDITIES/DEPROTONATION ENTHALPIES ..48
   1.6 REFERENCES 49

2 INTRA- AND INTERMOLECULAR HYDROGEN BONDING IN FORMOHYDROXAMIC ACID 58
   2.1 INTRODUCTION 58

- 2.2 RESULTS AND DISCUSSION 59
  - 2.2.1 CONFORMATIONAL STABILITY AND ISOMERISM 59
  - 2.2.2 INTRAMOLECULAR HYDROGEN BONDING.. 64
  - 2.2.3 INTERMOLECULAR HYDROGEN BONDING BETWEEN FORMOHYDROXAMIC ACID AND WATER 66
  - 2.2.4 CHARGE ANALYSIS 77
  - 2.2.5 INTERMOLECULAR HYDROGEN BONDING IN DIMERIC UNITS 78
  - 2.2.6 NATURAL BOND ORBITAL (NBO) ANALYSIS 100
- 2.3 CONCLUSIONS 110
- 2.4 REFERENCES 111

3 INTRA- AND INTERMOLECULAR HYDROGEN BONDING IN THIOFORMOHYDROXAMIC ACID 113

- 3.1 INTRODUCTION 113
- 3.2 RESULTS AND DISCUSSION 114
  - 3.2.1 CONFORMATIONAL STABILITY AND TAUTOMERISM 114
  - 3.2.2 INTRAMOLECULAR HYDROGEN BONDING...................................... 124
  - 3.2.3 INTERMOLECULAR HYDROGEN BONDING WITH WATER 128
- 3.3 CONCLUSIONS 145
- 3.4 REFERENCES 146

4 THE ROLE OF ISOMERISM AND EFFECT OF MEDIUM ON STABILITY OF ANIONS OF FORMO- AND THIOFORMO-HYDROXAMIC ACIDS 148

- 4.1 INTRODUCTION 148
- 4.2 RESULTS AND DISCUSSION 150
  - 4.2.1 DEPROTONATION ENTHALPIES OF FHA/TFHA IN GAS AND AQUEOUS PHASE 150
  - 4.2.2 RELATIVE STABILTY OF FHA/TFHA ANIONS................................... 153
  - 4.2.3 VARIATION IN GEOMETRICAL PARAMETERS 154
  - 4.2.4 ATOMIC CHARGES 157
  - 4.2.5 VARIATION IN ELECTRON DELOCALIZATIONS 158
  - 4.2.6 ISOMERISM IN FHA/TFHA ANIONS 161
  - 4.2.7 INTERMOLECULAR H-BONDING WITH WATER IN ANIONS OF FHA/TFHA 173
  - 4.2.8 SOLVENT EFFECT ON DEPROTONATION PROCESSES IN FHA/TFHA 181
- 4.3 CONCLUSIONS 186
- 4.4 REFERENCES 188

5 HYDROGEN BOND COOPERATIVITY IN DIMERS OF HYDROXAMIC ACIDS 189

- 5.1 INTRODUCTION 189
- 5.2 RESULTS AND DISCUSSIONS 190
  - 5.2.1 PROTON AFFINITY 190
  - 5.2.2 GAS PHASE ACIDITIES 209
  - 5.2.3 DIMERIZATION AND THE STABILIZATION ENERGIES 218
  - 5.2.4 HYDROGEN BOND COOPERATIVITY 240
  - 5.2.5 SOLUTION PHASE STUDIES 251
- 5.3 CONCLUSIONS 252
- 5.4 REFERENCES 253

6 HYDROGEN BONDING ABILITY AND ACID-BASE BEHAVIOUR OF FORMYLPHOSPHINOUS ACID: AN ISOSTERE OF HYDROXAMIC ACID 255

- 6.1 INTRODUCTION 255
- 6.2 RESULTS AND DISCUSSION 256
  - 6.2.1 RELATIVE STABILITIES AND ISOMERIC ANALYSIS 256
  - 6.2.2 INTRAMOLECULAR H-BONDING 287
  - 6.2.3 ELECTRON DELOCALIZATIONS IN ISOMERIC FORMS OF FPA 288
  - 6.2.4 DEPROTONATION ENTHALPIES AND STABILITIES OF ANIONS 292
  - 6.2.5 ATOMIC CHARGE ANALYSIS OF THE ANIONS 297
  - 6.2.6 ELECTRON DELOCALIZATIONS IN THE ANIONS 299
  - 6.2.7 PROTON AFFINITIES 301
  - 6.2.8 NBO ANALYSIS OF PROTONATED SPECIES 310
  - 6.2.9 HYDROGEN BONDING OF FPA WITH WATER 312
  - 6.2.10 HYDROGEN BONDING OF FPO WITH WATER 329
- 6.3 CONCLUSIONS 333
- 6.4 REFERENCES 335

7 HYDROGEN BONDING INTERACTIONS OF HYDROXAMIC ACIDS WITH AMINO ACID SIDE CHAIN GROUPS 336

- 7.1 INTRODUCTION 336
- 7.2 RESULTS AND DISCUSSION 337
  - 7.2.1 THE HYDROGEN BONDING OF HYDROXAMIC ACIDS WITH METHYLAMINE (MeNH$_2$) 338
  - 7.2.2 THE HYDROGEN BONDING OF HYDROXAMIC ACIDS WITH ACETIC ACID (AcOH) 349

7.2.3 THE HYDROGEN BONDING OF HYDROXAMIC ACIDS WITH
METHANOL (MeOH) 364

7.2.4 THE HYDROGEN BONDING OF HYDROXAMIC ACIDS WITH
METHANETHIOL (MeSH).. 374

7.2.5 HYDROGEN BONDING OF HYDROXAMIC ACIDS WITH
METHANESELENOL (MeSeH):... 383

7.3 CONCLUSIONS 391

7.4 REFERENCES 393

CONCLUSIONS **394**

Important List of Abbreviations

AIM	Atoms in molecules
AO	Atomic Orbital
B3LYP	Becke 3 Lee Yang Parr
BSSE	Basis Set Superposition Error
CI	Configuration Interaction
CP	Counterpoise
DFT	Density Functional Theory
GTO	Gaussian Type Orbital
GS	Ground State
GTF	Gaussian Type Function
HF	Hartree Fock Energy
L1	Theoretical Level B3LYP/6-31+G*
L2	Theoretical Level B3LYP/Aug-cc-pVDZ
L3	Theoretical Level MP2/6-31+G*
L4	Theoretical Level MP2/ Aug-cc-pVDZ
MO	Molecular Orbital
MP2	Moller-Plesset Second Order
NBO	Natural Bond Orbital
NPA	Natural Population Analysis
PA	Proton Affinity
PCM	Polarized Continuum Model
PES	Potential Energy Surface
RHF	Restricted Hartree Fock
RTS	Rotational Transition State
SCF	Self Consistent Field

STO	Slater Type Orbital
TS	Transition State
UHF	Unrestricted Hartree Fock
ZPVE	Zero point Vibrational Energy
CBS	Complete basis set

INTRODUCTION AND QUANTUM MECHANICAL BACKGROUND

1.1 INTRODUCTION:

Non covalent interactions are known to play important role in chemistry, physics and bio disciplines as these determine structure, stability and dynamics of the biological systems [1]. In biology, the selectivity and recognition of specific sites is achieved with non covalent interactions. These interactions reveal the properties of biological molecules and offer a glimpse into possible origin of these molecules. In chemistry, non covalent interactions are known to influence chemical reactions and the design of building blocks held together by non covalent interactions comprises an entire field, known as supramolecular chemistry.

Various types of non covalent interactions include

1. Pi-pi interactions: π-π interactions are caused by intermolecular overlapping of p-orbitals in π-conjugated systems e.g. π-π interactions in benzene dimer (**Ia**).
2. London or Dispersion forces: are the weakest temporary attractive forces that result when the electrons in two adjacent atoms occupy positions that make the atoms form temporary dipoles e.g. N_2 (**Ib**).
3. Dipole-Dipole forces: are attractive forces between the positive end of one polar molecule and the negative end of another polar molecule e.g. interaction between methanol and chloroform (**Ic**).

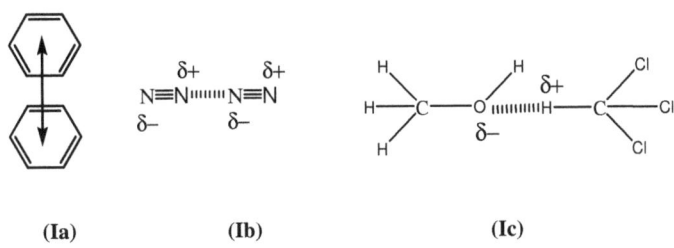

(**Ia**)   (**Ib**)   (**Ic**)

4. Cation pi interactions: Cation-π interactions are noncovalent molecular interactions between the face of an electron-rich π system (e.g. benzene, ethylene) with an adjacent cation (e.g. Li^+, Na^+) (**Id**).
5. Ion dipole interactions: Interactions between a charged species and a polar molecule e.g. hydration of NaCl in water (**Ie**).

6. Hydrogen bonding interactions: When a covalently bound hydrogen atom forms a second bond to another atom, the second bond is referred to as hydrogen bond e.g. liquid water **(If)**.

(Id) (Ie) (If)

1.1.1 HYDROGEN BONDING:

The original concept of hydrogen bonding (H-bonding) was first put forward by Latimer and Rodebush (1920) [2]. H-bond was quoted for the first time in 1923 in a chemistry book by G.N. Lewis. The definition by L. Pauling (1939) states that "under certain conditions an atom of hydrogen is attracted by rather strong forces to two atoms, instead of only one, so that it may be considered to be acting as a bond between them. This is called the hydrogen bond [3]." Earlier it was recognized that H-bond results when electronegative end of highly polar bonds, consisting of F, N or O atoms interacts with positive end H-X. More modern definitions allow other types of H-bond donor and acceptor groups, including weak donors such as C-H, P-H and As-H bonds or weak H-bond acceptors such as π-cloud of a benzene ring. The latest provisionally recommended definition by International Union of Pure and Applied Chemistry (IUPAC) task group is "The hydrogen bond is an attractive interaction between a hydrogen atom from a molecule or a molecular fragment X-H in which X is more electronegative than H, and an atom or a group of atoms in the same or a different molecule, in which there is evidence of bond formation."

The utility of H-bond comes from their strong directional preference which sets them apart from other intermolecular interactions and their intermediate strength, placing them between covalent and the pure van der Waals interactions [4]. Energetics of H-bond in comparison to thermal energy (~ 0.6 kcal/mol at 300 k) and energetics of covalent bonds (~ 100 kcal/mol) and van der Waals interactions (<0.5

kcal/mol) are recognized as a crucial parameter in evolution of life processes. A system of H-bonded particles appears to have right balance of stability and dynamics. In contrast, a system of covalently bound particles is relatively rigid and a system of particles interacting via van der Waals forces is extremely dynamic at room temperature. Predicting the properties of H-bonding is an important problem in science due to great role of this bond in biological phenomena, chemical reactions, relations within condensed phases and crystal engineering etc [5-7].

The geometrical characterization of the H-bond constituted with a donor A-H and an acceptor B may be described in terms of parameters as r, d, D and θ as shown in **II** [8]. Strength of a H-bond depends upon its length, linearity, nature of its microenvironment [9]. The strength of H-bond ranges from weak (4 kcal/mol), through strong (4-15 kcal/mol) to very strong (15-40 kcal/mol) [10]. H-bond strengths are much higher in gas phase than in water or polar solvents. The efficacy of H-bond depends substantially not only on donor and acceptor ability but also on ability of H-bond donor to polarize the acceptor and of the acceptor to become polarized.

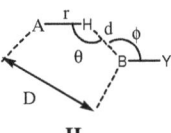

II

The molecular orbital treatment of H-bonding in terms of highest occupied molecular orbital (HOMO) of H-bond acceptor 'B' and lowest unoccupied molecular orbital (LUMO) of H-bond donor AH suggests that H-bond results when the total energy of H-bonded adduct is lower than the total energy of H-bond donor and H-bond acceptor and the small relative energy difference between the HOMO and LUMO favors the adduct formation. Thus for the H-bonded adduct formation B+AH ⇌ BHA, three possibilities exist. In Figure **1.1 a**, there is large energy difference between the HOMO and LUMO i.e. the energy of LUMO of AH is above the energy of HOMO of B that suggests poor matching between the orbitals and hence weak orbital interactions between the two is anticipated. In Figure **1.1 b**, where the energy of HOMO of B and LUMO of AH show a large difference and the energy of HOMO is lower than that of LUMO, no H-bonded adduct results as the situation favor complete transfer of proton from AH to B. For a good match between the orbital energies (of HOMO and LUMO) as shown in Figure **1.1 c**, strong interaction between the orbitals favors strongly H-bonded adduct [11].

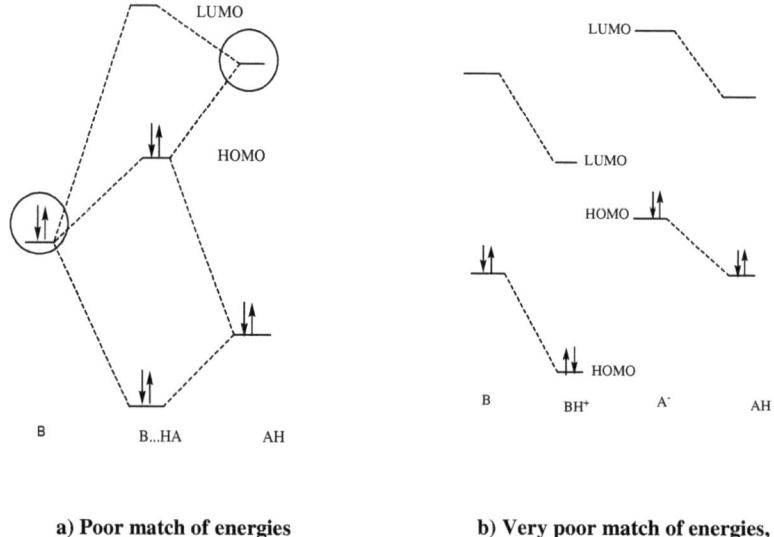

a) **Poor match of energies little or no H-bonding**

b) **Very poor match of energies, transfer of hydrogen ion**

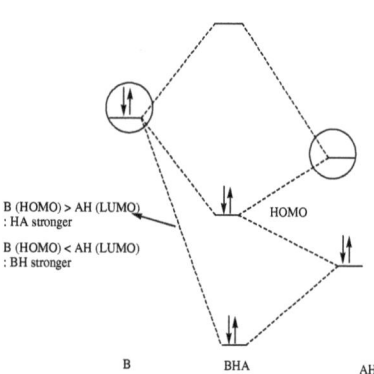

c) **Good match of energies, good H-bonding**

Figure 1.1: Relative positions of HOMO and LUMO of H-bond acceptor and donor respectively and the strength of H-bonding.

The most ubiquitous and perhaps the simplest example of a H-bond is between the water molecules. The anomalous behavior of water is assigned to the H-bonded network present in water and ice. Intramolecular H-bonds have significant influence on the properties such as charge distribution within molecules, relative stability of conformers and reactivity [12]. The preference for a specific stereochemistry in large organic molecules is sometimes determined by the intramolecular H-bonding interactions. The crystal structural properties are determined by intermolecular H-bonding whenever possible.

The structures of proteins, nucleic acids, α-helix, β-sheet of polypeptides and double helices of polynucleotides are also dictated by optimal stabilization offered by the H-bonds. The significant role of H-bonding interactions in solvation, diffusion through biological tissues of membranes and adsorption on the surfaces is well recognized [12-14]. H-bonding plays role in biological recognition, assembly processes, in electron transfer and radical scavenging. H-bonding enables enzymes to bind to their substrates; antibodies bind their antigen, proteins to bind their receptors. Catalytic action of enzymes is based on the fact that first the enzyme holds the substrate at active site through a variety of interactions such as H-bonding, dipole-dipole interaction, van der waals interactions etc [15-17]. This ability of H-bonding is recognized for drug protein interactions and hence helps in drug designing. H-bonding capabilities deeply influence the transport and ADME (Administration, distribution, metabolism and excretion) properties of a molecule as well as its specific interaction with biological receptors. Many quantitative structure activity relationship (QSAR) studies have been reported in which H-bonding interactions play a key role in modelling a particular target activity [18].

Several techniques like Nuclear magnetic resonance (NMR), Infrared spectroscopy (IR), and X-ray diffraction are employed to study the H-bonding [19]. Diffraction methods like X-ray, neutron diffraction are especially important to researchers to study weak H-bond interactions because they in contrast to strong H-bonds are generally not detected with ease by spectroscopic methods. Theoretical methods are also useful in the recognition of H-bonding interactions. They provide information about the structure of H-bonded complexes and benchmark values for energies of intermolecular interactions without complication effects of solid state or solution environment.

1.1.2 HYDROXAMIC ACIDS:

The hydroxamic acids (HAs) known since 1869 with discovery of oxalohydroxamic acid by Lossen, consist of a class of naturally occurring weak organic acids mainly obtained from microbial sources like fungi, bacteria and plant tissues etc [20]. Aspergillic acid (**III**), an N-hydroxylated pyrazine isolated from *Aspergillus flavus* (1966) was one of the first naturally occurring HA isolated [21]. HAs are recognized as compounds of pharmacological, toxicological and pathological importance playing key role in facilitating the proper functions of enzymes in electron and oxygen transport and other life sustaining processes [22].

Specific enzyme inhibition is a classical goal of drug design and the specificity of enzyme inhibitors must rest to a great deal on H-bonding between their tails and enzyme's protein around them [23]. HAs are collagenase inhibitors, antiinfectives and influenza virus polymerase inhibitors [24,25]. Horseradish peroxidase form spectroscopically distinct, reversible complexes with HA, hydrazides, amides etc. Binding of these compounds with enzyme depend upon polar and steric character of R, nucleophilicity of heteroatoms and H-bonding capacity of –COXY (X-Y=NH, OH, NH-NH$_2$ etc) [26].

HAs number among few compounds effective as nucleophillic reactivators of sarin-inactivated chymotrypsin or acetylcholine esterase [27,28]. They are potent and selective inhibitors of a range of enzymes including peroxidases [29], hydrolases [30], ureases [31], lipoxygenases [32], cyclooxygenases [33], peptide deformilases [34] and matrix metalloproteinases (MMPs) [35]. MMP are a class of enzymes performing functions like tissue repair and remodeling the elements of extracellular matrix. Their abnormal expression results in arthritis, tumor metastasis and central nervous system diseases etc [36-38]. HA is key functional group of therapeutic targets at matrix MMP involved in cancers and other drug targets associated with cardiovascular diseases, AIDS and Alzheimer's disease [32]. Piperidine sulfonamide aryl HA analogs act as MMP inhibitors. Some other potent HA based MMP inhibitors include marimastat, CGS-27023A (**IV**),

RO-32-3555 etc [39]. Succinyl, malonyl and glutaryl hydroxamates e.g. N-Sulfonyl amino acid hydroxamates are reported to be inhibitors of metalloenzymes and binding of a hydroxamate group to MMP involves the coordination of the two hydroxamate oxygens to the catalytic zinc ion and stable H-bonds between hydroxamate nitrogen and the carbonyl of the enzyme backbones as shown in **Scheme 1.1** [40,41].

Some of the hydroxamates also act as effective carbonic anhydrase inhibitors [42-44]. The HA such as acetohydroxamic acid (AHA) is well established as effective inhibitor of plant and bacterial urease and has been shown to effectively inhibit ureolytic activity and to lower blood ammonia levels in mice, rats, sheep, cows, dogs and men. HA derivatives of salicylic acid are cyclooxygenase (COX-1 and COX-2) inhibitors [33,45].

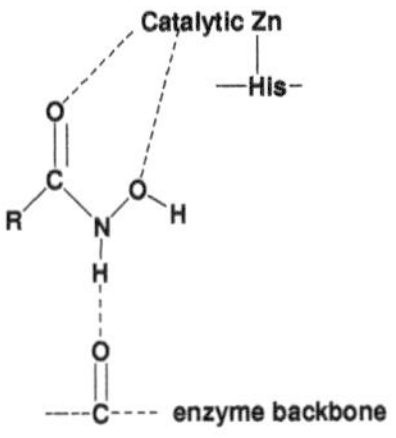

Scheme 1.1: Enzyme inhibitory action

Tepoxalin (**V**) is dual COX and peroxidase (POX) inhibitor [46]. The aspartic acid β hydroxamate exhibited antitumor activity in L5178Y leukemia, therapeutic effect on friend erythroleukemia and antiproliferative activity on friend virus infected erythropoeitic progenitor cells. HA derivatives of short chain fatty acids, butyryl and propionyl hydroxamate, subericbishydroxamic acid, suberoylanilide hydroxamic acid (SAHA), succinamide hydroxamic acids are potent inhibitors of histone deacetylase (HDAC), tumor cell proliferation and shown to induce fetal hemoglobin. Combined effect of certain HDAC inhibitors on erythropoiesis and on γ gene expression make them desirable targets for development of β chain hemoglobinopathie. HA derivatives can stimulate in vivo erythropoiesis and fetal globin production in a thalassemic murine model [47].

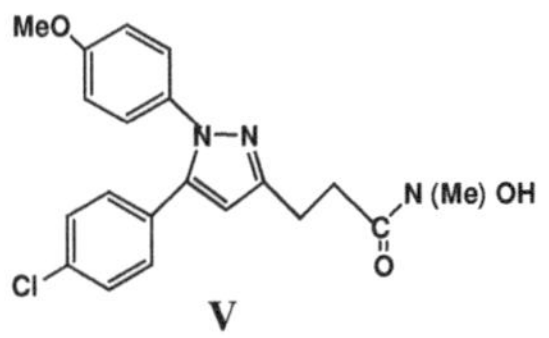

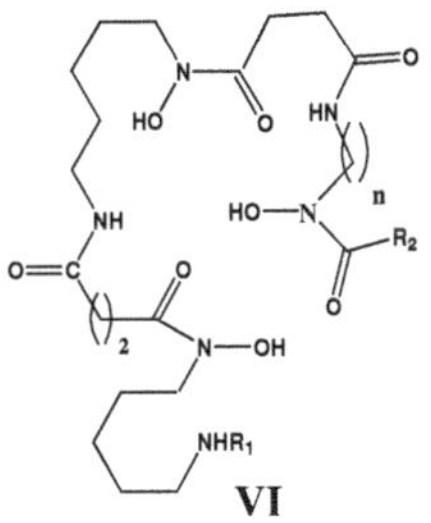

HA based chelation therapy is useful for treatment of cancer, metal poisoning, iron overload, malaria, allergic diseases and tuberculosis. Desferrioxamine (a natural trihydroxamic acid) (**VI**) is a drug for human use in case of iron overload associated with transfusional treatment of β-thalassaemia or Cooley's anemia [48].

Recently developed salicylhydroxamic acid is useful for treatment of thalassaemia major. Bufexamac (**VII**), a hydroxamic acid compound with anti-inflammatory properties is used for treatment of dermatological diseases [49]. Hydroxycarbamide is antineoplatic. Ibuproxam, oxametacin and bufexamac are anti-inflammatory and analgesic drugs. Adrafinil is adrenergic agonist and antidepressant. HAs are used in treatment of hepatic coma [50]. The acetylated salicyl hydroxamic acids are attractive aspirin substitute as they inhibit cyclooxegenase site of prostaglandin H synthetase-1(PGHS) by acetylation. Varinostat (Zolina^{TM}, SAHA) (**VIII**) is HA approved by the FDA for treatment of patients with cutaneous T-cell lymphoma [51].

Fosmidomycin (**IX**), first developed as an antibiotic was found to be reasonably effective antimalarial drug. Its anti-plasmodia properties are undoubtedly due to its HA group. The functionality has been employed as the first efficient antidote against nerve gas poisoning. Derivatives of 3, 4, 5 - trimethoxybenzohydroxamic acid act as mental tonic [52].

HAs have a wide spectrum of applications as being siderophores (iron chelators). Hydroxamate based siderophores for instance mycobactins, rhodotorulic acid (**X**) etc. have great therapeutic potentials [21]. HAs are chelating agents for many more metal ions in geochemical and other environments. HAs have also been ascribed a role in allelopathy (interaction between plants of same or different species). A host of HAs are active as pesticides, as plant growth promoters and as soil enhancers. HAs have inhibitory effect on plant growth e.g., DIBOA (2,4-dihydroxy-1,4-benzoxazin-3-one) (**XI**) and BOA (2-benzoxazolinone) (**XII**) were shown to have an inhibitory effect on root growth of cress (*Lepidum* *sativum*) and barnyardn grass (*Echinochloa crusgalli*) [53]. They impose a negative influence on survival and reproduction of aphids. Argandona et al. found inverse correlations between hydroxamic acid content and growth rate of the aphid *Metopolophium dirhodum* [54].

XI   **XII**

1.1.3 THIOHYDROXAMIC ACIDS:

Thiohydroxamic acids (THAs) are isosteres of HA where C=O is replaced by C=S. Sulfated S-glucosyl thiohydroxamates are biosynthesized by plants belonging to different families e.g. cruciferae, nastrium etc [55]. Using 'S' and 'O' atoms, THAs coordinate with metal ions like Fe^{+3}, Ni^{+2}, Cu^{+2} forming colored metal complexes and hence find applications in detection and quantitative determination of metals [56,57].

Tris chelates of THA and Fe^{3+} are believed to participate in bacterial iron transport systems [56]. O-acyl derivatives of THA are efficient precursors of C, N or P radicals [57,58]. A number of THAs found applications as biocidal compounds for bacteria, fungi, insects, mites and weeds etc. They are effective as antiperspirant and antihypertensive agents, as enzyme inhibitors, and as drugs for treatment of leukaemia. THAs have also been used to counteract the effect of war toxins and to alleviate paralysis [55]. Their other important applications include gravimetric and spectrophotometric determination of metals. Cyclic THA's such as 1-hydroxy-2(1H)-pyridinethione (pyrithione) (**XIII**) and 3-hydroxy-4-methyl-2(3H)-thiazolethione (HMTT) (**XIV**) find diverse applications as fungicides and alkoxy-radical precursors in synthetic procedures and mechanistic studies [59]. The complexes of zinc chloride, acetate **XIII**   **XIV** or oxide with thiohydroxamate group give enhanced micro biological activity. The presence of N-methylformothiohydroxamic acid commonly termed 'Thioformin' has been recognised in bacterial sources and is the only thiohydroxamic acid siderophore discovered so far [60]. The complexes of thioformin display antibiotic activity against

E.Coli NIHJ. Metal complexes of THA are deoxyribonucleic acid (DNA) cleavers, by accepting electrons from DNA they can trigger oxidative strand breaks. Pseudomonas florescens produces fluopsin C and fluopsin F which are THA complexes of Cu(II) and Fe(III) respectively. These antibiotics have high biological activity in relation to both gram positive and gram negative bacteria [61-63].

1.1.4 FORMYLPHOSPHINOUS ACID:

Phosphorus, the 11th most abundant element on earth crust is an essential element to all life being a structural and functional component of all organisms [64,65]. It is structural constituent in many cell components such as phosphoproteins, phospholipids in cell membrane and teeth bones. In some organisms it is present as intracellular polyphosphate storage granules. In the form of orthophosphate, it plays a key role in photosynthesis. It provides phosphate ester backbone of DNA and RNA and is thus crucial for transmission of chemical energy through Adenosine triphosphate (ATP) molecule [66]. Phosphinous acids are compounds of general formula H_2POH i.e. hydroxyphosphines. They are isosterically equivalent to well known hydroxylamine H_2NOH. Hydrophosphoryl compounds (HPCs) are attractive reagents in fine organic and organo element synthesis. In the last decade, their synthetic potential is utilized for synthesizing artificial analogues of oligonucleotides for antigenic therapy of certain diseases. Recently, HPCs have received new impetus because secondary phosphine oxides and their tautomers phosphinous acids have begun to be used as donor ligand in complexes with Pt, Pd, Ni, Rh which act as effective catalyst for cross coupling, hydroformylation, hydrophosphorylation etc [67,68]. Formohydroxamic acid (FHA) is a derivative of hydroxylamine with one of the hydrogen being substituted by formyl group (HC=O) while in case of phosphinous acids similar substitution results in the formation of the compound of formula H(C=O)PHOH called formylphosphinous acid (FPA). A number of literature reports on HAs are available, but virtually nothing is known about FPA except a few reports on parent phosphinous acids. Phosphinous acids usually oxidize easily to phosphinic acids. Only known example of thermally stable noncoordinated phosphinous acid is bis (trifluoromethyl)phosphinous acid $(CF_3)_2POH$ synthesized by Burg and Griffiths in 1960 [69]. Phosphinous acid (H_2POH) containing trivalent phosphorus are in tautomeric equilibrium with phosphine oxide (H_3PO) containing pentavalent phosphorus with the equilibrium almost completely shifted to the side of phosphine oxide [70,71]. Chatt and Heaton described that it is possible to stabilize phosphinous acid derivatives by coordination to transition metal complexes and thus shift the equilibrium to the side of phosphinous acids [72].

Theoretical and experimental study reveal that electron withdrawing effect of organic groups attached to phosphorus strongly influence equilibrium distribution between phosphine oxide and phosphinous acid tautomer [73,74]. According to Dubrovina and Borner transition metal complexes of phosphinous acids have catalytic activity in homogenous catalysis [75]. Phosphorus containing compounds like phosphinous acid group and its derivatives can modify the surface of nanoparticles in vapor phase. FPA has been reported to be a radical cation in mass spectrometry study by Heydorn et al. [76]. The chemistry of FPA is thus challenging and unexplored. A part of the present study is thus aimed at determining the properties of these model molecules. The study of such model systems can help in studying larger phosphorus containing systems e.g. Wittig olefination and phosphorylation of protein residues.

1.2 REVIEW OF LITERATURE:

Research on HAs was lacking till 1975, after which huge amount of information has accumulated with respect to their synthesis and biomedical applications. An electron diffraction study of FHA was published in 1976. It is most fundamental species among all HAs yet experimental data is available scarcely due to instability of FHA at room temperature, it decomposes violently above its melting point and the decomposition products include CO_2, NH_3, $HCONH_2$, $HCOOH$, NH_2OH along with FHA [77]. HAs like acetohydroxamic acid (AHA), benzohydroxamic acid (BHA), oxalodihydroxamic acid, glutarodihydroxamic acids etc. are stable. HAs are known to undergo Lossen rearrangement to yield isocyanates. The solution photolysis of HAs generally lead to amides/anilides [78].

The chemistry of HAs have been surrounded by a number of controversies because of the presence of keto-enol tautomerism and stereoisomerism in the molecules. IR spectra suggest that HAs in the solid state and in polar solvents exist in keto form (1Z, 1E) while in the non-polar solvents; the iminol form (2Z, 2E) is also present (**Scheme 1.2**) [79-81]. X-ray crystal structure study by Bracher and Small on AHA $0.5H_2O$ and FHA shows that in solid state they adopt Z-conformation while most other C- and N-substituted HA such as N-(4-cyanophenyl) acetohydroxamic acid, N-(3-cyanophenyl) acetohydroxamic acid) etc. adopt E-conformation due to steric effects [82,83]. Garcia et al. performed theoretical and experimental study on BHA and indicated Z form to be stable in gas phase, water and NMR measurements indicated presence of E form in acetone [84]. M. Saldyka and Z. Mielke in their publication in Chemical Physics Letter (2003) on the basis of matrix infrared isolation spectroscopy suggested that 1Z is the most stable form as it is stabilized intrinsically by intramolecular H-bond involving five membered chelate rings, besides they identified 1E isomer for the first time and reported the population ratio 1E/1Z to be 0.035 ± 0.009 [85]. Mielke et al. applied matrix isolation technique combined with FT-IR spectroscopy to study spectral characteristics of HCONHOH and its isotopic analogue HCONDOD and HCO^{15}NHOH and hence confirmed existence of keto form by presence of bands due to NH and C=O groups and lack of bands characteristic of C=N vibrations [77]. Saldyka et al. determined theoretical relative abundances of FHA isomers calculated from Gibbs free energies at 298.15K and reported these are 85%, 3.99%, 11% and 0.0062% for 1Z, 2Z, 1E and 2E forms respectively while results obtained by them experimentally in FHA/Ar matrixes by deposition of vapor over solid FHA indicate the relative abundances of 1Z, 2Z and 1E isomeric structures to be 94.3%, 2.5% and 3.1% for 1Z, 2Z and 1E respectively [86].

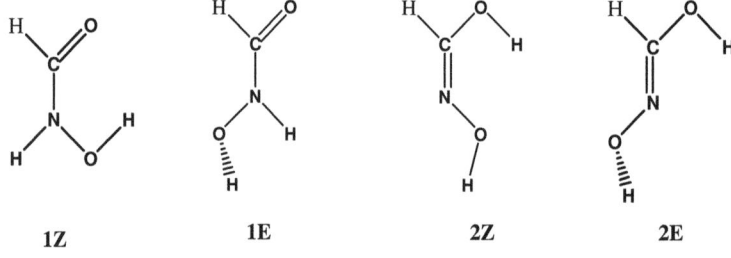

Scheme 1.2: Tautomers of HAs

Assignment of correct structure to HAs is challenging because several possible conformations strongly depend on concentration, temperature and nature of solvent etc [87]. Ab initio and NMR investigations suggest formation of imide structures both in undissociated HA and in their anions to be in minority. For AHA, 97% of neutral molecule and 94% of anion exist in Z-form but Z-imide structure is definitely present in metal chelates. Studies by several research groups determined the stability order of the various tautomeric forms of FHA to be 1Z > 1E > 2Z > 2E > dipolar form employing Density Functional Theory (DFT) methods. Wu and Ho determined the order of stability in FHA isomers to be 1Z > 2Z > 1E > 2E > dipolar form at the G2 theory [88]. Begona Garcia et al. in their ^{13}C NMR studies on phenyl benzohydroxamic acid (PBHA) advanced hypothesis of slower HZ→HE interconversion in methanol due to H-bonding with solvent [87]. R. Kakkar et al. investigated the barrier of interconversion of Z and E to be 18 kcal/mol which increases in aqueous solution. This rotational barrier decreases on methyl substitution [89]. Wu and Ho deduced the intramolecular proton transfer pathways which lead to interconversion of these five isomeric tautomers [88]. J.J. Ho et al. proposed concerted and stepwise mechanism (**Scheme 1.3**) for interconversion of five isomeric tautomers of FHA by studying intramolecular proton transfer between protonated FHA and water molecule theoretically and suggested the stepwise process to be the preferred one. They found that their barriers are sensitive to H-bond lengths [90].

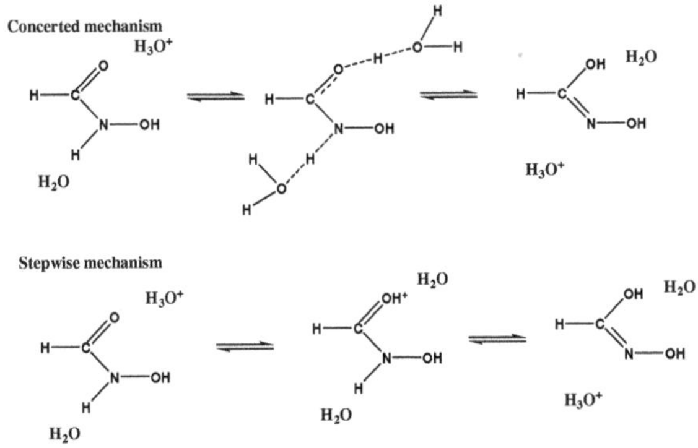

Scheme 1.3: Proton transfer between protonated FHA and water

The acid-base properties of biomolecules impact physiochemical considerations such as solubility, hydrophobicity and electrostatic interactions. Protonation plays an important role in proton transfer reactions, preparative chemistry, in living processes, in analysis by mass spectrometry and also has catalytic role in amide bond hydrolysis. Besides protonated molecules are important intermediates in reaction mechanisms. Deprotonation is important in evaluating reactivity, reaction mechanisms and structure of organic compounds. **Scheme 1.4** shows protonation and deprotonation sites of HAs. The HAs are suggested to exhibit two protonation sites (carbonyl oxygen and the oxime nitrogen). Carbonyl oxygen is most active protonation centre in solution and in the gas phase in FHA as suggested by several studies [91]. Caro et al. presented theoretical study of protonation processes in AHA in gas phase using MP2(FC)/cc-pVDZ and the molecular electrostatic potentials calculated by them show that the two oxygens and nitrogen to be the possible protonation sites. From the proton affinities determined, they concluded that protonation on C=O is favored on enthalpic grounds by 8.9 kcal/mol over the N-protonation [92].

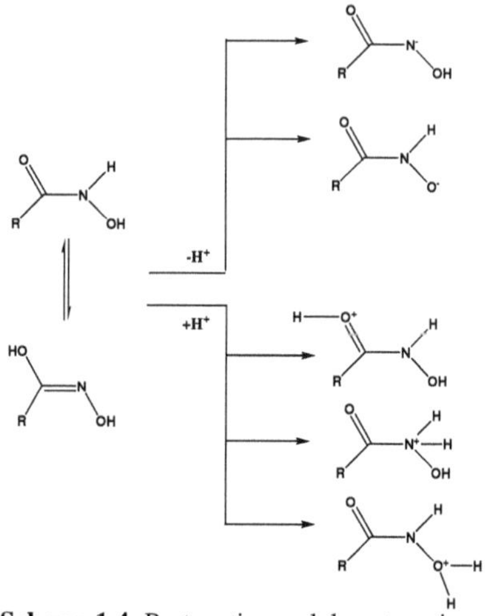

Scheme 1.4: Protonation and deprotonation

Remko et al. studied structure, gas phase acidities and vibrational frequencies of FHA and its sila derivatives and computed proton affinity (PA) of FHA to be -821.0 kJ/mol and acidity to be 1427 kJ/mol [93]. Ghosh et al. reported on protonation of 4-substituted BHA, 4-X ($C_6H_4CONHOH$) (X=H, OMe, Cl) in acidic medium of sulfuric acid, perchloric acid, hydrochloric acid at 25°C by UV spectrophotometric analysis and compared the results obtained by Hamett-Acidity function method, Bunnet-olsen method, Cox-Yates excess acidity function method etc. to rationalize difference between pK_{BH+} values determined by each method. The pK_{BH+} values are not influenced by the different mineral acids but show regular variation with substituents; an electron-withdrawing group (Cl) decreases and an electron-donating group (OMe) increases the pK_{BH+} values [94]. Garcia et al. also followed same techniques as that of Ghosh et al. to study protonation equilibria of N-phenylbenzohydroxamic acid, benzohydroxamic acid, salicylhydroxamic acid and N-p-tolylcinnamohydroxamic acids and found the change of pK_{BH+} to follow the sequence of catalytic efficiency of acids used in the order, HCl> H_2SO_4> $HClO_4$ [95].

HAs are weak acids with pka in the range ~ 9 which lies between pka of carboxylic acids (pka=4-5) and amides (pka~17). According to Exner et al. stronger acidity of HAs compared to amides is due to destabilizing inductive effect of hydroxyl group. They employed isodesmic reactions to study the acidity of both amides and HAs. They also concluded that high acidity of HAs is caused by the low energy of anions and not by high energy of acid molecules as in carboxylic acids [96]. Dynamic NMR and Nuclear Overhauser Effect spectroscopy (NOESY) (1D and 2D) experimental results for different HAs and their anions have shown that NH and OH compete for deprotonation sites however different titration experiments with strong bases have shown that only one proton can be removed, even at the highest basicity levels [84,97]. Many experimental studies based on IR and UV measurements in dioxane and aqueous alcohol solution as well as the theoretical studies support HAs to be N-acids [98]. N-ionization is also inferred by UV spectral changes as a function of pH, substituent effects, comparison with model compounds, NMR studies based on ^{17}O chemical shift changes and XPS in the solid state. Bordwell et al. performed a comparative study on acidities of several types of O, N and C acids including HAs and found that for both AHA and BHA, N-alkylation decreased acidity more than O-alkylation inferring parent acids to be OH acids in dimethyl sulfoxide (DMSO) [99]. Experimental studies by Exner et al. using FT-ICR (Fourier transform ion cyclotron) on AHA and derivatives yields N-anion and O-anion in equilibrium, with the equilibrium being dependent strongly on the reaction conditions like solvent and nature of R group. N-anion overweighs in gas phase and non-polar medium while O-anion is predominant in water [96]. Recent theoretical calculations by Ventura et al. predict FHA and AHA to behave as N-acids in gas phase and O-acids in solution phase [100]. Bagno et al. from heteronuclear relaxation time measurements indicated that AHA is predominantly O acid in aqueous phase while BHA is predominantly N-acid in aqueous phase [101]. Remko et al. calculated gas phase acidities of FHA and its sila derivatives R-M(=O)NHOH (R=H, CH_3, CF_3 and phenyl; M=C and Si) and concluded that thermodynamic stability of neutral and anionic species depend on substituent R. Change in acidities upon substitution of carbon by silicon was rationalized on the basis of electonegativity and charge capacity of these atoms and relative stability of individual species. According to them both formo- and silaformohydroxamic acid (HSiONHOH) are N-acids in gas phase and acidity increases in the order HSiONHOH < HCONHOH (R=H, CH_3, CF_3 and phenyl). Further the acidity order is CH_3M(=O)NHOH< HM (=O) NHOH) < Phe-M (=O) NHOH< CF_3M (=O) NHOH (M=C and Si) [102].

Wu and Ho suggested the strength of intramolecular H-bond and effect of dipole moment to be the two major factors governing the acidity of FHA [88]. Kevin Leung applied ab initio molecular dynamics to FHA anion tautomers and calculated hydration numbers. The cis-nitrogen deprotonated, cis- oxygen deprotonated and trans –oxygen deprotonated formohydroxamate tautomers form an average of 6.3, 6.9 and 6.0 H-bonds with water molecules in hydration shell respectively. Their predicted pair correlation functions and time dependence of hydration numbers suggested that water is highly structured around nominally negatively charged oxime oxygen in O-deprotonated tautomers but significantly less around the nitrogen atom in N-deprotonated species [103]. Yazal and Pang in their study showed that both zinc-coordinated AHA and N-methyl AHA exist in O-deprotonated Z keto form with their two oxygen atoms coordinated to zinc in proteins and zinc affinity of N-methyl AHA is 11 kcal/mol stronger than deprotonated AHA [104]. All the studies generated another controversy over the stability of anionic species and their mode of binding to metals.

Coordination chemistry of HAs is important in context of chemical biology since these bioligands (i) coordinate metal sites in Ni(II) containing metalloprotein urease and zinc (II) containing metalloprotein, MMP, carbonic anhydrase and tumor necrosis factor α-converting enzyme (TACE) (ii) feature as Fe(III) binding functional group in bacterially derived Fe(III) sequesteration molecules called siderophores. Each of (i) and (ii) couple into aspects of human medicine in terms of HA based inhibition of MMP and related enzymes and treatment of Fe(III) overload disease. HAs RC(O)N(R')OH (R=alkyl/aryl; R'=alkyl/aryl/H) coordinate to wide variety of transition metals predominantly as monoanionic hydroxamato or dianionic hydroximato O-O' bidentate ligand [105]. To determine the optimum HA structure for kinetic and thermodynamic stability of Fe(III) chelate, studies were undertaken by Crumbliss et al. that show that nature of C- and N-substituents influence the stability of metal-hydroxamato/imato complex whereby a more stable complex is formed when both the C- and N-substituents are electron donating, thereby increasing localized negative charge on coordinating oxygen atoms [106,107]. HAs act as a ligand toward many metal ions such as Al(III), Cr(III), Ga(III), Os(III), Eu(III), Be(II), Co(II), Ni(II), Cu(II), Zn(II), V(V), Pu(IV), Np(IV), U(VI), Mo(VI) and La(III) [108,109]. The strong affinity between most transition metal ions and HAs is supported by high magnitudes of overall stability constants [110].

HAs show diverse coordination modes which are characterized by X-ray crystallography and in solution. The deprotonation phenomena are related to chelating modes of HAs. They include (O, O), (N, O), (O-μ-O), N-bonding mode and NX (O, O) bridging modes. The (O, O) coordination mode (**Scheme 1.5 a**) is reported in majority of HA metal complexes. The (N, O) coordination mode (**Scheme 1.5 b**) is suggested in solution studies of aminohydroxamic acids [111]. The {O, μ-O} binding mode (**Scheme 1.5 c**) is observed in a series of dinuclear complexes used as models for urease activity [112]. N-bonding mode (**Scheme 1.5 d**) was observed in Ni(II) complex with glycine hydroxamic acid by Brown et al [113]. The (N X), O, O bridging mode (**Scheme 1.5 e**) is observed in metallacrown compounds.

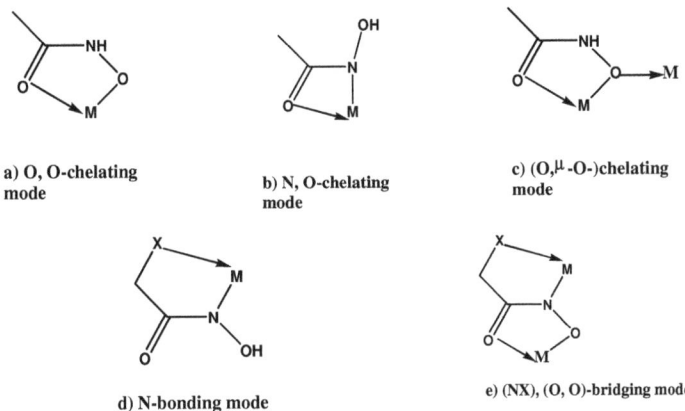

Scheme 1.5: Coordination modes of the hydroxamic group in metal complexes

Farkas et al. suggested that aminohydroxamic acids might be candidate as sequestering agents for some toxic metal ions e.g. Pb (II). Under physiological pH, complex formation of Pb(II) with THAs/HAs compete effectively with hydrolytic process of this toxic metal [114]. Raymond et al. synthesized a number of bis-thiohydroxamic acid with linking group of varying lengths and rigidities that are suggested as potential chelating agents for Pb^{+2} [115]. The mechanism and the kinetics of the binding of phenylbenzohydroxamic acids with Ni(II) has been studied in aqueous solution by stopped flow technique and the behavior of phenylbenzohydroxamic acids is studied in organic solvents by NMR spectroscopy [87]. First row transition metals usually occur in metallohydrolases in dinuclear active sites with bridging carboxylate as in urease Ni (II), arginase Mn(II) [116-118]. These dinuclear based enzymes are inhibited by HAs by their unique (O,O) coordination mode for which it is suggested that deprotonated 'OH' group bridges two metal centres and carbonyl oxygen bonded to one metal centre in a similar manner as is observed in the AHA inhibited C319A variant of *Klebsiella aerogens urease* [119]. Brown et al. applied DFT studies on pseudotetrahedral chelates of Zn-acetate and Zn-hydroxamate (ZnL$_2$) and their hydrate ZnCl$_2$.nH$_2$O to model zinc hydrolases and their inhibition by HAs and thereby concluded hydrates to be more stable than corresponding chelates in all cases [30]. R. kakkar et al. performed DFT calculations on a number of square planar hydroxamate chelates of several divalent metal ions in order to determine respective affinities for some biologically important ligands and concluded that both electrostatic and covalent contribution introduce variation in binding affinities of various metal ions [120]. With a view that antibiotic activity of THA complexes involves chelation with metals, R. Kakkar et al. prepared N-p-(ethylbenzene) thiohydroxamic acid chelates with several metal ions and by employing theoretical calculations concluded the geometry of complexes to be square planar. The electronic transitions do not involve metal d-orbitals, instead reorganization of charge within the ligand [121].

Two reviews on basic medicinal chemistry and pharmacology of HA derivatives were published by Williamson et al. and Lou et al. [32,122]. Anti inflammatory activity of HAs is also connected with their inhibition of S-lipoxygenase, an enzyme involved in the biosynthesis of leukotrienes mediators of inflammatory and allergic disorders. Wang et al. (1977) investigated mutagenic and antibacterial activities of HAs [123]. A suite of sulfonamide and sulfone hydroxamic acid TACE inhibitors (TNF-α converting enzyme) have shown promise in the ability to decrease the release of pro-inflammatory cytokines. J.J.-W. Duan et al. have reported discovery of β-benzamido hydroxamic acids as potent, selective and orally bioavailable TACE inhibitors [124]. Thiazole-5- hydroxamic acid has been designed and synthesized as a novel histone deacetylase (HDAC) inhibitor [125]. Aminosuberoyl hydroxamic acids are also a potent new class of HDAC inhibitors. The simple hydroxamates like acetohydroxamide, benzohydroxamide and hydroxyurea are weakly effective at inhibiting the growth of plasmodia (Holland et al.) with typical IC50 values in the micromolar range, as were a series of aromatic based HA synthesized by the Holland group [126]. Inhibitory effect of HAs on melanogenesis investigated by H.S. Rho et al. and results suggested that HA derivatives inhibited melanin synthesis via deactivation of tyrosinase [127]. Kochany (1991) published a review article which summarizes structures, reactions, photochemistry and applications of HAs with particular emphasis on environmental aspects [78]. Pande et al. investigated ten HAs as candidate for H-bond donors and acceptors in drug design and described the measurement of H-bond donor and H-bond acceptor strength of these HAs by measuring their log P (hydrophobicity) (O/W) values. H-bonding parameters computed by them are valuable for medicinal chemists [52].

Mielke et al. with a goal to identify NH or OH acting as H-bond donor for carbon monoxide (CO) molecule, studied interaction of FHA with CO employing FTIR matrix isolation technique and theoretical methods [128]. In a separate report, they employed IR matrix isolation studies on AHA complexes with HF and HCl [129]. Mielke et al. optimized and spectroscopically characterized two isomeric 1:1 complexes between FHA and N_2, the N_2 molecule interact with NH or OH group of FHA and N-H...N or O-H...N hydrogen bonds are formed [130]. Gejji et al. presented a study on electronic structure and vibrational characteristics of binary complexes of AHA...HX (X=F, Cl, Br) in order to understand its H-bonding ability [131].

Garcia et al. reported self aggregation of salicylhydroxamic acid and p-hydroxy benzohydroxamic acid (**XV**) by UV and ^{1}H NMR and suggested that dimers exist in even highly diluted solutions which points towards strong H-bonding between HA monomeric units [132]. Protons of HA group are strongly downfielded due to formation of H-bonded aggregates in the solution. NMR results by Garcia et al. on phenylbenzohydroxamic acid in acetone confirm Z-isomer to dimerise and dimer population was small in methanol and negligible in water as these protic solvents form H-bonds with the monomer [87]. Garcia et al. in another study reported existence of E-E, Z-Z and E-Z dimers of BHA in acetone at low temperature (-80°C) evidenced by variation of chemical shifts and appearance of crosspeaks in Correlation spectroscopy (COSY) spectra [133]. Proniewicz et al. on basis of NMR and Raman spectral studies of 2-(hydroxyimino) propanohydroxamic acid suggested that it forms dimers in solid state and clusters in DMSO [134]. Theoretical and experimental investigations performed by Ciofini revealed that the formation of dimers of N-phenylbenzohydroxamic acid is possible in gas phase with stability of dimers decreases from gas phase to aprotic solvents to protic solvents due to formation of H-bonds with solvent molecules [135].

THE PRESENT STUDY:

The importance of HAs in biochemical and physical processes motivates one to explore the characteristics of these compounds. The X=C-NH- (X=O,S) group of HAs and THAs serve as a model for amides, thiopurines and thiopyrimidines to study H-bonded interactions in DNA (Deoxyribonucleic acid) and RNA (Ribonucleic acid). The present study explores the stability, molecular properties, intramolecular H-bonding of FHA and its isosteres thioformohydroxamic acid (TFHA) and formylphosphinous acid (FPA) in neutral, anionic and cationic forms and intermolecular H-bonding with water, through dimerization and with amino acid side chain functionalities (-SH, -OH, -SeH, -CH$_3$COOH, -CH$_3$NH$_2$). Knowledge of the H-bonding parameters is valuable in the field of toxicology, pharmacology and environmental sciences.

1.3 QUANTUM MECHANICAL BACKGROUND:

The observation that the energy of the microscopic particles is quantized led to the birth of new era of science "quantum mechanics". The principles of quantum mechanics provide quantitative description of motion of electrons and nuclei, forming molecules that allow the determination of all the chemical properties of a system. The applications of quantum chemistry are to describe the behavior of atoms and molecules as pertaining to their reactivity.

1.3.1 THE SCHRODINGER EQUATION:

The Schrödinger equation describing the wave function of a collection of particles can be written in its generalized form as follows

$$\hat{H}\Psi(q,t) = E\Psi(q,t) \quad (1.3.1.1)$$

where $\hat{H}$ is Hamiltonian operator, $\Psi(q,t)$ is the wave function which depends on q (spatial coordinates) and t (time), and E represents energy of the system and is a scalar quantity. The wave functions can be written as product of spatial function and a time function

$$\Psi(q,t) = \psi(q)\,\psi(t)$$

where $\Psi(q)$ is spatial part of wave function dependent upon position coordinates and $\Psi(t)$ is a function of time alone. This separation is valid in general for most problems of interest. For most of the average properties, the time independent Schrödinger equation is used [136].

$$\hat{H}\Psi(q) = E\Psi(q)$$

The molecular Hamiltonian is given as

$$\hat{H} = -\frac{\hbar^2}{2m_e}\sum_i \nabla_i^2 - \frac{\hbar^2}{2}\sum_A \frac{1}{m_A}\nabla_A^2 - \sum_A\sum_i \frac{Z_A e^{'2}}{r_{iA}} + \sum_A\sum_{B>A} \frac{Z_A Z_B e^{'2}}{R_{AB}} + \sum_j\sum_{i>j} \frac{e^{'2}}{r_{ij}} \quad (1.3.1.2)$$

where A and B refer to nuclei and i and j refer to electrons. The first term is the operator for the kinetic energy of the electrons, the second term accounts for the operator for the kinetic energy of the nuclei. The third term is the potential energy of the attractions between the electrons and nuclei, r_{iA} being the distance between the electron i and nucleus A. The fourth term is the potential energy of the repulsions between the nuclei, R_{AB} being the distance beween nuclei A and B with atomic number Z_A and Z_B. The last term is the potential energy of the repulsions between the electrons, r_{ij} being the distance between electrons i^{th} and j^{th}. An exact solution to the Schrödinger equation is possible only for trivial molecular systems, several simplifications are applied for solving the equation for larger molecules. For the practical purposes, some approximations must be made to solve the Schrödinger equation.

1.3.2 THE BORN OPPENHEIMER (BO) APPROXIMATION:

This approximation proposed by Born and Oppenheimer (1927) for calculation of molecular wave functions is used to simplify the Schrödinger wave equation and is indispensible in quantum chemistry [137]. It is also known as adiabatic approximation. The full Hamiltonian operator for the molecular system is written as 1.3.1.2. Since the mass of a typical nucleus is 1000 times greater than that of electron thus electron distribution depends on position of nuclei and not on its velocity. BO approximation allows separation of wave function into its electronic and nuclear components.

$$\Psi_{Total} = \Psi_{Electr} \cdot \Psi_{Nuclear}$$

BO is the assumption that electronic motion and nuclear motion can be separated and the motion of nuclei and electron are independent of one other. Using

Born oppenheimer approximation, the electronic Hamiltonian $\hat{H}^{electronic}$ which separates kinetic energy term for the nuclei

$$\hat{H}^{electronic} = -\frac{1}{2}\sum_{i}^{electrons}\nabla_i^2 - \sum_{A}^{nuclei}\sum_{i}^{electrons}\frac{Z_A}{r_{Ai}} + \sum_{i>j}^{electrons}\sum_{j}^{electrons}\frac{1}{r_{ij}} \qquad (1.3.2.1)$$

1.3.3 THE VARIATIONAL PRINCIPLE:

The variation principle is another important simplification for the solution of Schrodinger equation. The ground state energy E_0 of some arbitrary system having ground state wave function Ψ_0 is given by

$$E_0 = \frac{\int \psi_0^* \hat{H} \psi_0 d\tau}{\int \psi_0^* \psi_0 d\tau} \qquad (1.3.3.1)$$

where $d\tau$ represents the appropriate volume element. With any trial wave function ϕ, the energy is evaluated as

$$E_\phi = \frac{\int \phi^* \hat{H} \phi d\tau}{\int \phi^* \phi d\tau} \qquad (1.3.3.2)$$

The variational principle states that

$$E_\phi \geq E_0$$

It is stated in words as "The expectation value of the Hamiltonian, Ĥ, calculated using a trial wave function, ϕ, E_ϕ is never lower in value than the true ground state energy, E_0, which is expectation value of Ĥ calculated using the true ground state wavefunction ψ" [138].

1.3.4 THE HARTREE FOCK (HF) SELF CONSISTENT FIELD APPROACH:

The HF-SCF method is an approximate method for the determination of the ground-state (GS) wave function and GS energy of a quantum mechanical many body system [139]. This approach assumes that any one electron moves in a potential that is spherical average due to other electrons and the nuclei. Spherically averaged potential is expressed as single charge centered on nucleus. The Schrodinger equation is then numerically solved for that electron in spherically averaged potential. HF results in separation of electron motions resulting in ordering of electrons in MOs. The many electron wave function Ψ for an N-electron molecule is written in terms of one electron space wavefunctions f_i and spin functions α and β

$$\Psi = f_1(1)\alpha f_1(2)\beta...f_{N/2}(N)\beta \quad (1.3.4.1)$$

The potential from nuclei is set by initial configuration of the molecule and potential from other electrons are determined from initial approximate wavefunctions resulting in HF Hamiltonian, $\hat{F}(1)$.

Molecular orbital wave functions, f_i, are eigenfunctions of Hartree-Fock Hamiltonian operator, $\hat{F}(1)$.

$$\hat{F}(1) = -\frac{1}{2}\nabla_i^2 - \sum_A^{nuclei}\frac{Z_A}{r_{iA}} + \sum_{j=1}^{N/2}[2\hat{J}_j(1) - \hat{K}_j(1)] \quad (1.3.4.2)$$

$$\hat{H}_{(1)}^{core} \equiv -\frac{1}{2}\nabla_i^2 - \sum_A\frac{Z_A}{r_{ia}} \quad (1.3.4.3)$$

First two terms in equation 1.3.4.2 correspond to kinetic energy operator of the electron and attraction between one electron and the nuclei of the molecule and two together constitute the core Hamiltonian. The next term $\hat{J}_j(1)$ is *Coulomb operator*.

$$\hat{J}_j(1)f(1) = f(1)\int |f_j(2)|^2 \frac{1}{r_{12}} d\tau_2 \quad (1.3.4.4)$$

The Coulomb operator accounts for the smeared-out electron potential with an electron density of $|f_j(2)|^2$. The last term in equation 1.3.4.2 is the *exchange operator*

$$\hat{K}_j(1)f(1) = f_j(1)\int \frac{f_j^*(2)f_i(2)}{r_{12}} d\tau_2 \quad (1.3.4.5)$$

where f is an arbitrary function and the integrals are definite integrals over all space. Exchange operator accounts the effects of spin correlation. Schrodinger equation is now solved for one electron $f_i(1)$

$$\hat{F}(1)f_i(1) = \varepsilon_i f_i(1) \quad (1.3.4.6)$$

Term ε_i correspond to orbital energy of electron ascribed by $f_i(1)$.

The energy of molecule in terms of the Hartree-Fock approach E^{HF}, is determined as follows.

$$E^{HF} = 2\sum_{i}^{\frac{N}{2}} \varepsilon_i - \sum_{i}^{\frac{N}{2}}\sum_{j}^{\frac{n}{2}}(2J_{ij} - K_{ij}) + \sum_{A>B}^{nuclei}\sum_{B}^{nuclei}\frac{Z_A Z_B}{R_{AB}} \qquad (1.3.4.7)$$

First summation in equation 1.3.4.7 is over all the orbital energies of occupied MOs. The terms J_{ij} and K_{ij} are determined by operating coulomb operator (equation 1.3.4.4) and exchange operator (equation 1.3.4.5) on $fi(1)$ and multiplying result by $fi*(1)$ and integrating overall space. The last summation term in equation 1.3.4.7 refers to inter-nuclear repulsion potential for a particular nuclear configuration.

The spatial one-electron wave function $f_i(n)$ is represented as linear combination of atom - centered functions (i.e. atomic orbitals) φ_{ik} called the linear combination of atomic orbitals (LCAO) approximation. The functions φ_{ik} constitute a basis set. The index k refers to specific atomic orbital wave function and the index i refers to its contribution to a specific molecular orbital.

$$f_i(n) = \sum_{k}^{N'} c_{ik}\varphi_{ik} \qquad (1.3.4.8)$$

The coefficients c_{ik} correspond to contribution of each atomic orbital (AO) to the corresponding MO.

The energy of given electron in a MO of a molecule ε_i is calculated as function of coefficients for that molecular orbital, c_{ik}. These equations are called *Roothaan Hall equations.*

$$\sum_{k} c_{ik} \hat{F} \varphi_k = \varepsilon_i \sum_{k} c_{ik}\varphi_k \qquad (1.3.4.9)$$

To calculate $\hat{F}$, an initial guess to coefficients for other molecular orbitals f_j must be made. Multiply 1.3.4.9 by φ_j^* (where j =1,2,3,....N') and integrating yields the following expression.

$$\sum_{k} c_{ik}(F_{jk} - \varepsilon_i S_{jk}) = 0 \qquad (1.3.4.10)$$

The terms F_{jk} are called the *Fock matrix elements*

$$F_{jk} = \left\langle \varphi_j \middle| \hat{F} \middle| \varphi_i \right\rangle \qquad (1.3.4.11)$$

The terms S_{jk} are the *overlap matrix elements*

$$S_{jk} = \left\langle \varphi_j \middle| \varphi_k \right\rangle \qquad (1.3.4.12)$$

Variational theory is used to optimize the coefficients by taking derivative of ε_i with respect to each coefficient and setting it equal to zero.

$$\det(F_{jk} - \varepsilon_i S_{jk}) = 0 \tag{1.3.4.13}$$

The optimized coefficients from equation 1.3.4.13 are compared to initial guess for the coefficients. This iterative procedure is called a self consistent field.

Restricted Hartree Fock (RHF) wave function is used generally for closed-shell states and Unrestricted Hartree fock (UHF) is molecular orbital method for open shell molecules. A Hartree-Fock wave function in which electrons whose spins are paired occupy the same spatial orbital is called a restricted Hartree-Fock (RHF) wave function while UHF is method for open shell molecules where number of electrons of each spin is not equal. UHF uses different molecular orbitals for the α and β electrons. Open shell systems where some of the electrons are not paired, can be dealt by one of the two methods

Restricted Open-shell Hartree Fock Method (ROHF)
Unrestricted Hartree Fock Method (UHF)

1.3.5 BASIS FUNCTIONS AND BASIS SETS:

The mathematical description of the orbital within a system can be carried out using several basis functions, is called as basis set [140-147]. Larger basis sets approximate the orbital more accurately by imposing few restrictions on the location of electrons in space. Slater suggested the following expression for basis functions that are known as Slater type orbitals (STO)

$$STO_{nlm}(r,\theta,\phi) = \frac{\left(\frac{2\zeta}{a_o}\right)^{n+\frac{1}{2}}}{[(2n)!]^{\frac{1}{2}}} r^{n-1} e^{-\zeta r/a_o} Y_l^m(\theta,\phi) \tag{1.3.5.1}$$

ζ is orbital exponent, a_0 is Bohr's radius, n, l and m are quantum numbers, Y_l^m is spherical harmonics. Slater orbitals corresponded to a set of functions which decayed exponentially with distance from the nuclei. S.F. Boys came up with an alternative when he developed the Gaussian Type function (GTF) equation:

$$g_{ijk} = N x_a^i y_a^j z_a^k e^{-\alpha r_a^2} \tag{1.3.5.2}$$

N is normalization constant.

α is positive orbital exponent.

i, j, k are non-negative integers.

When i + j + k = 0, i.e. i=0, j=0, k=0, the GTF is called an s-type Gaussian.

With i + j + k = 1, the function describes p-type Gaussian which contains the factor x_a, y_a, z_a. Similarly with i + j + k = 2, a d-type Gaussian results. Besides the GTF is better over STO in that molecular integrals are calculated fast and it can be differentiated trivially any number of times in comparison to STO which is difficult to differentiate analytically and is time consuming to calculate molecular integrals.

An alternative to Cartesian Gaussians is spherical Gaussians, whose real form is

$$Nr_a^{n-1}e^{-\alpha r_a^2}(Y_l^{m*} \pm Y_l^m)/2^{1/2} \qquad (1.3.5.3)$$

GTFs do not have radial nodes; however, radial nodes can be obtained by *combining* different GTFs. In Gaussian-basis-set terminology, instead of using the individual Gaussian functions as basis functions, each basis function is taken as a linear combination of a small number of Gaussians, according to

$$\chi_r = \sum_u d_{ur} g_u \qquad (1.3.5.4)$$

Where the g_u's' are Cartesian Gaussians centered on same atom and having the same i, j, k values as one other, but different α's. The coefficients d_{ur} are constants that are held fixed during the calculation. In equation 1.3.5.4, χ_r is called a contracted Gaussian-type function (CGTF) and g_u's are called primitive Gaussians.

The terminology for the basis set defines nature and number of basis functions used to approximate the total electronic wave function during theoretical calculations. The minimal basis set is STO-nG, where n is an integer representing the number of primitive Gaussian functions comprising a single basis function that approximates one STO for each AO/MO.

Orbitals are distorted (polarized) upon molecule formation. A split valence basis set uses one contracted Gaussian function for inner shell AO while two or more contracted Gaussian functions are used to describe the valence AO. Split valence basis sets allow the change in size of the orbitals but not the shape, it can be solved in the form of polarized basis set by adding orbitals with angular momentum beyond what is required for the description of each atom in the valence shell. Pople et al. developed several split valence basis sets using the Gaussian functions. For example in 3-21G basis set, each inner shell AO is represented by a single CGTF that is linear combination of three primitive Gaussians. For each valence shell AO, there are two basis functions one of which is CGTF (Linear combination of two Gaussian primitives) and one is single Gaussian. In order to allow the shape of molecular orbitals to vary, polarization functions are added. For example additions of 2p functions to 1s function of hydrogen atom; the function $C_1 1s + C_2 2p_X + C_3 2p_Y + C_4 2p_Z$ is obtained as a part of molecular orbital. The function will be polarized in the direction determined by the coefficients C2, C3 and C4. Polarized basis set adds d-functions to first and second row atoms and f-functions to transition metal elements indicated by 6-31G* or (d) in the basis set notation. The 6-31G* basis set is valence double zeta polarized basis set that adds to the 6-31G set six d-type CGTF on each of atoms Li through Ca and ten f-type functions on each of atoms Sc through Zn. One asterisk (*) at the end of a basis set denotes that d functions are added to heavy atoms. Two asterisks (**) means that p functions are added to hydrogen atom in addition to the d functions on heavy atoms for example 6-31G** basis set adds to the 6-31G* set a set of three p-type Gaussian polarization functions on each hydrogen and helium atom. Diffuse functions are large-size versions of s- and p-type functions and allow orbitals to occupy larger regions of space. They are important for systems where electrons are relatively far from the nucleus: molecules with lone pairs, anions, other systems with significant negative charge, systems in excited states, system with low ionization potential, desciption of absolute acidities etc. Diffuse basis sets are represented by the '+' signs. One '+' means diffuse functions are added to heavy atoms while '++' signals that diffuse functions are added to hydrogen atoms as well e.g. 3-21+G is formed from 3-21G by addition of four highly diffuse functions (s, p_x, p_y, p_z) on each non hydrogen atom. Dunning and coworkers developed the CGTF basis sets cc-pVDZ, cc-pVTZ, cc-pVQZ and cc-pV5Z (collectively denoted cc-pVXZ). The 'cc-p', stands for 'correlation consistent polarized' and the 'V' indicate they are valence only basis sets example is aug-cc-pVDZ - Augmented versions of the preceding basis sets with added diffuse functions [148].

1.3.6 ELECTRON CORRELATION:

Coulomb hole is the region in which the probability of finding another electron is small. The motions of the electrons are correlated to each other and designated as electron correlation. Hence electron correlation refers to the interaction between electrons in a quantum system whose electronic structure is being considered.

Within HF system, the antisymmetric wave function is approximated by a single Slater determinant. Exact wave functions, however can't be generally expressed as single determinant as electron correlation can't be accounted for, thus leading to total electronic energy different from the exact solution of the non-relativistic Schrödinger equation within the Born-Oppenheimer approximation. Therefore the HF limit is always above this exact energy. The difference is called *correlation energy*, a term coined by Löwdin. The correlation energy is defined as

$$E_{corr} = E_{exact} - E_{HF} \qquad (1.3.6.1)$$

Because the Hartree-Fock energy is always above the exact energy, the correlation energy is always negative,

$$E_{corr} < 0 \qquad (1.3.6.2)$$

Several methods have emerged in order to provide for electron correlation.
Two methods deal with electron correlation

    i)   Configuration interaction [149]
    ii)   Moller-Plesset Perturbation Theory (MPn) [150,151]

CONFIGURATION INTERACTION (CI)/ CONFIGURATION MIXING (CM):

It is a post HF linear variational method for including electron correlation and solving the non relativistic Schrodinger equation within the BO approximation for a quantum chemical multielectron system. *Configuration Interaction* means the mixing of different electronic configurations. CI proceeds by constructing other determinants by replacing one or more occupied orbitals within the HF determinant with virtual (unoccupied) orbitals. CI uses a variational wave function that is linear combination of Configuration state functions (CSFs) built from spin orbitals,

$$\Psi = \sum_{I=0} C_I \Phi_I^{SO} = C_0 \Phi_0^{SO} + C_1 \Phi_1^{SO} + ... \qquad (1.3.6.3)$$

Where ψ is electronic ground state of the system and SO stands for spin orbitals. If the expansion includes all possible CSFs, then it is full configuration interaction procedure. The first term in the above expansion is HF determinant. Other CSFs can be characterized by number of spin orbitals that are exchanged with virtual orbitals from HF determinant. The configuration functions in a CI calculation are classified as singly excited, doubly excited, triply excited, ----, according to whether 1, 2, 3, ---- electrons are excited from occupied to unoccupied orbitals.

MOLLER PLESSET (MP) THEORY:

Perturbation theory methods provide technique to deal with systems of many interacting particles. The Moller-Plesset method uses perturbation theory to correct for electron correlation in a many electron-systems. The MP method has advantage in being faster than CI computations; however the disadvantage is that it is not Variational. Non variational result is not, in general, an upper bound of true ground state energy. In MP method, the zero order Hamiltonian $\hat{H}^{(0)}$ is defined as sum of N-electron Hatree-Fock Hamiltonians, $\hat{H}_i^{HF}$, as given in equation 1.3.6.4

$$\hat{H}^{(0)} = \sum_{i=1}^{N} \hat{H}_i^{HF} \tag{1.3.6.4}$$

The first order perturbation is the difference between the zero order Hamiltonian in equation 1.3.6.4 and electronic Hamiltonian in equation 1.3.2.1

$$\hat{H}^{electronic} - H^{(0)} \tag{1.3.6.5}$$

The Hartree-Fock ground state wavefunction ψ^{HF}, is an eigen function of Hartree-Fock Hamiltonian $H^{(0)}$, with eigen value $E^{(0)}$. The HF energy associated with normalized ground–state HF wavefunction is the following expectation value.

$$\begin{aligned} E_{HF} &= \langle \Psi^{HF} | H^{(electronic)} | \Psi^{HF} \rangle \\ &= \langle \Psi^{HF} | H^{(0)} | \Psi^{HF} \rangle + \langle \Psi^{HF} | \hat{H}^{(1)} | \Psi^{HF} \rangle \\ &= E^{(0)} + E^{(1)} \end{aligned} \tag{1.3.6.6}$$

Hence, the HF energy is the sum of the zero and first order energy. The first correction to the ground-state energy of the system as the result of electron correlation is given by second order perturbation theory.

$$E_0^{(2)} = \sum_{J \neq 0} \frac{\langle \Psi_J^{HF} | \hat{H}^{(1)} | \Psi_0^{HF} \rangle \langle \Psi_0^{HF} | \hat{H}^{(1)} | \Psi_J^{HF} \rangle}{E_0^{(0)} - E_J} \tag{1.3.6.7}$$

A Moller–Plesset computation to a second order energy correction is called an MP2 computation and higher order energy corrections are called MP3, MP4 and so on.

1.3.7 DENSITY FUNCTIONAL THEORY (DFT):

DFT is a quantum mechanical theory to investigate the electronic structures of many body systems, in particular atoms, molecules and the condensed phases. DFT computes the electron correlation via general functional of the electron density. It is ultimately derived from Thomas-Fermi- Dirac model, and from Slater's work. Such methods owe their origin to the Hohenberg- Kohn (H-K) theorem [152,153]. The first Hohenberg-Kohn theorem asserts that the electron density of any system determines all ground-state properties of the system, that is, $E = E[\rho_0]$, where ρ_0 is the ground-state density of the system. The second H-K theorem states that the trial density function ρ_{tr} for a given system that satisfies $\int \rho_{tr} dr = n$ (total number of electrons), gives the energy functional $E[\rho_{tr}] \geq E_0$ (the true ground state energy). Kohn and Sham devised a method for evaluating electron density using a fictitious system of non interacting electrons. Following their work, DFT functional partitions the electronic energy into several components which are computed separately: the kinetic energy (E^T), the potential energy of electron-nuclear interaction (E^V), the Coulomb repulsion (E^J), and an exchange correlation term (E^{XC}) accounting for the remainder of the electron-electron interaction.

$$E = E^T + E^V + E^J + E^{XC} \qquad (1.3.7.1)$$

E^J is given by following expression:

$$E^J = \frac{1}{2} \iint \rho(\vec{r}_1)(\Delta r_{12})^{-1} \rho(\vec{r}_2) d\vec{r}_1 d\vec{r}_2 \qquad (1.3.7.2)$$

The exchange correlation energy functional E^{XC} is defined by

$$E_{XC}[\rho] \equiv \Delta \overline{T}[\rho] + \Delta \overline{V}_{ee}[\rho] \qquad (1.3.7.3)$$

where ΔT and ΔV are the differences in average ground state kinetic energies and potential energies of the molecules and reference system of the non-interacting electrons respectively with electron density equal to that in the molecule and

$$\Delta \overline{V}_{ee}[\rho] \equiv \overline{V}_{ee}[\rho] - \frac{1}{2} \iint \frac{\rho(r_1)\rho(r_2)}{r_{12}} dr_1 dr_2 \qquad (1.3.7.4)$$

The immediate and most obvious advantage of DFT is the inclusion of electron correlation. Although this alteration imposes an additional computational step, these integrals are computationally easily handled. Various approximate functions have been used in molecular DFT calculations, for example, local density approximation (LDA).

In a molecule, the positive charge is not distributed uniformly, but is located at the nucleus. Hence, ρ varies rapidly in a molecule and LDA approximation is applied. But the results are not in agreement with the experimental results, thus improved version of LDA called local spin density approximation (LSDA) was proposed. The LSDA uses different orbital for electrons with different spins.

Some commonly used exchange functional are Perdew and Wang's 1986 functional (which contains no empirical parameters) designated as PW86, Becke's 1988 functionals denoted B88, and Perdew and Wang's 1991 exchange functional PW91. The PW86 functionals and B88 functionals work about equally well in predicting molecular properties. The others are LYP (Lee-Yang-Parr) functional, P86, B96. Most popular hybrid method is B3LYP [154-157]. It uses Becke's three parameters exchange functional with Lee-Yang-Parr (LYP) correlation functional. This method is known to give better results for geometries than at MP2 level and often energetic comparable to QCISD(T) level. Another familiar hybrid method is B3PW91 [158], which uses Perdew correlation functional instead of LYP in B3LYP method.

1.4 COMPUTATIONAL DETAILS:

Gaussian 98W package, the windows version of Gaussian 98 suite of programs is used for performing ab initio molecular orbital (MO) and density functional theory (DFT) calculations [159]. DFT calculations have been performed using B3LYP three parameter density functional, including Becke's gradient exchange correction [154] and the Lee-Yang-Parr correlation functional [155]. As these molecules possess several lone pair of electrons, inclusion of diffuse functions in the basis sets and electron correlation is highly recommended. To study the effect of electron correlation on the geometries and energies, full geometry optimizations are performed using B3LYP/6-31+G* and MP2/6-31+G* levels without any symmetry constraints. The Pople's split valence basis set 6-311++G** and Dunning electron correlation basis set Aug-cc-pVDZ are also employed to improve the results wherever possible. Frequencies are computed analytically in order to characterize each stationary point as a minimum or transition state as first order saddle point and also to determine the zero point vibrational energies (ZPVE). The ZPVE values have been scaled by a factor of 0.9806, 0.9670, 0.9787 and 0.9675 respectively for B3LYP/6-31+G*, MP2/6-31+G*, B3LYP/Aug-cc-pVDZ and MP2/Aug-cc-pVDZ levels respectively [160,161].

1.5 METHODOLOGY TO CALCULATE VARIOUS PROPERTIES:

1.5.1 CALCULATION OF STABILIZATION ENERGIES:

A supersystem refers to a system formed by non covalent interaction between two or more molecular entities e.g. H-bond systems. The stabilization energy (S.E.) of such a system is calculated as difference between energy of supersystem and sum of the energies of its subsystems using expression 1.5.1.1

$$S.E. = E_{ab} - \{E_a + E_b\} \qquad (1.5.1.1)$$

where E_{ab} is electronic energy of super system, E_a and E_b refer to energies of subsystems. Aggregation is accompanied by structural distortions and the distortion energies which estimate the relaxation of monomers on complexation, are calculated using equation 1.5.1.2

$$E_{Dis} = (E_a + E_b) - (E_{a,Dis} + E_{b,Dis}) \qquad (1.5.1.2)$$

where E_a and E_b are energies of individual monomeric forms in gas phase and $E_{a,Dis}$ and $E_{b,Dis}$ are single point energies obtained for the distorted isolated monomer geometry upon complexation [162].

1.5.2 CORRECTION FOR BASIS SET INCOMPLETENESS:

Stabilization energy for the molecular clusters is affected by serious fact that different basis sets are used for evaluation of energies of the supersystem and its subsystems. In a supermolecule consisting of two molecules basis functions from one molecule can help compensate for the basis set incompleteness on the other molecule and vice versa. This effect is known as basis set superposition error (BSSE) [163]. Origin of BSSE lies in possibility that the unused basis functions of second unit in associated complex may augment the basis set of first unit, thereby lowering its energy compared to calculation of this unit alone. An approximate way of assessing BSSE is counterpoise (CP) correction method introduced by Boys and Bernardi in which the BSSE can be estimated as the difference between monomer energies with the regular basis and the energies calculated with the full set of basis functions for whole supermolecule. The CP method calculates each of the units with just the basis functions of other (without the nuclei or electron) using so called "ghost orbitals". BSSE is zero in the case of complete basis set.

1.5.3 ATOMIC CHARGES AND ORBITAL INTERACTIONS:

Natural Bond Orbital Analysis (NBO) developed by Weinhold, originated as a technique for studying hybridisation and covalency effects in polyatomic wave functions and has been extensively used to explore the various electronic transitions [164-166]. NBO program makes extensive provision for energetic analysis of NBO interactions. NBO program is capable of doing natural population analysis (NPA), natural bond orbitals (NBOs), natural localized MO (NLMOs), natural hybrid orbitals (NHOs), energetic analysis of wavefunction in terms of interactions between NBOs, localized analysis of molecular dipole moment in terms of NLMO and NBO bond moments and their interactions etc. This program can be used extensively to study various kinds of second order interactions present in the molecules- hence is a tool to understand different electronic interactions in the molecules. NBO is based on method for optimally transforming a given wavefunction into localized form.

The various natural localized sets can be considered to result from a sequence of transformation of input atomic orbital basis set (χ_i).

Input basis → NAOs → NHOs → NBOs → NLMOs.

Each natural localized set forms a complete orthonormal set of one electron functions for expanding delocalized MOs. The overlap associated with NAOs can be used to estimate the strength of orbital interactions. The optimal condensation of occupancy in natural localized orbital leads to partitioning into high and low occupancy orbital types. The natural minimal basis functions make little contribution to molecular properties. Each pair of valence hybrids h_A and h_B in the NHO basis set give rise to bond σ_{AB} and antibond σ^*_{AB} in the NBO basis.

$$\sigma_{AB} = c_A h_A + c_B h_B \quad (1.5.3.1)$$

$$\sigma^*_{AB} = c_A h_A - c_B h_B \quad (1.5.3.2)$$

NBO analysis transforms the delocalized many electron wavefunction into optimized electron pair bonding subunits in a set of Lewis type (σ bonding and lone pairs) and non Lewis type (such as Rydberg and σ^* antibonding orbitals). The interactions among the later two groups of orbitals can be used as measure of electronic delocalization within the systems under investigation and is investigated by 2nd order perturbation theory. Since these interactions lead to loss of occupancy from the localized NBOs of the idealized Lewis structure into the empty non-Lewis orbitals, they are referred to as 'delocalization' corrections to the zeroth-order natural Lewis structure. For each donor NBO (i) and acceptor NBO (j), the stabilization energy $E^{(2)}$ associated with delocalization i→j is estimated as

$$E^{(2)} = \Delta E_{ij} = q_i \frac{F^2(i,j)}{E_j - E_i} \qquad (1.5.3.3)$$

where q_i is donor orbital occupancy

E_j, E_i are diagonal elements (orbital energies)

$F_{i,j}$ is off diagonal elements associated with NBO fock matrix [167].

NBO has become a reliable tool for H-bond analysis and is used to derive information on the changes in charge densities in the proton donor, as well as changes in bonding and the antibonding space. H-bonds involve charge transfer from proton acceptor to proton donor. Improper/non conventional H-bonding on the other hand is a complicated two step process. Charge transfer from lone pair of electron donor is mainly directed to antibonding orbitals in remote part of complex, which causes elongation of bonds in that part of the complex. The primary effect is accompanied by secondary effect of structural reorganization of the proton donor leading to contraction of A-H bond and blue shift in A-H stretch frequency [168].

1.5.4 INTERACTIONS WITH MEDIUM BASED ON DIELECTRIC CONTINUUM MODEL:

There has been revolutionary progress in the theories for calculating molecular properties in solution. Among various methods, self consistent reaction field (SCRF) methods based on dielectric continuum model for solvent have been widely used in interpreting solvent effects on solute electronic structure. In these methods, solute occupies some kind of cavity surrounded by dielectric continuum of solvent. A dipole/multipole in solute include electric (reaction) field in solvent medium, which in turn will act to electrostatically stabilize the solute. SCRF models differ in definition of cavity and reacton field. Simplest reaction field model requires two parameters, solvent dielectric constant and cavity radius. In Onsagar's reaction field model, solute is placed in spherical cavity of radius a_0 within the solvent [169-173]. However spherical nature of cavity does not give realistic picture for solvated molecule.

Tomasi's Polarization continuum model (PCM) [174,175] defines cavity as set of spherical overlapping atoms with the appropriate van der Waals radii. A modification of PCM model is IPCM- Isodensity polarization continuum model that defines cavity as an isosurface of total electron density of the solute. It uses two parameters as the dipole model: dielectric constant and isodensity value. IPCM does not take into account the coupling of isosurface and electron density. It is achieved in self-consistent polarized model (SCI-PCM). SCRF methods however can't treat a microscopic solvation structure because solvation free energy is estimated in relation to dielectric constant, which are macroscopic properties of solvents.

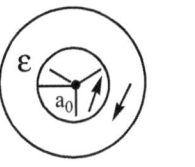 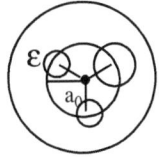

SCRF=DIPOLE   SCRF=PCM

The stabilization energy associated with the dimerization process in medium can be considered to be sum of $\Delta\Delta G_{Assoc}$ and $(S.E.)_g$ and are evaluated using expression 1.5.4.1. $\Delta\Delta G_{Assoc}$ is the difference between free energy of solvation of the dimer and the isolated monomers and calculated using expression 1.5.4.2.

$$(S.E.)_{aq} = (S.E.)_g + \Delta\Delta G_{Assoc} \qquad (1.5.4.1)$$

$$(\Delta\Delta G_{Assoc}) = (\Delta G_{sol})_{dimer} - [(\Delta G_{sol})_{monomer1} + (\Delta G_{sol})_{monomer2}] \qquad (1.5.4.2)$$

Effects of dielectric of medium on the free energy accompanying deprotonation of HAs have been evaluated in chapter 4. ΔG is free energy change for deprotonation processes using isolated conformers in water medium and calculated by relation below [91]

$$\Delta G = \Delta G^{iso} - \Delta G_s(HA) + \Delta G_s(A^-) + \Delta G_s(H^+) \qquad (1.5.4.3)$$

Where ΔG^{iso} is the free energy change of dissociation processes in gas phase and is estimated within ideal gas model. The term A⁻ denotes anion of HAs formed after deprotonation. ΔG_S is solvation free energy of isolated conformers and anions.

ΔG' refers to free energy of deprotonation when water aggregates of HAs and their anion water aggregates are placed in medium and is calculated by using relation

$$\Delta G' = \Delta G^{iso} - \Delta G_S^=(HA) + \Delta G_S^=(A^-) + \Delta G_S^=(H^+) \qquad (1.5.4.4)$$

where $\Delta G_S^=$ is solvation free energy of water aggregates of isolated conformers and anion water aggregates in solution phase. $\Delta G_S^=(H^+)$ has been used as 276.52 at MP2/Aug-cc-pVDZ as suggested by Senent et al [91].

The solution pK_a values have been calculated using expression devised by Liptak et al. [176]

$$pK_a = \{G(A^-_{gas}) - G(HA_{gas}) + \Delta G_S(A^-) - \Delta G_S(HA) - 269.0\}/1.3644 \quad (1.5.4.5)$$

The value for $G_{gas}(H^+)$ is -6.28 kcal/mol and experimental value for $(\Delta G)_s (H^+)$ is -264.61 kcal/mol. When the ΔG_{gas} reference state is converted from 1 atm to 1M, the value of $(\Delta G)_s (H^+)$ becomes -262.72 kcal/mol and $G_{gas}(H^+)$ changes to -6.3 kcal/mol.

1.5.5 PROTON AFFINITIES:

The proton affinity of base B is defined as the negative of the enthalpy change (ΔH) of the process in equation [177].

$$B + H^+ \rightarrow BH^+ \quad (1.5.5.1)$$

$$\Delta H = \Delta H_f(BH^+) - \Delta H_f(B) - \Delta H_f(H^+) \quad (1.5.5.2)$$

$$PA = -\Delta H = -[(E_{e(BH+)} + ZPVE_{(BH+)} + H_{vib(BH+)}) - (E_{e(B)} + ZPVE_{(B)} + H_{vib(B)})] + 5/2RT \quad (1.5.5.3)$$

where E_e is the electronic energy, ZPVE is scaled zero point energy, H_{vib} is the vibrational enthalpy correction scaled by 0.8945 at the MP2/6-31+G* and 0.9989 at the B3LYP/6-31+G* theoretical level and the term 5/2RT includes ΔnRT [$\Delta(PV)$ for the above reaction and translational energy of the proton].

1.5.6 GAS PHASE ACIDITIES/DEPROTONATION ENTHALPIES:

The gas-phase acidity is defined as the enthalpy change of deprotonation (ΔH^{298}) for process [178].

$$AH(g) \rightarrow A^-(g) + H^+(g) \quad (1.5.6.1)$$

The enthalpy of deprotonation, ΔH^{298}, was computed using equations below, where

$$\Delta H = \Delta E^{298} + \Delta(PV) \quad (1.5.6.2)$$

$$\Delta E^{298} = E^{298}(A^-) + 3/2RT - E^{298}(AH) \quad (1.5.6.3)$$

$E^{298}(AH)$ and $E^{298}(A^-)$ stand for the total energies of the most stable conformation of acids and their anions, including the thermal energy correction at T = 298.15 K.

1.6 REFERENCES:

1. K.E. Riley, P. Hobza, J. Phys. Chem. A 111 (2007) 8257.
2. W.M. Latimer, W.H. Rodebush, J. Am. Chem. Soc. 42 (1920) 1431.
3. S.J. Grabowski, Annu. Rep. Prog. Chem. Sect. C 102 (2006) 131.
4. T. Steiner, G.R. Desiraju, J. Chem. Soc. Chem. Commun. (1998) 891.
5. G.A. Jaffrey, W. Saenger, Hydrogen Bonding in Biological Structures, Springer, Berlin, 1991.
6. J.D. Dunitz, X-ray Analysis and the Structure of Organic Molecules, Comell University Press, Ithaca, 1979.
7. G.R. Desiraju, Crystal Engineering. The Design of Organic Solids, Elsevier, Amsterdam, 1989 (pp. 142-165).
8. G.R. Desiraju, T. Steiner, The Weak Hydrogen Bond In Structural Chemistry and Biology, Oxford University Press, 1999.
9. W.W. Cleland, P.A. Frey, J.A. Gerlt, J. Biol. Chem. 273 (1998) 25529.
10. R. Parthasarathi, V. Subramanian, N. Satyamurthy, J. Phys. Chem. A 110 (2006) 3349.
11. G.L. Miessler, D.A. Tarr, Inorganic chemistry, Pearson Education, Inc., 2004 (pp 178-182).
12. F.T.T. Huque, J.A. Platts, Org. Biomol. Chem. 1 (2003) 1419.
13. D.E. Leahy, J.J. Morris, P.J. Taylor, P.R. Wait, J. Chem. Soc. Perkin Trans. 2 (1992) 705.
14. S. Spange, A. Reuter, Langmuir 15 (1999) 141.
15. L. Senthilkumar, T.K. Ghanty, S.K. Ghosh, J. Phys. Chem. A 109 (2005) 7575.
16. W.D. Arnold, E. Oldfield, J. Am. Chem. Soc. 122 (2000) 12835.
17. K. Osvald, N.R. Kathryn, J.B. Russell, J. Phys. Chem. A 105 (2001) 6552.
18. T. Fujita, T. Nishioka, M. Nakajima, J. Med. Chem. 20 (1977) 1071.
19. P. Hobza, Z. Havlas, Theor. Chem. Acc. 108 (2002) 325.
20. H. Lossen, Liebigs Ann. Chem. 150 (1869) 314.
21. M.J. Miller, Chem. Rev. 89 (1989) 1563.
22. E.B. Paniago, S. Carvalho, Ciência e Cultura 40 (1988) 629.
23. T. Tekeste, H. Vahrenkamp, Inorg. Chim. Acta 360 (2007) 1523.
24. B.W. Clare, A. Scozzafava, C.T. Supuran, J. Med. Chem. 44 (2001) 2253.

25. H.R. Onishi, B.A. Pelak, L.S. Gerckens, L.L. Silver, F.M. Kahan, M.H. Chen, A.A. Patchett, S.M. Galloway, S.A. Hyland, M.S. Anderson, C.R.H. Raetz, Science 274 (1996) 980.
26. G.R. Schonbaum, J. Biol. Chem. 248 (1973) 502.
27. W. Cohen, B.F. Erlanger, J. Am. Chem. Soc. 82 (1960) 3928.
28. G.T. Gilbert, T. Wagner-Jauregg, G.M. Steinbery, Arch. Biochem. Biophys. 93 (1961) 469.
29. C. Indiani, E. Santoni, M. Becucci, A. Boffi, K. Fukuyama, G. Smulevich, Biochemistry 42 (2003) 14066.
30. D.A. Brown, L.P. Cuffe, N.J. Fitzpatrick, A.T. Ryan, Inorg. Chem. 43 (2004) 297.
31. Z. Amtul, A.U. Rahman, R.A. Siddiqui, M.I. Choudhary, Curr. Med. Chem. 9 (2002) 1323.
32. E.M.F. Muri, M. J. Nieto, R.D. Sindelar, J.S. Williamson, Curr. Med. Chem. 9 (2002) 1631.
33. C.M. Dooley, M. Devocelle, B. Mcloughlin, K.B. Nolan, D.J. Fitzgerald, C.T. Sharkey, Mol. Pharmacol. 63 (2003) 450.
34. D. Chen, C. Hackbarth, Z.J. Ni, C. Wu, W. Wang, R. Jain, Y. He, K. Bracken, B. Weidmann, D.V. Patel, J. Trias, R.J. White, Z. Yuan, Antimicrob. Agents. Chemother. 48 (2004) 250.
35. W.P. Steward, A.L. Thomas, Expert. Opin. Investig. Drugs 9 (2000) 2913.
36. M.G. Natchus, R.G. Bookland, B. De, N.G. Almstead, S. Pikul, M.J. Janusz, S.A. Heitmeyer, E.B. Hookfin, L.C. Hsieh, M.E. Dowty, C.R. Dietsch, V.S. Patel, S.M. Garver, F. Gu, M.E. Pokross, G.E. Mieling, T.R. Baker, D.J. Foltz, S.X. Peng, D.M. Bornes, M.J. Strojnowski, Y.O. Taiwo, J. Med. Chem. 43 (2000) 4948.
37. D.T. Puerta, S.M. Cohen, Inorg. Chem. 41 (2002) 5075.
38. D.T. Puerta, M.O. Griffin, J.A. Lewis, D. Romero-Perez, R. Garcia, F.J. Villarreal, S.M. Cohen, J. Biol. Inorg. Chem. 11 (2006) 131.
39. S. Kumaran, S.P. Gupta, J. Enzym. Inhib. Chem. 22 (2007) 23.
40. M. Barbarić, S. Uršic, V. Pilepić, B. Zorc, A.H. Brundić, A. Nagl, M. Grdiša, K. Pavelic, R. Snoeck, G. Andrei, J. Balzarini, E.D. Clercq, M. Mintas, J. Med. Chem. 48 (2005) 884.

41. E. Ganea, M. Trifan, A.C. Laslo, G. Putina, C. Cristescu, Biochem. Soc. Trans. 35 (2007) 689.
42. L.R. Scolnick, A.M. Clements, J. Liao, L. Crenshaw, Hellberg, J. May, T.R. Dean, D.W. Christianson, J. Am. Chem. Soc. 9 (1997) 850.
43. A. Scozzafava, C.T. Supuran, J. Med. Chem. 43 (2000) 3677.
44. E. Kimura, Acc. Chem. Res. 34 (2001) 171.
45. P.J. Loll, C.T. Sharkey, S.J. O'Connor, C.M. Dooley, E. O'Brien, M. Devocelle, K.B. Nolan, B.S. Zelinsky, D.J. Fitz-gerald, Mol. Pharmacol. 60 (2001) 1407.
46. S.C.C. Tam, D.H.S. Lee, E.Y. Wang, D.G. Munroe, C.Y. Lau, J. Biol. Chem. 270 (1995) 13498.
47. H. Cao, M. Jung, G. Statmatoyannopoulos, Exp. Hematol. 33 (2005) 1443.
48. W.F. Anderson, Inorganic Chemistry in Biology and Medicine, Matrell, A.E. Ed, American Chemical Society: Washington, DC, 1973 (Chapter 15).
49. F. Leoni, A. Zaliani, G. Bertolini, G. Porro, P. Pagani, P. Pozzi, G. Dona, G. Fossati, S. Sozzani, T. Azam, R. Bufler, G. Fantuzzi, I. Goncharav, S.H. Kim, B.J. Pomerantz, L.L. Reznikov, B. Siegmund, C.A. Dinarello, P. Mascagni, Proc. Natl. Aca. Sci. 99 (2002) 2995.
50. W.N. Fishbein, C.L. Strecter, J.E. Daly, J. Pharmacol. Exp. Ther. 186 (1973) 173.
51. S. Grant, C. Easley, P. Kirkpatrick, Nat. rev. Drug Disc. 6 (2007) 21.
52. H. Vanjari, R. Pande, J. Pharm. Biomed. Anal. 33 (2003) 783.
53. J.A. Klun, W.D. Guthrie, A.R. Hallauer, W.A. Russell, Crop. Sci. 10 (1970) 87.
54. V.H. Argandona, J.G. Luza, H.M. Niemeyer, L.J. Corcuera, Phytochemistry 19 (1980) 1665.
55. R. Kakkar, A. Dua, S. Zaidi, Org. Biomol. Chem. 5 (2007) 547.
56. A. Mitchell, K.S. Murray, P.J. Newman, P.E. Clark, Aust. J. Chem. 30 (1977) 2439.
57. A. Chimiak, W. Przychodzen, J. Rachon, Heteroat. Chem. 13 (2002) 169.
58. D.H.R. Barton, R.A.V. Embse, Tetrahedron 54 (1998) 12475.
59. A.D. Bond, W. Jones, J. Phys. Org. Chem. 13 (2000) 395.
60. G. Winkelmann, V.D. Helm, J.B. Neilands, Iron Transport in microbes, Plants and Animals, VCH, Weinheim, New York, 1987.

61. Y. Egawa, K. Umino, Y. Ito, T. Okuda, J. Antibiot. 24 (1971) 124.
62. S. Itoh, K. Inuzuka, T. Suzuki, J. Antibiot. 23 (1970) 542.
63. K. Shirahata, D. Deguchi, T. Hayashi, I. Matsubara, T. Suzuki, J. Antibiot. 23 (1970) 546.
64. W.A. Bridger, J.F. Henderson, Cell Adenosine Triphosphate Physiology, Wiley, Newyork, 1983.
65. A.L. Lehninger, D.L. Nelson, M.M. Cox, Principles of Biochemistry, Worth Publishers: New York, 1993.
66. A. Paytan, K. McLaughlin, Chem. Rev. 107 (2007) 563.
67. L.-B. Han, N. Choi, M. Tanaka, Organometallics 15 (1996) 3259.
68. C.Y. Li, J. Organomet. Chem. 653 (2002) 63.
69. J.E. Griffiths, A.B. Burg, J. Am. Chem. Soc. 82 (1960) 1507.
70. B. Hoge, P. Garcia, H. Willner, H. Oberhammer, Chem. Eur. J. 12 (2006) 3567.
71. D.E.C. Corbridge, Phosphorus: An Outline of its Chemistry, Biochemistry and Uses, Elsevier, Amsterdam, 1995.
72. J. Chatt, B.T. Heaton, J. Chem. Soc. A (1968) 2745.
73. B. Hoge, C. Thösen, T. Herrmann, P. Panne, I. Patenburg, J. Fluorine Chem. 125 (2004) 831.
74. B. Hoge, W. Wiebe, S. Hettl, S. Neufeind, C. Thösen, J. Organomet. Chem. 690 (2005) 2382.
75. N.V. Dubrovina, A. Borner, Angew. Chem. 116 (2004) 6007.
76. L.N. Heydorn, P.C. Burgers, P.J.A. Ruttink, J.K. Terlouw, Int. J. Mass Spectrom. 228 (2003) 759.
77. M. Saldyka, Z. Mielke, J. Phys. Chem. A 106 (2002) 3714.
78. E. Lipczyńska-Kochany, Chem. Rev. 91 (1991) 477.
79. D.A. Brown, R.A. Coogan, N.J. Fitzpatrick, W.K. Glass, D.E. Abukshima, M. Algrén, K. Smolander, T.T. Pakkanen, T.A. Pakkanen, M. Peräkylä, J. Chem. Soc. Perkin Trans. 2 (1996) 2673.
80. D. Hadži, D. Provoršek, Spectrochim. Acta 10 (1957) 38.
81. O. Exner, K, M. Horák, Collect. Czech. Chem. Commun. 24 (1959) 968.
82. B.H. Bracher, R.W.H. Small, Acta Crystallogr. B 26 (1970) 1705.
83. R.R. Mooherla, D.R. Powell, C.L. Barnes, D. van der Helm, Acta Crystallogr. C39 (1983) 868.

84. B. García, S. Ibeas, J.M. Leal, F. Secco, M. Venturini, M.L. Senent, A. Niño, A. Muñoz, Inorg. Chem. 44 (2005) 2908.
85. M. Saldyka, Z. Mielke, Chem. Phys. Lett. 371 (2003) 713.
86. M. Saldyka, Z. Mielke, Vib. Spectrosc. 45 (2007) 46.
87. B. García, S. Ibeas, A. Muñoz, J.M. Leal, C. Ghinami, F. Secco, M. Venturini, Inorg. Chem. 42 (2003) 5434.
88. D.-H. Wu, J.J. Ho, J. Phys. Chem. A 102 (1998) 3582.
89. R. Kakkar, R. Grover, P. Chadha, Org. Biomol. Chem. 1 (2003) 2200.
90. S.-J. Yen, C.-Y. Lin, J.-J. Ho, J. Phys. Chem. A 104 (2000) 11771.
91. N. Mora-Diez, M.L. Senent, B. García, Chem. Phys. 324 (2006) 350.
92. C. Muñoz-Caro, A. Niño, M.L. Senent, J.M. Leal, S. Ibeas, J. Org. Chem. 65 (2000) 405.
93. M. Remko, J. Šefčíková, J. Mol. Struct. (Theochem) 528 (2000) 287.
94. K.K. Ghosh, P. Tamrakar, S.K. Rajput, J. Org. Chem. 64 (1999) 3053.
95. B. García, S. Ibeas, F.J. Hoyuelos, J.M. Leal, F. Secco, M. Venturini, J. Org. Chem. 66 (2001) 7986.
96. S. Böhm, O. Exner, Org. Biomol. Chem. 1 (2003) 1176.
97. B. Garcia, F. Secco, S. Ibeas, A. Muñoz, F.J. Hoyuelos, J.M. Leal, M.L. Senent, M. Venturini, J. Org. Chem. 72 (2007) 7832.
98. Exner, B. Kakác, Collect. Czech. Chem. Commun. 28 (1963) 1656.
99. F.G. Bordwell, H.E. Fried, D.L. Hughes, T.-Y. Lynch, A.V. Satish, Y.E. Whang, J. Org. Chem. 55 (1990) 3330.
100. O.N. Ventura, J.B. Rama, L. Turi, J.J. Dannenberg, J. Am. Chem. Soc. 115 (1993) 5754.
101. A. Bagno, C. Comuzzi, G. Scorrano, J. Am. Chem. Soc. 116 (1994) 916.
102. M. Remko, J. Phys. Chem. A 106 (2002) 5005.
103. K. Leung, S.B. Rempe, J. Am. Chem. Soc. 126 (2004) 344.
104. J.E. Yazal, Y.-P. Pang, J. Phys. Chem. B 104 (2000) 6499.
105. R. Codd, Coord. Chem. Rev. 252 (2008) 1387.
106. C.P. Brink, A.L. Crumbliss, Inorg. Chem. 23 (1984) 4708.
107. B. Monzyk, A.L. Crumbliss, J. Am. Chem. Soc. 101 (1979) 6203.
108. M. Gasper, R. Grazina, A. Bodor, E. Farkas, M.A. Santos, J. Chem. Soc. Dalton Trans. 5 (1999) 799.

109. E. Farkas, K. Megyeri, I. Somsak, L. Kovacs, J. Inorg. Biochem. 70 (1998) 41.
110. B. Kurzak, H. Kozlowski, E. Farkas, Coord. Chem. Rev. 114 (1992) 169.
111. E. Farkas, D.A. Brown, R. Cittaro, W.K. Glass, J. Chem. Soc. Dalton Trans. (1993) 3903.
112. A.J. Stemmler, J.M. Kampf, M.L. Kirk, V.L. Pecoraro, J. Am. Chem. Soc. 117 (1995) 6368.
113. D.A. Brown, A.L. Roche, T.A. Pakkanen, T.T. Pakkanen, K. Smolander, J. Chem. Soc. Chem. Commun. (1982) 676.
114. D. Bátka, E. Farkas, J. Inorg. Biochem. 100 (2006) 27.
115. S. Rupprecht, K. Langemann, T. Lügger, J.M. McCormick, K.N. Raymond, Inorg. Chim. Acta 243 (1996) 79.
116. E. Jabri, M.B. Carr, R.P. Hausinger, P.A. Karplus, Science 268 (1995) 998.
117. S.L. Roderick, B.W. Matthews, Biochemistry 32 (1993) 3907.
118. Z.F. Kanyo, L.R. Scolnick, D.E. Ash, D.W. Christianson, Nature 383 (1996) 554.
119. M.A. Pearson, L.O. Michel, R.P. Hausinger, P.A. Karplus, Biochemistry 36 (1997) 8164.
120. R. Kakkar, R. Grover, P. Gahlot, J. Mol. Struct. (Theochem) 767 (2006) 175.
121. R. Kakkar, R. Sharma, A. Dua, Spectrochim. Acta Part A 62 (2005) 819.
122. B. Lou, K. Yang, Mini. Rev. Med. Chem. 3 (2003) 609.
123. C.Y. Wang, L.H. Lee, Antimicrob. Agents Chemother. 11 (1977) 753.
124. J.J. –W. Duan, L. Chen, Z. Lu, C.-B. Xue, R.-Q. Liu, M.B. Covington, M. Qian, Z.R. Wasserman, K. Vaddi, D.D. Christ, J.M. Trzaskos, R.C. Newton, C.P. Decicco, Biorg. Med. Chem. Lett. 18 (2008) 241.
125. S.-K. Anandan, J.S. Ward, R.D. Brokx, T. Denny, M.R. Bray, D.V. Patel, X.-Y. Xiao, Biorg. Med. Chem. Lett. 17 (2007) 5995.
126. K.P. Holland, H.L. Elford, V. Bracchi, C.G. Annis, S.M. Schuster, D. Chakrabarti, Antimicrob. Agents Chemother. 42 (1998) 2456.
127. H.S. Rho, H.S. Baek, S.M. Ahn, J.W. Yoo, D.H. Kim, H.G. Kim, Bull. Korean Chem. Soc. 30 (2009) 475.
128. M. Saldyka, Z. Mielke, Chem. Phys. 300 (2004) 209.
129. M. Saldyka, Z. Mielke, J. Mol. Struct. (Theochem) 692 (2004) 163.
130. M. Saldyka, Z. Mielke, J. Mol. Struct. (Theochem) 708 (2004) 183.

131. K.A. Joshi, S.P. Gejji, Chem. Phys. Lett. 415 (2005) 110.
132. B. García, F. Secco, S. Ibeas, A. Muñoz, F.J. Hoyuelos, J.M. Leal, M.L. Senent, M. Venturini, J. Org. Chem. 72 (2007) 7832.
133. B. García, S. Ibeas, J.M. Leal, F. Secco, M. Venturini, M.L. Senent, A. Niño, C. Muñoz, Inorg. Chem. 44 (2005) 2908.
134. A. Kaczor, L.M. Proniewicz, Spectochim. Acta A 62 (2005) 1023.
135. I. Ciofini, Magn. Reson. Chem. 42 (2004) S48.
136. E. Schrodinger, Phys. Rev. 28 (1926) 1049.
137. M. Born, R. Oppenheimer, Ann. Phsik 84 (1927) 457.
138. R.G. Parr, W. Yang, Density Functional Theory of Atoms and Molecules, Oxford University Press, 1989.
139. M. Mueller, Fundamentals of quantum Chemistry: Molecular Spectroscopy and Modern Electronic Structure Computations, Kluwer Academic/Plenum Publishers, Newyork, 2001.
140. E. Clementi, Modern Techniques in Computational Chemistry, ESCOM: Leiden 1990.
141. P.M.W. Gill Adv. Quantum Chem. 25 (1994) 141.
142. G. Frenking, I. Antes, M. Boehme, S. Dapprich, A.W. Ehlers, V. Jonas, A. Neuhaus, M. Otto, R. Stegmann, A. Veldkamp, S.F. Vyboishchikov, Rev. Comput. Chem. 8 (1996) 63.
143. V. Bonifacic, S. Huzinaga, J. Chem. Phys. 60 (1974) 2779.
144. Y. Sakai, E. Miyoshi, M. Klobukowski, S. Huzinaga, J. Comput. Chem. 8 (1987) 226.
145. Y. Sakai, E. Miyoshi, M. Klobukowski, S. Huzinaga, J. Comput. Chem. 8 (1987) 256.
146. S. Huzinaga, L. Seijo, Z. Barandiaran, M. Klobukowski, J. Chem. Phy. 86 (1987) 2132.
147. Z. Barandiaran, L. Seijo, S. Huzinaga, J. Chem. Phy. 93 (1990) 5843.
148. I.N. Levine, Quantum Chemistry, Engelwood Cliffs, New jersey, 1991.
149. G.W. Trucks, M.J. Frisch, Analytical Second Derivatives of Excited States: Configuration Interaction Singles Theory and Application , 1998.
150. C. Moller, M.S. Plesset, Phys. Rev. 46 (1934) 618.
151. R. Krishnan, M.J. Frisch, J.A. Pople, J. Chem. Phys. 72 (1980) 4244.
152. P. Hohenberg, B. Kohn, Phys. Rev. B 136 (1964) B864.

153. W. Kohn, L.J. Sham, Phys. Rev. A 140 (1965), A1133.
154. A.D. Becke, J. Chem. Phys. 98 (1993) 5648.
155. C. Lee, W. Yang, R.G. Parr, Phys. Rev. B 37 (1988) 785.
156. S.H. Vosko, L. Wilk, M. Nusair, Can. J. Phys. 58 (1980) 1200.
157. P.J. Stephens, F.J. Delwin, C.F. Chabalowski, M.J. Frisch, J. Phys. Chem. 98 (1994) 11623.
158. J.P. Perdew, Y. Wang, Phys. Rev. B 45 (1992) 13244.
159. M.J. Frisch, G.W. Trucks, H.B. Schlegel, G.E. Scuseria, M.A. Robb, J.R. Cheeseman, V.G. Zakrzewski, J.A. Montgomery, Jr. R.E. Stratmann, J.C. Burant, S. Dapprich, J.M. Millam, A.D. Daniels, K.N. Kudin, M.C. Strain, O. Farkas, J. Tomasi, V. Barone, M. Cossi, R. Cammi, B. Mennucci, C. Pomelli, C. Adamo, S. Clifford, J. Ochterski, G.A. Petersson, P.Y. Ayala, Q. Cui, K. Morokuma, N. Rega, P. Salvador, J.J. Dannenberg, D.K. Malick, A.D. Rabuck, K. Raghavachari, J.B. Foresman, J. Cioslowski, J.V. Oritz, A.G. Baboul, B.B. Stefanov, G. Liu, A. Liashenko, P. Piskorz, I. Komaromi, R. Gomperts, R.L. Martin, D.J. Fox, T. Keith, M.A. Al-Laham, C.Y. Peng, A. Nanayakkara, M. Challacombe, P.M.W. Gill, B. Johnson, W. Chen, M.W. Wong, J.L. Andres, C. Gonzalez, M. Head-Gordon, E.S. Replogle, J.A. Pople, Gaussian 98, Revision-A.11.2, Gaussian Inc. Pittsburgh, PA, 2001.
160. A.P. Scott, L. Radom, J. Phys. Chem. 100 (1996) 16502.
161. P. Sinha, S.E. Boesch, C. Gu, R.A. Wheeler, A.K. Wilson, J. Phys. Chem. A 108 (2004) 9213.
162. C. Aleman, J. Phys. Chem. A 105 (2001) 6717.
163. F. Jensen, Introduction to Computational Chemistry, Wiley, Newyork, 2007.
164. J.P. Foster, F. Weinhold, J. Am. Chem. Soc. 102 (1980) 7211.
165. A.E. Reed, F. Weinhold, L.A. Curtiss, Chem. Rev. 88 (1988) 899.
166. A.E. Reed, P.V.R. Schleyer, J. Am. Chem. Soc. 112 (1990) 1434.
167. L. Senthilkumar, P. Kolandaivel, J. Mol. Struct. 791 (2006) 149.
168. P. Hobza, Z. Havlas, Chem. Rev. 100 (2000) 4253.
169. L. Onsanger, J. Am. Chem. Soc. 58 (1936) 1486.
170. J.G. Kirkwood, J. Chem. Phys. 2 (1934) 351.
171. M.W. Wong, M.J. Frisch, K.B. Wiberg, J. Am. Chem. Soc. 113 (1991) 4776.
172. M.W. Wong, M.J. Frisch, K.B. Wiberg, J. Am. Chem. Soc. 114 (1992) 523.
173. M.W. Wong, M.J. Frisch, K.B. Wiberg, J. Am. Chem. Soc. 114 (1992) 1645.

174. S. Mierus, E. Scocco, J. Tomasi, J. Chem. Phys. 55 (1981) 117.
175. J. Tomasi, M. Persico, Chem. Rev. 94 (1994) 2027.
176. M.D. Liptak, K.C. Gross, P.G. Seybold, S. Feldgus, G.C. Shields, J. Am. Chem. Soc. 124 (2002) 6421.
177. P.V. Bharatam, P. Iqbal, A. Malde, R. Tiwari, J. Phys. Chem. A 108 (2004) 10509.
178. M. Remko, K.R. Liedl, B.M. Rode, J. Chem. Soc. Perkin Trans. 2 (1996) 1743.

INTRA- AND INTERMOLECULAR HYDROGEN BONDING IN FORMOHYDROXAMIC ACID

2.1 INTRODUCTION:

Hydrogen bonding (H-bonding) is regarded as an important interaction in determining the shapes, properties and functions of many biomolecules [1-4]. Prediction of the properties of intra and intermolecular H-bonding is one of the important problems in science [5]. Intramolecular H-bond exists in many biomolecules such as hormones, coenzymes, proteins etc. OH...O, NH...O, NH...N are most common of the several intramolecular H-bonds observed in biological systems [6]. Intramolecular H-bonding, though weaker than its intermolecular counterpart, can exert a pronounced effect upon the properties such as charge distribution within molecule, relative stability, reactivity, thermal and photochemical equilibria of configurational isomers etc [7]. Intermolecular H-bonding in a chemical system changes the number, mass, shape and electronic structure of the participants while intramolecular H-bonds affect only the electronic structure of the molecule and the effect on physical properties is negligible. The intermolecular H-bond interactions between the molecules of same type decide the crystal structural properties [8]. The intermolecular H-bonding with water is particularly important because of water being the most commonly used solvent and its high abundance in biological systems. Water has ability to form variety of hydrogen bonds due to its double donor and double acceptor capacity. Many structural features that are necessary for the biological functions of the biomolecules depend on the interactions with the surrounding water and among the molecules themselves.

The hydroxamic acids (HAs) consist of a class of compounds having numerous applications such as tumor inhibitors, antimicrobial agents, antituberculosis agents, as key pharmacophore in many chemotherapeutic agents, pesticides and as promoter of plant growth etc [9-15]. They possess three atoms with lone pair of electrons, having ability not only to form intermolecular H-bonds with the conventional H-bond donors like O-H and N-H bonds but also intramolecular H-bonds. When in aqueous solution, intermolecular H-bond interactions between water and HAs compete with intramolecular H-bond interactions, then the tautomerism in solution becomes solvent dependent [16]. Theoretical solvent models are unable to include specific interactions like H-bond. Quantum mechanical calculations can provide information about the hydrogen bonded complexes and the strength of binding interactions. In order to understand the intra and intermolecular H-bonding in HA molecules of biological interest, the aggregation of the formohydroxamic acid (FHA) molecule with water and dimerization possibilities among various isomeric forms of FHA are studied. The understanding of nature of forces governing the self aggregation and aggregation with water molecules shall be helpful for a better physical-chemical description of such systems.

2.2 RESULTS AND DISCUSSION:
2.2.1 CONFORMATIONAL STABILITY AND ISOMERISM:

FHA has undergone structural analysis by a number of research groups and it has been concluded that it exists in several isomeric forms [17,18]. Full geometry optimization of the molecule resulted in seven minima on the potential energy surface (Figure 2.1). Out of these, two correspond to cis and trans arrangement of C(=O)NO unit of the keto tautomer (**1E**, **1Z** respectively), four correspond to enol tautomers namely **2E**, **2Z**, **2E2**, **2Z2** and one zwitter ion conformation (**3**) as shown in Figure 2.1. The geometrical parameters of all these seven conformations at B3LYP/Aug-cc-pVDZ [**L2**] and MP2/Aug-cc-pVDZ [**L4**] levels are listed in Table 2.1. The various enolic isomeric forms as well as zwitter ion form is planar due to presence of C=N while keto forms are non planar since both the C-N bond and C-O bonds in them have partial double bond character.

The relative energy differences among these forms are given in the Table 2.2. The order of stability of all conformations of FHA in gas phase is found to be **1Z** > **1E** > **2Z** > **2E** > **2Z2** > **2E2** > (**3**) at B3LYP/Aug-cc-pVDZ theoretical level and is the same as reported by D. M. Stinchcomb et al. and R. Kakkar et al. [17,18]. The order of gas phase stability found using ab initio methods employing [**L4**] theoretical level is **1Z** > **2Z** > **1E** > **2E** > **2Z2** > **2E2** > (**3**) (Table 2.2) and it agrees well with Wu and Ho reports [19]. The higher stability of **1Z** form has been assigned by Saldyka et al. to the presence of intramolecular H-bonding, evidence for which has been obtained from IR studies of the argon matrix of the FHA [9]. In a separate IR study in argon matrix by them, the presence of **1E** isomer has been reported with the population ratio of **1E** to **1Z** isomer to be 0.035±0.009 in the gas phase at room temperature [20].

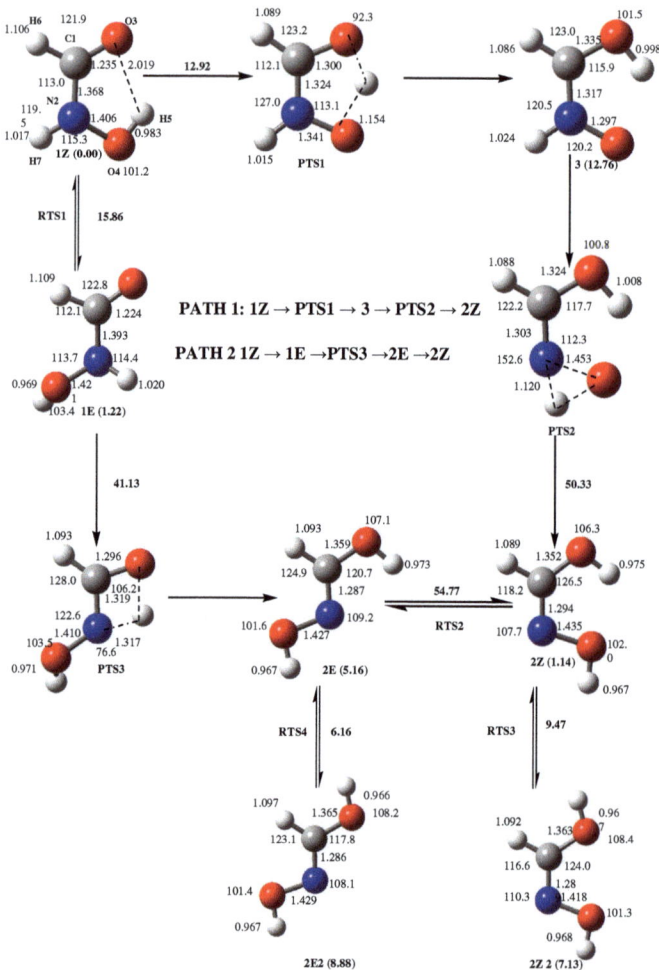

Figure 2.1: Interrelationship among isomeric forms of formohydroxamic acid.

Table 2.1: Geometrical parameters of isomeric forms of FHA at B3LYP/Aug-cc-pVDZ [**L2**] and (MP2/Aug-cc-pVDZ) [**L4**] theoretical levels. All the bond distances are in angstrom (Å), angles and dihedrals are in degrees (°).

Parameters	1Z	1E	Parameters	2Z	2E	2Z2	2E2	Parameters	3
N2-C1	1.357	1.381	N2-C1	1.280	1.274	1.275	1.272	N2-C1	1.303
	(1.368)	(1.393)		(1.294)	(1.287)	(1.289)	(1.286)		(1.317)
O3-C1	1.226	1.214	O3-C1	1.345	1.351	1.357	1.358	O3-C1	1.326
	(1.235)	(1.224)		(1.352)	(1.359)	(1.363)	(1.365)		(1.335)
O4-N2	1.397	1.410	O4-N2	1.426	1.415	1.407	1.415	O4-N2	1.307
	(1.406)	(1.421)		(1.435)	(1.427)	(1.418)	(1.429)		(1.297)
H5-O4	0.981	0.968	H5-O3	0.972	0.971	0.965	0.964	H5-O3	0.992
	(0.983)	(0.969)		(0.975)	(0.973)	(0.967)	(0.966)		(0.998)
H6-C1	1.106	1.109	H6-C1	1.088	1.092	1.091	1.096	H6-C1	1.084
	(1.106)	(1.109)		(1.089)	(1.093)	(1.092)	(1.097)		(1.086)
H7-N2	1.013	1.018	H7-O4	0.964	0.965	0.965	0.965	H7-N2	1.021
	(1.017)	(1.020)		(0.967)	(0.967)	(0.968)	(0.967)		(1.024)
O3-C1-N2	121.9	122.9	O3-C1-N2	126.8	121.0	124.6	118.4	O3-C1-N2	117.0
	(121.9)	(122.8)		(126.5)	(120.7)	(124.0)	(117.8)		(115.9)
O4-N2-C1	116.8	115.8	O4-N2-C1	108.8	110.6	111.9	109.7	O4-N2-C1	120.6
	(115.3)	(113.7)		(107.7)	(109.2)	(110.3)	(108.1)		(120.5)
H5-O4-N2	101.8	104.7	H5-O3-C1	107.5	108.1	109.5	109.2	H5-O3-C1	103.0

Contd......

Parameters	1Z	1E	Parameters	2Z	2E	2Z2	2E2	Parameters	3
	(101.2)	(103.4)		(106.3)	(107.1)	(108.4)	(108.2)		(101.5)
H6-C1-N2	113.2	112.3	H6-C1-N2	118.3	125.0	116.7	123.2	H6-C1-N2	123.0
	(113.0)	(112.1)		(118.2)	(124.9)	(116.6)	(123.1)		(123.0)
H7-N2-C1	122.2	116.2	H7-O4-N2	102.9	102.6	102.3	102.4	H7-N2-C1	120.4
	(119.5)	(114.4)		(102.0)	(101.6)	(101.3)	(101.4)		(120.0)
O4-N2-C1-O3	-10.0	159.7	O4-N2-C1-O3	0.0	180.0	-0.0	180.0	O4-N2-C1-O3	0.1
	(-12.0)	(157.7)		(0.0)	(180.0)	(0.0)	(180.0)		(0.0)
H5-O4-N2-C1	4.3	117.5	H5-O3-C1-N2	0.0	0.0	180.0	180.0	H5-O3-C1-N2	-0.2
	(4.5)	(119.0)		(0.0)	(0.0)	(180.0)	(180.0)		(-0.0)
H6-C1-N2-O4	172.6	-23.4	H6-C1-N2-O4	180.0	-0.0	180.0	-0.0	H6-C1-N2-O4	179.9
	(171.4)	(-25.9)		(180.0)	(-0.0)	(180.0)	(0.0)		(180.0)
H7-N2-C1-O3	-154.0	25.7	H7-O4-N2-C1	179.9	180.0	180.0	180.0	H7-N2-C1-O3	179.9
	(-148.1)	(29.7)		(180.0)	(180.0)	(180.0)	(180.0)		(180.0)

Table 2.2: Relative energies (kcal/mol) of tautomeric forms of FHA at B3LYP/Aug-cc-pVDZ **[L2]** and MP2/Aug-cc-pVDZ **[L4]** theoretical levels.

Species	L2	L4
1Z	0.00	0.00
1E	1.47	1.22
2Z	2.68	1.14
2E	5.83	5.16
2Z2	8.11	7.13
2E2	9.58	8.88
3	11.99	12.76

Keto ↔ enol tautomerism in FHA has been studied by a number of research groups and it has been recognized that H-bonding interactions play an important role in the stability of tautomers and proton transfer in case of hydroxamic acids. Figure 2.1 show various pathways for the interconversion of different isomeric forms of FHA. **1Z**, **1E**; **2Z**, **2E** and **2Z2**, **2E2** pairs are rotamers of each other about C-N bond. Though **1Z** and **1E** tautomeric forms differ by only 1.22 kcal/mol at **[L4]** theoretical level, the **1Z → 1E** transformation through C-N rotational transition state **(RTS1)** requires energy barrier of 15.86 kcal/mol and the similar energy barrier for **2Z → 2E (RTS2)** interconversion is 54.77 kcal/mol at **[L4]** theoretical level. The significant difference in C-N rotation barriers of keto and iminol forms i.e. approximately 40 kcal/mol, arises due to the difference in π character of C-N bond in the two cases. **2Z2** and **2E2** are C-O rotamers of **2Z** and **2E** forms respectively. The **2Z → 2Z2 (RTS3)** and **2E → 2E2 (RTS4)** require barriers of 9.47 and 6.16 kcal/mol respectively at **[L4]** theoretical level. The gas phase transformation of keto forms to iminol forms can be rationalized through two probable pathways. As can be seen from Figure 2.1, the paths involve the transformation **Path 1: 1Z → TS1 → 3 → TS2 → 2Z** and **Path 2: 1Z → 1E → TS3 → 2E → 2Z**. The barriers associated with different rotational transition states (RTS) and proton transfer transition states (PTS) leading to tautomerization are listed in Table 2.3 and it shows path 2 is favorable over path 1 in the gas phase.

Table 2.3: Activation barriers (kcal/mol) for interconversion among isomeric forms of FHA at MP2/Aug-cc-pVDZ **[L4]** theoretical level.

Species	Transition State	Activation barriers
RTS1	1Z → 1E	15.86
RTS2	2Z → 2E	54.77
RTS3	2Z → 2Z2	9.47
RTS4	2E → 2E2	6.16
PTS1	1Z → 3	12.92
PTS2	3 → 2Z	50.33
PTS3	1E → 2E	41.13

2.2.2 INTRAMOLECULAR HYDROGEN BONDING:

Several intramolecular H-bonding interactions are probable in all the isomeric forms due to the presence of multiple H-bond donor groups A-H (A=N, O) and H-bond acceptor sites B (B=O, N). All those atoms among which intramolecular H-bonding could be expected to occur were analyzed for intra atomic separations and the angles that the bridging hydrogen makes. The atomic separations between different H-bond donors and H-bond acceptors have been scrutinized and the values that are less than the sum of van der Waals radii of interacting atoms are recorded in Table 2.4 along with the A-H...B angle. The sum of van der Waals radii is also included in the Table 2.4 for comparison. The shorter the distance from the sum of van der Waals radii and closer the angle to 180°; stronger is the H-bond interaction. The difference in the non covalently bonded distances from the sum of the van der Waals radii (Δr) are also reported in the same table. The results suggest that though in many cases, H...B distance is far less than the sum of their van der Waals radii but H-bonding interactions are still expected to be negligible because of the acute angle, the bridging hydrogen makes.

The strong intramolecular H-bonding interactions in **1Z** tautomer of FHA are indicated by large value of Δr (0.581Å) for H5...O3 and O4-H5...O3 angle of 119.3°. A new bond path connecting H5...O3 and a ring critical point between carbonyl oxygen and hydroxyl hydrogen is located using Atoms in Molecules (AIM) analysis. Four other nonbonded distances in the **1Z** conformation as reported in the table are also much less than the sum of van der Waals radii for the respective atoms but the H-bond interactions in these are expected to be negligible because of the acute angle. With acute angle, the repulsive interactions increase and dominate over the attractive forces. Several nonbonded distances in **1E** are also much less than the sum of van der Waals radii of the respective atoms. But no stabilization is expected out of these as the A-H...B angle is much less than 180° required for strong H-bond interactions. **1Z** has H5...O3 distance of 2.019 Å and angle O4-H5...O3 to be 119.3° suggesting relatively stronger H-bonding interactions in **1Z** than **2Z** where H5...O4 and O3-H5...O4 values are 2.007 Å and 112.9° respectively.

Table 2.4: Important intramolecular atomic separations in angstrom (Å), angles at bridging hydrogen in degrees (°) in the various isomers of FHA at MP2/Aug-cc-pVDZ [L4] theoretical level.

Species	Atomic Separations (r)		Δr*	Angles at bridging hydrogen	
1Z	H7…O4	2.007	0.593	O2-H7…O4	40.9
	H5…N2	1.865	0.875	O4-H5…N2	47.7
	H5…O3	2.019	0.581	O4-H5…O3	119.3
	H6…N2	2.068	0.672	C1-H6…N2	37.5
	H6…O3	2.077	0.523	C1-H6…O3	29.2
1E	H5…N2	1.897	0.843	O4-H5…N2	47.8
	H6…O4	2.453	0.147	C1-H6…O4	71.8
	H6…N2	2.081	0.659	C1-H6…N2	38.3
	H7…O3	2.531	0.069	N2-H7…O3	65.3
	H7…O4	2.016	0.584	N2-H7…O4	41.4
	H6…O3	2.069	0.531	C1-H6…O3	29.0
2Z	H5…O4	2.007	0.593	O3-H5…O4	112.9
	H6…N2	2.047	0.693	C1-H6…N2	33.9
	H6…O3	2.067	0.533	C1-H6…O3	36.2
	H7…N2	1.890	0.850	O4-H7…N2	48.0
2E	H6…O4	2.424	0.176	C1-H6…O4	65.8
	H5…N2	2.309	0.431	O3-H5…N2	77.3
	H6…O3	2.066	0.534	C1-H6…O3	36.8
	H7…N2	1.878	0.862	O4-H7…N2	48.1
2Z2	H6…N2	2.028	0.712	C1-H6…N2	34.6
	H7…N2	1.868	0.872	O4-H7…N2	48.1
	H6…O3	2.028	0.572	C1-H6…O3	34.0
2E2	H6…N2	2.098	0.642	C1-H6…N2	30.9
	H7…N2	1.877	0.863	O4-H7…N2	48.3
	H6…O4	2.372	0.228	C1-H6…O4	67.5
	H6…O3	2.126	0.474	C1-H6…O3	34.1

(Sum of van der Waals radii, $r_{VW} = r_O + r_H = 2.6$ Å and $r_H + r_N = 2.74$ Å). $\Delta r^* = r_{VW} - r$

2.2.3 INTERMOLECULAR HYDROGEN BONDING BETWEEN FORMOHYDROXAMIC ACID AND WATER:

Though the dominance of **1Z** conformation for FHA in the gas phase is well recognized but the equilibrium between the keto tautomeric **Z-E** forms of acetohydroxamic acid has been suggested in the solution phase by Lindberg et al. and Brown et al [21,22]. Garcia et al. concluded on the basis of NMR spectra of benzohydroxamic acid (BHA) in acetone that BHA is present mainly as the **E** conformation, the **Z**-conformation being only 25% of the total population [23]. This conclusion suggests that intermolecular interactions become important in the solution phase in deciding tautomeric equilibrium of HAs. In order to understand the nature of these intermolecular H-bonding interactions, aggregation of all isomers with one water molecule has been analyzed for all the conformations. The relative energy difference between **2Z** and **2Z2** is 5.99 kcal/mol and between **2E** and **2E2** is 3.72 kcal/mol at **[L4]** theoretical level. As **2Z2** and **2E2** conformers are much higher in energy in comparison to their respective **2Z** and **2E** conformers, only the latter conformers are selected for studying the aggregation with water molecule. The aggregates of single water molecule with each of the four conformations (**1Z, 1E, 2Z, 2E**) have been optimized for different relative positions of water at **[L2]** and **[L4]** theoretical levels with respect to the HA and fifteen aggregates that are observed to be minima on the potential energy surface are shown in Figure 2.2. The geometrical parameters of these adducts at **[L4]** level are listed in Tables 2.5-2.8. The variations in different bond lengths and angles of the isomeric forms on water adduct formation are shown in Tables 2.9-2.10. The bond of FHA which acts as H-bond donor towards water shows elongation for the conventional O-H and N-H bond donors but for C-H being unconventional H-bond donor there is either contraction of C-H bond length or no change upon aggregation with water. In all aggregates where carbonyl acts as H-bond acceptor, there is elongation of carbonyl bond accompanied by contraction of adjacent C-N bond.

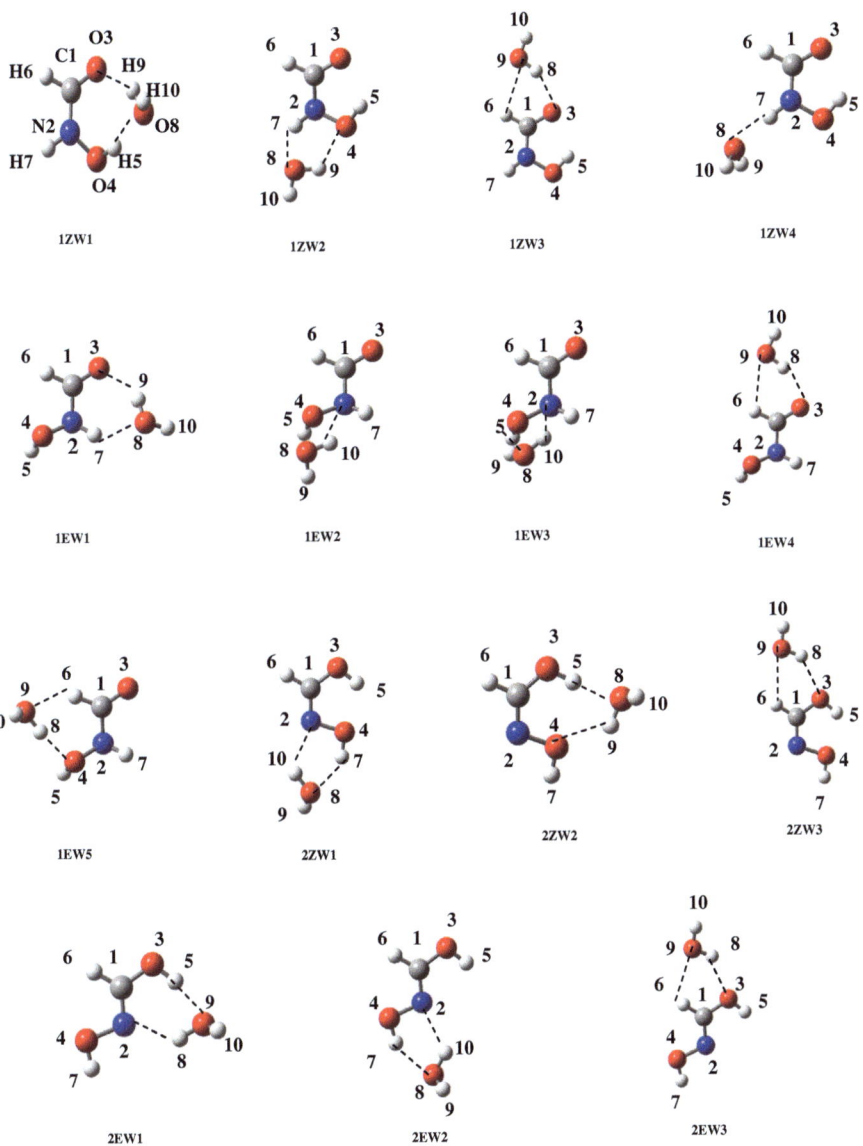

Figure 2.2: The optimized structures of the hydrogen bonded complexes of FHA-H$_2$O

Table 2.5: Important geometrical parameters of FHA-H$_2$O adducts at MP2/Aug-cc-pVDZ **[L4]** theoretical level. All bond distances are in angstrom (Å), angles and dihedrals are in degrees (°).

Parameters	1ZW1	Parameters	1ZW2	Parameters	1ZW3	Parameters	1ZW4
N2-C1	1.355	N2-C1	1.366	N2-C1	1.355	N2-C1	1.354
O3-C1	1.239	O3-C1	1.237	O3-C1	1.243	O3-C1	1.242
O4-N2	1.390	O4-N2	1.413	O4-N2	1.401	O4-N2	1.404
H5-O4	0.990	H5-O4	0.985	H5-O4	0.982	H5-O4	0.984
H6-C1	1.109	H6-C1	1.105	H6-C1	1.104	H6-C1	1.106
H7-N2	1.013	H7-N2	1.023	H7-N2	1.015	H7-N2	1.024
O8-H5	1.743	O8-H7	2.046	H8-O3	1.969	O8-H7	1.895
H9-O8	0.981	H9-O8	0.971	O8-H7	0.974	H9-O8	0.967
H10-O8	0.966	H10-O8	0.965	H10-O9	0.964	H10-O8	0.967
O3-C1-N2	125.9	O3-C1-N2	122.1	O3-C1-N2	121.4	O3-C1-N2	122.1
O4-N2-C1	122.7	O4-N2-C1	114.1	O4-N2-C1	116.8	O4-N2-C1	115.8
H5-O4-N2	104.6	H5-O4-N2	101.1	H5-O4-N2	101.3	H5-O4-N2	100.6
H6-C1-N2	110.8	H6-C1-N2	113.1	H6-C1-N2	114.2	H6-C1-N2	113.1
H7-N2-C1	121.3	H7-N2-C1	120.3	H7-N2-C1	121.7	H7-N2-C1	124.1
O8-H5-O4	159.8	O8-H7-N2	130.9	H8-O3-C1	97.4	O8-H7-N2	175.6
H9-O8-H5	87.1	H9-O8-H7	83.8	O9-H8-O3	147.0	H9-O8-H7	113.9
H10-O8-H9	105.2	H10-O8-H7	141.0	H10-O9-H8	104.9	H10-O8-H7	116.6
O4-N2-C1-O3	-13.6	O4-N2-C1-O3	11.7	O4-N2-C1-O3	-10.3	O4-N2-C1-O3	-8.9
H5-O4-N2-C1	59.1	H5-O4-N2-C1	-4.0	H5-O4-N2-C1	3.6	H5-O4-N2-C1	4.3
H6-C1-N2-O4	169.9	H6-C1-N2-O4	188.6	H6-C1-N2-O4	172.5	H6-C1-N2-O4	173.6
H7-N2-C1-O3	188.0	H7-N2-C1-O3	144.6	H7-N2-C1-O3	206.1	H7-N2-C1-O3	-155.5
O8-H5-O4-N2	-65.5	O8-H7-N2-C1	234.8	H8-O3-C1-N2	-179.6	O8-H7-N2-C1	149.9
H9-O8-H5-O3	-6.6	H9-O8-H7-N2	-8.0	O9-H8-O3-C1	-2.1	H9-O8-H7-N2	-45.4
H10-O8-H9-O3	255.1	H10-O8-H7-N2	245.7	H10-O9-H8-O3	178.2	H10-O8-H7-N2	76.3

Table 2.6: Important geometrical parameters of FHA- H$_2$O adducts at MP2/Aug-cc-pVDZ[L4] theoretical level. All bond distances are in angstrom (Å), angles and dihedrals are in degrees (°).

Parameters	1EW1	Parameters	1EW2	1EW3	Parameters	1EW4	Parameters	1EW5
N2-C1	1.372	N2-C1	1.394	1.392	N2-C1	1.381	N2-C1	1.400
O3-C1	1.235	O3-C1	1.225	1.225	O3-C1	1.231	O3-C1	1.223
O4-N2	1.414	O4-N2	1.418	1.417	O4-N2	1.416	O4-N2	1.426
H5-O4	0.960	H5-O4	0.978	0.979	H5-O4	0.969	H5-O4	0.970
H6-C1	1.107	H6-C1	1.109	1.109	H6-C1	1.107	H6-C1	1.107
H7-N2	1.025	H7-N2	1.021	1.021	H7-N2	1.020	H7-N2	1.021
O8-H7	2.019	O8-H5	1.879	1.875	H8-O3	1.951	H8-O4	2.021
H9-O8	0.978	H9-O8	0.966	0.966	O9-H8	0.974	O9-H8	0.970
H10-O8	0.965	H10-O8	0.969	0.969	H10-O9	0.964	H10-O9	0.965
O3-C1-N2	123.5	O3-C1-N2	122.9	123.1	O3-C1-N2	122.4	O3-C1-N2	121.7
O4-N2-C1	114.9	O4-N2-C1	114.0	114.1	O4-N2-C1	114.7	O4-N2-C1	114.3
H5-O4-N2	103.8	H5-O4-N2	102.0	102.0	H5-O4-N2	103.7	H5-O4-N2	103.4
H6-C1-N2	112.7	H6-C1-N2	112.0	112.0	H6-C1-N2	113.2	H6-C1-N2	113.0
H7-N2-C1	117.4	H7-N2-C1	113.9	114.3	H7-N2-C1	116.0	H7-N2-C1	114.1
O8-H7-N2	138.0	O8-H5-O4	150.2	150.9	H8-O3-C1	99.0	H8-O4-N2	117.0
H9-O8-H7	83.9	H9-O8-H5	130.5	129.5	O9-H8-O3	150.9	O9-H8-O4	154.0
H10-O8-H9	105.6	H10-O8-H5	92.1	92.9	H10-O9-H8	104.7	H10-O9-H8	104.9
O4-N2-C1-O3	161.9	O4-N2-C1-O3	158.6	159.1	O4-N2-C1-O3	159.4	O4-N2-C1-O3	157.0
H5-O4-N2-C1	116.4	H5-O4-N2-C1	116.2	118.9	H5-O4-N2-C1	116.2	H5-O4-N2-C1	116.2
H6-C1-N2-O4	-21.2	H6-C1-N2-O4	-25.0	-24.5	H6-C1-N2-O4	-24.1	H6-C1-N2-O4	-26.7
H7-N2-C1-O3	24.8	H7-N2-C1-O3	30.5	30.2	H7-N2-C1-O3	27.5	H7-N2-C1-O3	30.0
O8-H7-N2-O4	207.6	O8-H5-O4-N2	-9.2	2.8	H8-O3-C1-N2	-186.8	H8-O4-N2-C1	-4.5
H9-O8-H7-O3	-6.0	H9-O8-H5-O4	127.1	-124.1	O9-H8-O4-C1	-0.7	O9-H8-O4-H6	32.6
H10-O8-H9-O3	156.4	H10-O8-H5-O4	15.2	-12.0	H10-O9-H8-O3	178.8	H10-O9-H8-O4	-177.5

Table 2.7: Important geometrical parameters of FHA- H$_2$O adducts at MP2/Aug-cc-pVDZ [L4] theoretical level. All bond distances are in angstrom (Å), angles and dihedrals are in degrees (°).

Parameters	2ZW1	Parameters	2ZW2	Parameters	2ZW3
N2-C1	1.294	N2-C1	1.295	N2-C1	1.293
O3-C1	1.350	O3-C1	1.341	O3-C1	1.359
O4-N2	1.425	O4-N2	1.441	O4-N2	1.435
H5-O3	0.977	H5-O3	0.988	H5-O3	0.975
H6-C1	1.089	H6-C1	1.091	H6-C1	1.089
H7-O4	0.979	H7-O4	0.967	H7-O4	0.967
O8-H7	1.896	O8-H5	1.779	H8-O3	2.119
H9-O8	0.966	H9-O8	0.972	O9-H8	0.969
H10-O8	0.974	H10-O8	0.966	H10-O9	0.965
O3-C1-N2	125.8	O3-C1-N2	130.9	O3-C1-N2	125.6
O4-N2-C1	108.5	O4-N2-C1	110.6	O4-N2-C1	107.8
H5-O3-C1	105.8	H5-O3-C1	112.8	H5-O3-C1	106.4
H6-C1-N2	114.7	H6-C1-N2	115.0	H6-C1-N2	119.3
H7-O4-N2	101.0	H7-O4-N2	100.7	H7-O4-N2	102.0
O8-H7-O4	144.9	O8-H5-O3	171.2	H8-O3-C1	101.1
H9-O8-H7	123.2	H9-O8-H5	86.2	O9-H8-O3	133.5
H10-O8-H7	81.2	H10-O8-H5	115.6	H10-O9-H8	104.8
O4-N2-C1-O3	-0.1	O4-N2-C1-O3	-0.3	O4-N2-C1-O3	-0.0
H5-O3-C1-N2	-0.4	H5-O3-C1-N2	-2.4	H5-O3-C1-N2	-0.0
H6-C1-N2-O4	179.9	H6-C1-N2-O4	179.6	H6-C1-N2-O4	180.0
H7-O4-N2-C1	178.9	H7-O4-N2-C1	183.7	H7-O4-N2-C1	180.0
O8-H7-O4-N2	-4.0	O8-H5-O3-C1	219.4	H8-O3-C1-N2	180.3
H9-O8-H7-O4	108.2	H9-O8-H5-O4	-9.5	O9-H8-O3-C1	-0.3
H10-O8-H7-N2	3.3	H10-O8-H5-O3	29.4	H10-O9-H8-O3	180.2

Table 2.8: Important geometrical parameters of FHA- H$_2$O adducts at MP2/ Aug-cc-pVDZ [L4] theoretical level. All bond distances are in angstrom (Å), angles and dihedrals are in degrees (°).

Parameters	2EW1	Parameters	2EW2	Parameters	2EW3
N2-C1	1.294	N2-C1	1.287	N2-C1	1.286
O3-C1	1.345	O3-C1	1.359	O3-C1	1.366
O4-N2	1.430	O4-N2	1.419	O4-N2	1.425
H5-O3	0.989	H5-O3	0.972	H5-O3	0.972
H6-C1	1.095	H6-C1	1.093	H6-C1	1.093
H7-O4	0.967	H7-O4	0.977	H7-O4	0.967
H8-N2	2.022	O8-H7	1.952	H8-O3	2.113
O9-H8	0.978	H9-O8	0.966	O9-H8	0.969
H10-O9	0.966	H10-O8	0.973	H10-O9	0.965
O3-C1-N2	122.3	O3-C1-N2	121.2	O3-C1-N2	119.6
O4-N2-C1	108.9	O4-N2-C1	110.1	O4-N2-C1	109.3
H5-O3-C1	108.3	H5-O3-C1	107.4	H5-O3-C1	107.2
H6-C1-N2	122.8	H6-C1-N2	124.5	H6-C1-N2	126.2
H7-O4-N2	102.0	H7-O4-N2	100.9	H7-O4-N2	101.7
H8-N2-C1	107.1	O8-H7-O4	144.3	H8-O3-C1	101.6
O9-H8-N2	136.6	H9-O8-H7	122.1	O9-H8-O3	134.5
H10-O9-H8	105.3	H10-O8-H7	78.5	H10-O9-H8	104.8
O4-N2-C1-O3	179.3	O4-N2C1O3	180.0	O4-N2-C1-O3	180.1
H5-O3-C1-N2	-0.2	H5-O3-C1-N2	0.7	H5-O3-C1-N2	-0.2
H6-C1-N2-O4	-0.6	H6-C1-N2-O4	0.0	H6-C1-N2-O4	0.1
H7-O4-N2-C1	173.9	H7-O4-N2-C1	178.7	H7-O4-N2-C1	180.1
H8-N2-C1-O3	1.3	O8-H7-O4-N2	-1.9	H8-O3-C1-N2	188.4
O9-H8-N2-H5	-5.8	H9-O8-H7-O4	106.0	O9-H8-O3-C1	-1.7
H10-O9-H8-O3	121.1	H10-O8-H7-N2	4.3	H10-O9-H8-O3	181.9

Table 2.9: Variations in geometrical parameters of FHA on aggregation with water at MP2/Aug-cc-pVDZ [L4] theoretical level.

Parameters	1ZW1	1ZW2	1ZW3	1ZW4	1EW1	1EW2	1EW3	1EW4	1EW5
N2-C1	-0.013	-0.002	-0.013	-0.014	-0.021	0.001	-0.001	-0.012	0.007
O3-C1	0.004	0.002	0.008	0.007	0.011	0.001	0.001	0.007	-0.001
O4-N2	-0.016	0.007	-0.005	-0.002	-0.007	-0.003	-0.004	-0.005	0.005
O3-H5	0.007	0.002	-0.001	0.001	-0.009	0.009	0.010	0.000	0.001
H6-C1	0.003	-0.001	-0.002	0.000	-0.002	0.000	0.000	-0.002	-0.002
H7-N2	-0.004	0.006	-0.002	0.007	0.005	0.001	0.001	0.000	0.001
O4-C1-N2	4.0	0.2	-0.5	0.2	0.7	0.1	0.3	-0.4	-1.1
O4-N2-C1	7.4	-1.2	1.5	0.5	1.2	0.3	0.4	1.0	0.6
H5-O4-N2	3.4	0.1	0.1	-0.6	0.4	-1.4	-1.4	0.3	0.0
H6-C1-N2	-2.2	0.1	1.2	0.1	0.6	-0.1	-0.1	1.1	0.9
H7-N2-C1	1.8	0.8	2.2	4.6	3.0	0.5	-0.1	1.6	-0.3

Table 2.10: Variations in geometrical parameters of FHA on aggregation with water at MP2/Aug-cc-pVDZ [L4] theoretical level.

Parameters	2ZW1	2ZW2	2ZW3	2EW1	2EW2	2EW3
N2-C1	0.000	0.001	-0.001	0.007	0.000	-0.001
O3-C1	-0.002	-0.011	0.007	-0.082	0.000	0.007
O4-N2	-0.010	0.006	0.000	0.003	-0.008	-0.002
O3-H5	0.002	0.013	0.000	0.016	-0.001	-0.01
H6-C1	0.000	0.002	0.000	0.002	0.000	0.000
O4-H7	0.012	0.000	0.002	0.000	0.010	0.000
O3-C1-N2	-0.7	4.4	-0.9	1.6	0.5	-1.1
O4-N2-C1	0.8	2.9	0.1	-0.3	0.9	0.1
H5-O3-C1	-0.5	6.5	0.1	1.2	0.3	0.1
H6-C1-N2	-3.5	3.2	1.1	-2.1	-0.4	1.3
H7-O4-N2	-1.0	-1.3	0.0	0.4	-0.7	0.1

The presence of H-bonds in the FHA-H$_2$O adducts has been concluded from the analysis of geometrical parameters. The Table 2.11 displays the important H-bonded distances between the atoms of FHA-H$_2$O aggregates whose distances are less than sum of their respective van der Waals radii. The angle and dihedral angle associated with H-bonding interactions are other important geometrical parameters that determine the strength of H-bonding and thus the requisite angles are also included in the table. The table also incorporates the stabilization energies arising as a result of aggregation at **[L4]** theoretical level. In almost all the aggregates, water molecule forms two hydrogen bonds, one as H-bond donor and one as H-bond acceptor. Only in **1ZW4**, one H-bond formation is seen. The analysis of geometrical parameters of FHA-H$_2$O aggregates indicates that in almost all the water aggregates of **1Z** except **1ZW1**, the H5…O3 intramolecular interaction which is the dominant interaction present in **1Z** conformation remains intact.

1ZW1 adduct forms two homonuclear O-H…O H-bonds with the water molecule and this aggregate has the maximum stabilization (11.30 kcal/mol at MP2/Aug-cc-pVDZ) associated with the aggregation. As this aggregation involves breakage of intramolecular H5…O3 H-bond interaction, hence it clearly indicates that intermolecular interactions between water and **1Z** conformation are stronger than intramolecular interactions present in **1Z**. Variations in the intramolecular distances on aggregate formation are smaller and are anticipated as the result of intermolecular H-bond formation with the water molecule.

Table 2.12 lists the geometrical parameters for intramolecular H-bonding upon aggregation. The intramolecular H5…O3 distance is increased to 2.704 Å in **1ZW1** from 2.019 Å distance in **1Z** conformation at **[L4]** theoretical level indicating loss of intramolecular H-bonding interaction upon aggregate formation. **2Z** and its aggregates are also stabilized by weak intramolecular H-bonding. Intramolecular H-bonding remains intact in **1ZW2, 1ZW3, 1ZW4** of **1Z** and **2ZW1, 2ZW2, 2ZW3, 2ZW4** of **2Z**.

The sum of angles around nitrogen in **1Z, 1E** and their water adducts at **[L4]** theoretical level are also incorporated in Table 2.11. All the adducts show pyramidal character at nitrogen with larger deviations from planarity for **1E** and its adducts as compared to **1Z**.

Table 2.11 Important intermolecular H-bonding parameters (hydrogen bond distances[a], hydrogen bond angles[b]), charges on hydrogen bond acceptor and hydrogen at MP2/Aug-cc-pVDZ [L4] theoretical level along with stabilization energies S.E.[c,d] and distortion energies E_{Dis}[e] and sum of angles around nitrogen in various hydrogen bonded aggregates of FHA-H$_2$O.

Species	Hydrogen bond distances[a]	Hydrogen bond angles[b]		Atomic charges		S.E.[c] [S.E.][d]	E_{Dis}[e]	Sum of angles around N
1ZW1	1.862	O8-H9...O3	151.5	q$_H$(q$_O$)	0.538 (-0.790)	11.30	2.89	356.8
	1.743	O4-H5...O8	159.8	q$_H$(q$_O$)	0.548 (-1.017)	[10.88]		
1ZW2	2.046	N2-H7...O8	130.9	q$_H$(q$_O$)	0.444(-1.001)	6.60	0.17	343.7
	2.185	O8-H9...O4	124.5	q$_H$(q$_O$)	0.512(-0.633)	[7.35]		
1ZW3	1.969	O9-H8...O3	147.0	q$_H$(q$_O$)	0.515(-0.794)	5.66	0.21	350.5
	2.566	C1-H6...O9	104.0	q$_H$(q$_O$)	0.174(-1.011)	[5.89]		
1ZW4	1.895	N2-H7...O8	175.6	q$_H$(q$_O$)	0.455(-0.980)	6.25 [6.84]	0.26	352.1
1EW1	2.019	N2-H7...O8	138.0	q$_H$(q$_O$)	0.450(-1.021)	8.41	0.54	345.6
	1.946	O8-H9...O3	145.7	q$_H$(q$_O$)	0.527(-0.761)	[8.38]		
1EW2	1.879	O4-H5...O8	150.2	q$_H$(q$_O$)	0.534(-0.992)	6.72	0.15	338.4
	2.503	O8-H10...N2	107.7	q$_H$(q$_N$)	0.506(-0.471)	[7.26]		
1EW3	1.875	O4-H5...O8	150.9	q$_H$(q$_O$)	0.534(-0.991)	6.77	0.14	339.2
	2.529	O8-H10...N2	106.4	q$_H$(q$_N$)	0.506(-0.469)	[7.26]		
1EW4	1.951	O9-H8...O3	150.9	q$_H$(q$_O$)	0.515(-0.743)	5.39	0.17	341.9

Contd.....

Species	Hydrogen bond distances[a]		Hydrogen bond angles[b]		Atomic charges		S.E.[c] [S.E.][d]	E_{Dis}[e]	Sum of angles around N
1EW5	H6...O9	2.656	C1-H6...O9	102.2	$q_H(q_O)$	0.162(-1.010)	[5.37]		
	H8...O4	2.021	O9-H8...O4	154.0	$q_O(q_H)$	0.504(-0.629)	3.33	0.13	337.8
	H6...O9	2.530	C1-H6...O9	133.7	$q_H(q_O)$	0.159(-0.996)	[3.46]		
2ZW1	H10...N2	2.165	O8-H10...N2	123.3	$q_N(q_H)$	0.518(-0.341)	7.75	0.23	-
	H7...O8	1.896	O4-H7...O8	144.9	$q_H(q_O)$	0.544(-0.999)	[7.92]		
2ZW2	H5...O8	1.779	O3-H5...O8	171.2	$q_H(q_O)$	0.557(-0.992)	7.48	1.80	-
	H9...O4	2.021	O8-H9...O4	133.7	$q_O(q_H)$	0.521(-0.698)	[7.32]		
2ZW3	H8...O3	2.119	O9-H8...O3	133.5	$q_O(q_H)$	0.504(-0.806)	3.53	0.09	-
	H6...O9	2.513	C1-H6...O9	113.5	$q_H(q_O)$	0.209(-0.993)	[3.96]		
2EW1	H5...O9	1.801	O3-H5...O9	159.7	$q_H(q_O)$	0.553(-1.009)	9.16	0.51	-
	H8...N2	2.022	O9-H8...N2	136.6	$q_N(q_H)$	0.526(-0.390)	[8.71]		
2EW2	H10...N2	2.157	O8-H10...N2	126.2	$q_N(q_H)$	0.514(-0.365)	6.54	0.16	-
	H7...O8	1.952	O4-H7...O8	144.3	$q_H(q_O)$	0.530(-0.998)	[6.59]		
2EW3	H6...O9	2.532	C1-H6...O9	113.6	$q_H(q_O)$	0.203(-0.992)	3.33	0.09	-
	H8...O3	2.113	O9-H8...O3	134.5	$q_O(q_H)$	0.503(-0.794)	[3.70]		

a- in angstrom (Å)
b- in degrees (°)
c- at MP2/Aug-cc-pVDZ (kcal/mol)
d- at MP2/6-31+G* (kcal/mol)
e- in kcal/mol

Table 2.12: Hydrogen bond distances (Å), angles (°) for intramolecular H-bond in **1Z, 2Z** of FHA and their aggregates at MP2/Aug-cc-pVDZ **[L4]** theoretical level.

Species	H5...O3	O4-H5...O3	Species	H5...O4	O3-H5...O4
1Z	2.019	119.3	**2Z**	2.007	112.9
1ZW1	2.704	88.1	**2ZW1**	1.994	113.7
1ZW2	1.983	120.2	**2ZW2**	2.311	102.4
1ZW3	2.036	118.5	**2ZW3**	1.988	113.2
1ZW4	2.007	119.9	**2ZW4**	2.018	112.6

Though in **1ZW2, 1ZW3, 1ZW4** aggregates the intramolecular H-bond remains intact, yet these have comparatively smaller stabilizations relative to that in **1ZW1** upon aggregation with water. This can be rationalized in terms of difference of nature of H-bond acceptor and donor atoms and hence its strength in forming FHA-H_2O aggregate. The aggregates **1ZW3, 1EW4, 1EW5, 2ZW3, 2EW3** contain at least one of the two H-bonds as an unconventional C-H...O interaction and therefore smaller stabilization energies are anticipated. The intramolecular H-bond in **1ZW2** upon interaction with water gets strengthened as compared to the parent **1Z** molecule as can be seen from decreased H-bond H5...O3 separation from 2.019 Å to 1.983 Å. The relative stability based upon the most stable structure of aggregate of each conformation is reported in the Table 2.11. The stabilization energies for all the aggregates are in the range 3.33 to 11.30 kcal/mol. The aggregates **1ZW1, 1EW1, 2ZW2** and **2EW1** are the most stabilized aggregate structures for **1Z, 1E, 2Z** and **2E** isomeric forms of FHA respectively at **[L4]** theoretical level. The carbonyl oxygen is H-bond acceptor to OH of water in **1ZW1** and **1EW1** while the second H-bond involves OH of **1Z** and NH of **1E** as H-bond donor to oxygen of water. The high stabilization resulting from interaction with water also favors **1Z** conformer over others.

The stabilization energy associated with FHA-H_2O aggregate formation has three contributing components (i) deformation (distortion) energy (ii) interaction energy due to H-bonding with water (iii) loss of intramolecular interaction energy. Though the third component is difficult to separate from the deformation energy component but is important when water molecule acts as a shield for intramolecular

H-bonding without deformation of the molecule. The distortion energy also contributes to the total stabilization energy. The FHA-H_2O system undergoes significant distortion for some adducts. The distortion energies have been evaluated as the difference in energies of the monomer conformation in the gas phase (in relaxed state) and its energy in the aggregate (in distorted state) and are also reported in Table 2.11.

2.2.4 CHARGE ANALYSIS:

Table 2.13 records natural population analysis (NPA) charges on all atoms of FHA in its different isomeric forms and their aggregates with water. From the charge analysis, the C=O is found to be most polar bond in FHA as the C1 and O3 are highly charged in all the structures. Upon adduct formation, charges on all atoms of FHA undergo variations but largest change occurs on atoms participating in intermolecular H-bond formation i.e. in all aggregates the negative charge on H-bond acceptor atom increases and simultaneously the positive charge on H-bond donor atom decreases. The high positive and negative charges on the atoms involved in formation of H-bond clearly suggest that electrostatics also play an important role towards interaction energy.

Table 2.13: Atomic charges (NPA) on atoms of FHA and FHA-H_2O aggregates at MP2/Aug-cc-pVDZ **[L4]** theoretical level.

Species	C1	N2	O3	O4	H5	H6	H7
1Z	0.696	-0.423	-0.750	-0.603	0.522	0.148	0.410
1ZW1	0.712	-0.421	-0.790	-0.615	0.548	0.139	0.413
1ZW2	0.699	-0.437	-0.758	-0.633	0.528	0.150	0.444
1ZW3	0.696	-0.411	-0.794	-0.596	0.523	0.174	0.419
1ZW4	0.691	-0.443	-0.782	-0.613	0.523	0.147	0.455
1E	0.719	-0.449	-0.703	-0.605	0.497	0.138	0.403
1EW1	0.730	-0.460	-0.761	-0.597	0.496	0.145	0.449
1EW2	0.721	-0.471	-0.709	-0.630	0.534	0.140	0.401
1EW3	0.719	-0.469	-0.711	-0.629	0.534	0.137	0.403
1EW4	0.725	-0.441	-0.743	-0.597	0.498	0.162	0.407
1EW5	0.709	-0.445	-0.698	-0.629	0.506	0.159	0.403
2Z	0.535	-0.300	-0.782	-0.683	0.534	0.187	0.509
2ZW1	0.546	-0.341	-0.783	-0.699	0.537	0.185	0.544
2ZW2	0.554	-0.317	-0.808	-0.698	0.557	0.181	0.506
2ZW3	0.527	-0.293	-0.806	-0.686	0.540	0.209	0.510
2ZW4	0.555	-0.347	-0.775	-0.672	0.533	0.206	0.509
2E	0.541	-0.328	-0.771	-0.630	0.511	0.181	0.497
2EW1	0.578	-0.390	-0.799	-0.627	0.553	0.180	0.494

Contd.....

Species	C1	N2	O3	O4	H5	H6	H7
2EW2	0.552	-0.365	-0.770	-0.645	0.508	0.183	0.530
2EW3	0.531	-0.325	-0.794	-0.626	0.517	0.203	0.496
2EW4	0.548	-0.338	-0.767	-0.661	0.510	0.206	0.504

2.2.5 INTERMOLECULAR HYDROGEN BONDING IN DIMERIC UNITS:

The intermolecular interactions are important for FHA with multiple H-bond donor and H-bond acceptor sites. Intermolecular H-bonds can lead to dimer formation and may influence in a cooperative fashion the relative strength of intramolecular H-bonds. The search for minimum energy dimeric structures of different tautomeric forms of FHA has lead to twenty five structures at MP2/6-31+G* [L3] theoretical level. The optimized dimeric structures are shown in Figure 2.3 and their geometrical parameters are recorded in Tables 2.14-2.20.

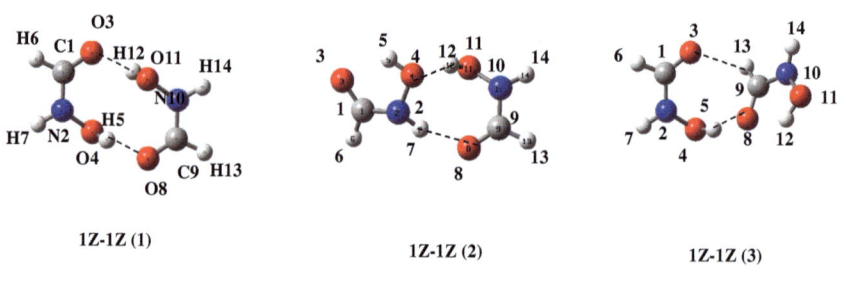

1Z-1Z (1)   1Z-1Z (2)   1Z-1Z (3)

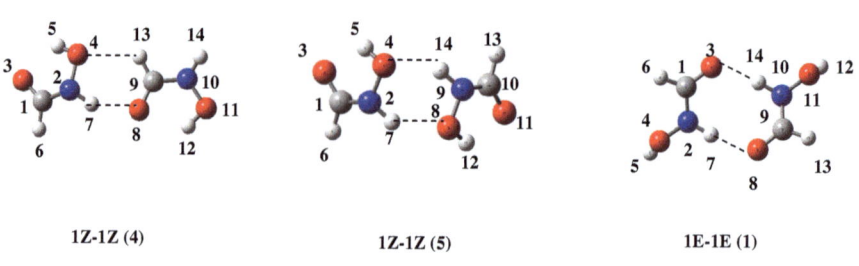

1Z-1Z (4)   1Z-1Z (5)   1E-1E (1)

Contd.....

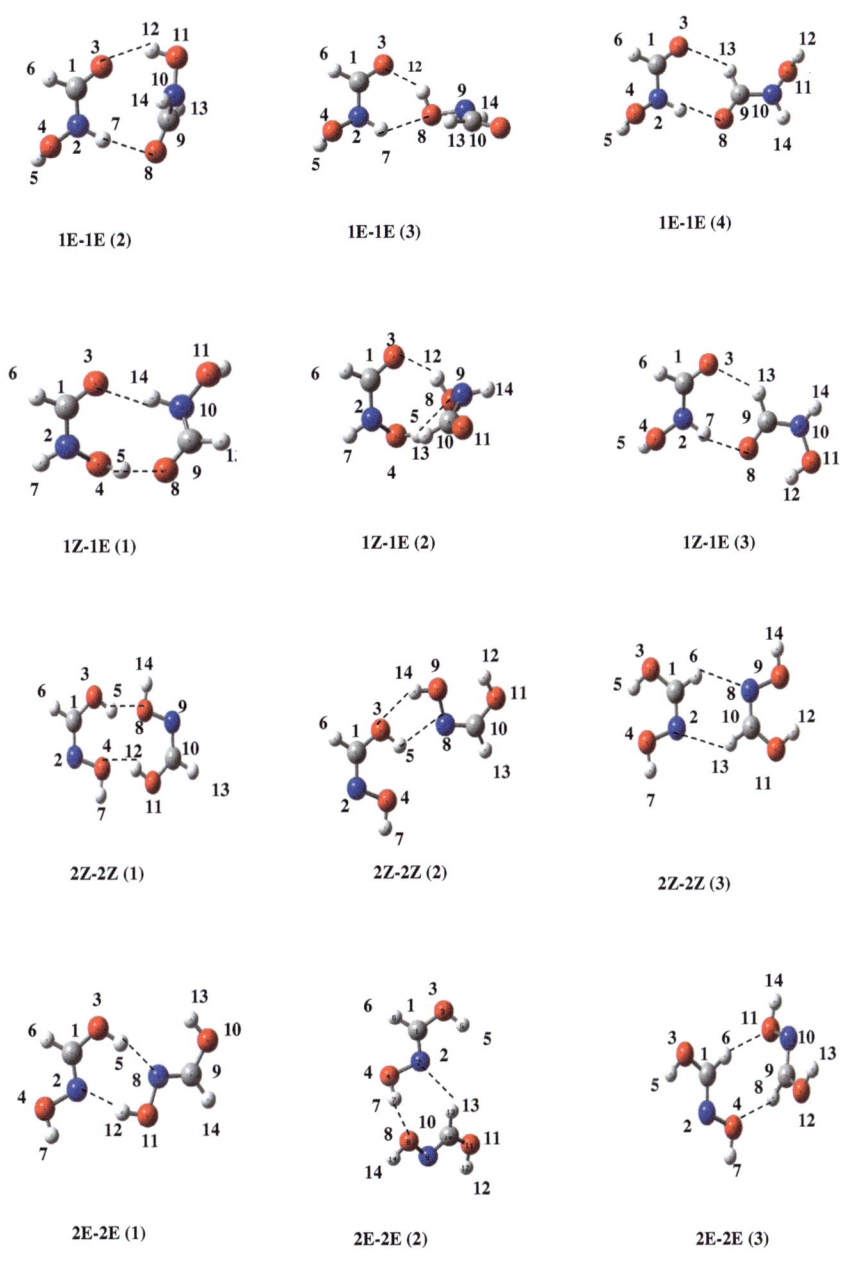

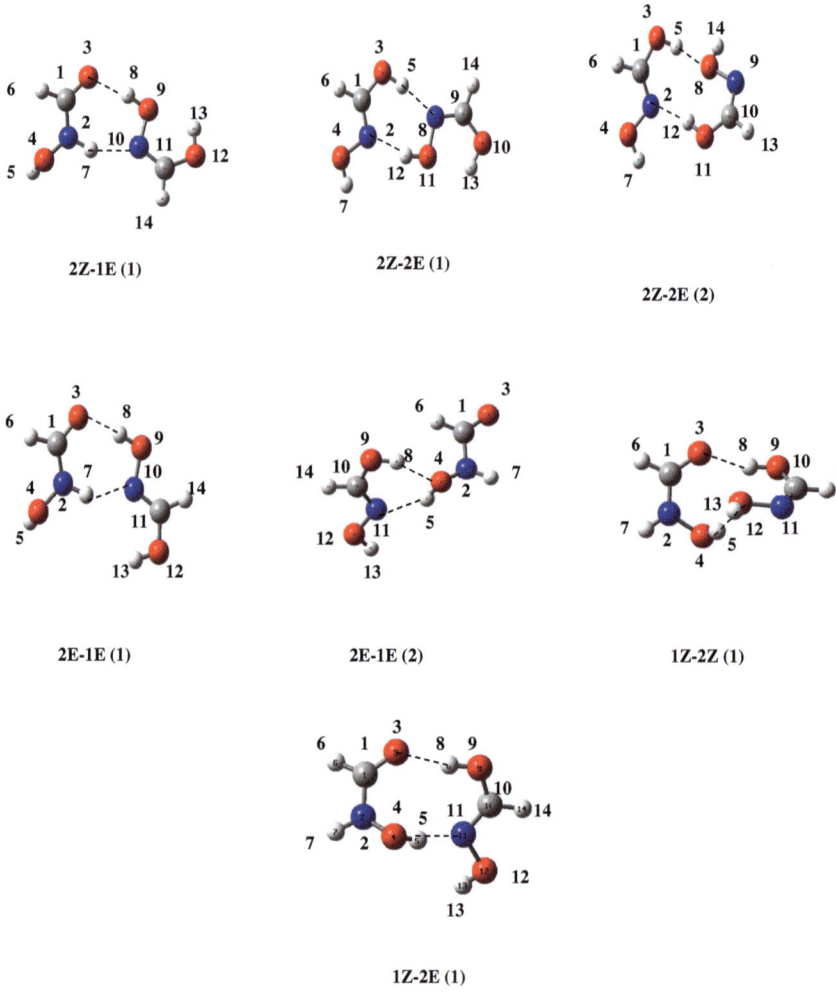

Figure 2.3: The structures of various hydrogen-bonded dimeric complexes formed among FHA tautomers considered in the present work.

Table 2.14: Geometrical parameters of FHA dimers at MP2/6-31+G* [L3] theoretical level. All the bond distances are in angstrom (Å), angles and dihedrals are in degrees (°).

Parameters	1Z-1Z (1)	Parameters	1Z-1Z (2)	Parameters	1Z-1Z (3)	Parameters	1Z-1Z (4)
N2-C1	1.359	N2-C1	1.357	N2-C1	1.361	N2-C1	1.354
O3-C1	1.235	O3-C1	1.240	O3-C1	1.232	O3-C1	1.240
O4-N2	1.394	O4-N2	1.420	O4-N2	1.397	O4-N2	1.415
H5-O4	0.992	H5-O4	0.994	H5-O4	0.993	H5-O4	0.990
H6-C1	1.102	H6-C1	1.098	H6-C1	1.104	H6-C1	1.099
H7-N2	1.017	H7-N2	1.029	H7-N2	1.015	H7-N2	1.026
O8-H5	1.808	O8-H5	1.872	O8-H5	1.771	O8-H7	1.907
C9-O8	1.235	C9-O8	1.235	C9-O8	1.252	C9-O8	1.248
N10-C9	1.359	N10-C9	1.357	N10-C9	1.341	N10-C9	1.344
O11-N10	1.395	O11-N10	1.397	O11-N10	1.401	O11-N10	1.401
H12-O11	0.992	H12-O11	0.990	H12-O11	0.989	H12-O11	0.988
H13-C9	1.102	H13-C9	1.102	H13-C9	1.096	H13-C9	1.096
H14-N10	1.017	H14-N10	1.015	H14-N10	1.013	H14-N10	1.014
O3-C1-N2	125.8	O3-C1-N2	121.9	O3-C1-N2	125.6	O3-C1-N2	122.2
O4-N2-C1	118.4	O4-N2-C1	113.8	O4-N2-C1	119.7	O4-N2-C1	115.0
H5-O4-N2	106.8	H5-O4-N2	101.1	H5-O4-N2	104.5	H5-O4-N2	101.4
H6-C1-N2	111.7	H6-C1-N2	113.6	H6-C1-N2	111.1	H6-C1-N2	113.5
H7-N2-C1	118.7	H7-N2-C1	123.5	H7-N2-C1	119.6	H7-N2-C1	123.6
O8-H5-O4	153.4	O8-H5-O4	163.4	O8-H5-O4	168.1	O8-H7-N2	154.7
C9-O8-H7	114.8	C9-O8-H7	123.7	C9-O8-H5	111.2	C9-O8-H7	109.8
N10-C9-O8	125.8	N10-C9-O8	125.5	N10-C9-O8	120.4	N10-C9-O8	121.0
O11-N10-C9	118.4	O11-N10-C9	119.6	O11-N10-C9	118.2	O11-N10-C9	117.9
H12-O11-N10	106.8	H12-O11-N10	104.9	H12-O11-N10	101.6	H12-O11-N10	101.8
H13-C9-N10	111.7	H13-C9-N10	111.7	H13-C9-O8	123.6	H13-C9-O8	123.8
H14-N10-C9	118.7	H14-N10-C9	120.3	H14-N10-C9	124.2	H14-N10-C9	123.8
O4-N2-C1-O3	-19.9	O4-N2-C1-O3	-11.9	O4-N2-C1-O3	-18.6	O4-N2-C1-O3	-11.6

Contd.....

Parameters	1Z-1Z (1)	Parameters	1Z-1Z (2)	Parameters	1Z-1Z (3)	Parameters	1Z-1Z (4)
H5-O4-N2-C1	103.9	H5-O4-N2-C1	10.6	H5-O4-N2-C1	83.2	H5-O4-N2-C1	7.5
H6-C1-N2-O4	164.0	H6-C1-N2-O4	171.9	H6-C1-N2-O4	166.3	H6-C1-N2-O4	172.0
H7-N2-C1-O3	-161.6	H7-N2-C1-O3	-149.2	H7-N2-C1-O3	197.2	H7-N2-C1-O3	-149.7
O8-H5-O4-N2	195.9	O8-H5-O4-N2	17.1	O8-H5-O4-N2	276.1	O8-H7-N2-O4	3.2
C9-O8-H7-N2	59.7	C9-O8-H7-N2	-22.3	C9-O8-H5-O4	9.0	C9-O8-H7-N2	-3.0
N10-C9-O8-H5	33.6	N10-C9-O8-H5	-25.7	N10-C9-O8-H5	184.8	N10-C9-O8-H7	181.8
O11-N10-C9-O8	19.9	O11-N10-C9-O8	-18.3	O11-N10-C9-O8	9.5	O11-N10-C9-O8	-9.9
H12-O11-N10-O3	-5.6	H12-O11-N10-O4	13.0	H12-O11-N10-C9	-5.0	H12-O11-N10-C9	5.6
H13-C9-N10-O11	196.0	H13-C9-N10-O11	166.2	H13-C9-O8-O3	-11.1	H13-C9-O8-O4	-1.9
H14-N10-C9-O8	161.5	H14-N10-C9-O8	196.0	H14-N10-C9-H13	-23.1	H14-N10-C9-O8	201.3

Table 2.15: Geometrical parameters of FHA dimers at MP2/6-31+G* [L3] theoretical level. All the bond distances are in angstrom (Å), angles and dihedrals are in degrees (°).

Parameters	1Z-1Z (5)	Parameters	1E-1E (1)	Parameters	1E-1E (2)
N2-C1	1.361	N2-C1	1.359	N2-C1	1.359
O3-C1	1.237	O3-C1	1.239	O3-C1	1.238
O4-N2	1.418	O4-N2	1.415	O4-N2	1.411
H5-O4	0.992	H5-O4	0.977	H5-O4	0.977
H6-C1	1.098	H6-C1	1.099	H6-C1	1.100
H7-N2	1.023	H7-N2	1.031	H7-N2	1.026
O8-H7	2.013	O8-H7	1.903	O8-H7	2.002
N9-O8	1.418	C9-O8	1.239	C9-O8	1.240
C10-N9	1.360	N10-C9	1.359	N10-C9	1.356
O11-C10	1.237	O11-N10	1.415	O11-N10	1.401
H12-O8	0.992	H12-O11	0.977	H12-O11	0.986
H13-C10	1.098	H13-C9	1.099	H13-C9	1.100
H14-N9	1.023	H14-N10	1.031	H14-N10	1.018
O3-C1-N2	121.9	O3-C1-N2	123.5	O3-C1-N2	123.9

Contd......

Parameters	1Z-1Z (5)	Parameters	1E-1E (1)	Parameters	1E-1E (2)
O4-N2-C1	114.2	O4-N2-C1	114.9	O4-N2-C1	115.3
H5-O4-N2	101.4	H5-O4-N2	103.9	H5-O4-N2	104.1
H6-C1-N2	113.6	H6-C1-N2	113.5	H6-C1-N2	112.9
H7-N2-C1	122.3	H7-N2-C1	119.0	H7-N2-C1	120.8
O8-H7-N2	145.0	O8-H7-N2	169.1	O8-H7-N2	164.0
N9-O8-H7	103.9	C9-O8-H7	124.7	C9-O8-H7	100.5
C10-N9-O8	114.2	N10-C9-O8	123.5	N10-C9-O8	123.4
O11-C10-N9	121.9	O11-N10-C9	114.9	O11-N10-C9	118.0
H12-O8-N9	101.4	H12-O11-N10	103.9	H12-O11-N10	103.5
H13-C10-N9	113.6	H13-C9-N10	113.5	H13-C9-N10	112.8
H14-N9-O8	109.2	H14-N10-C9	119.0	H14-N10-C9	119.4
O4-N2-C1-O3	-12.3	O4-N2-C1-O3	163.7	O4-N2-C1-O3	164.8
H5-O4-N2-C1	9.2	H5-O4-N2-C1	118.5	H5-O4-N2-C1	116.9
H6-C1-N2-O4	171.7	H6-C1-N2-O4	-19.3	H6-C1-N2-O4	-18.1
H7-N2-C1-O3	-147.6	H7-N2-C1-O3	24.9	H7-N2-C1-O3	22.8
O8-H7-N2-O4	-5.3	O8-H7-N2-O4	204.7	O8-H7-N2-O4	193.9
N9-O8-H7-N2	-13.9	C9-O8-H7-N2	0.7	C9-O8-H7-N2	8.7
C10-N9-O8-H7	232.7	N10-C9-O8-H7	-14.4	N10-C9-O8-H7	79.8
O11-C10-N9-O8	-12.2	O11-N10-C9-O8	163.6	O11-N10-C9-O8	201.2
H12-O8-N9-C10	9.2	H12-O11-N10-C9	118.6	H12-O11-N10-O3	1.0
H13-C10-N9-O8	171.7	H13-C9-N10-O11	-19.3	H13-C9-N10-O11	23.0
H14-N9-O8-O4	2.3	H14-N10-C9-O3	2.1	H14-N10-C9-H13	169.4

Table 2.16: Geometrical parameters of FHA dimers at MP2/6-31+G* [L3] theoretical level. All the bond distances are in angstrom (Å), angles and dihedrals are in degrees (°).

Parameters	1E-1E (3)	Parameters	1E-1E (4)	Parameters	2Z-2Z (1)
N2-C1	1.369	N2-C1	1.371	N2-C1	1.290
O3-C1	1.235	O3-C1	1.233	O3-C1	1.342
O4-N2	1.415	O4-N2	1.420	O4-N2	1.442
H5-O4	0.977	H5-O4	0.976	H5-O3	0.987
H6-C1	1.099	H6-C1	1.101	H6-C1	1.084
H7-N2	1.023	H7-N2	1.029	H7-O4	0.976
O8-H7	2.155	O8-H7	1.936	O8-H5	1.941
N9-O8	1.416	N9-O8	1.235	N9-O8	1.442
C10-N9	1.382	C10-N9	1.374	C10-N9	1.290
O11-C10	1.226	O11-N10	1.416	O11-C10	1.342
H12-O8	0.988	H12-O11	0.976	H12-O11	0.987
H13-C10	1.102	H13-C9	1.098	H13-C10	1.084
H14-N9	1.019	H14-N10	1.019	H14-O8	0.976
O3-C1-N2	123.3	O3-C1-N2	123.1	O3-C1-N2	128.3
O4-N2-C1	114.7	O4-N2-C1	114.1	O4-N2-C1	108.5
H5-O4-N2	103.9	H5-O4-N2	103.6	H5-O3-C1	110.1
H6-C1-N2	113.2	H6-C1-N2	113.1	H6-C1-N2	117.1
H7-N2-C1	117.8	H7-N2-C1	116.8	H7-O4-N2	101.7
O8-H7-N2	132.9	O8-H7-N2	164.9	O8-H5-O3	140.2
N9-O8-H7	150.5	C9-O8-H7	113.2	N9-O8-H5	142.9
C10-N9-O8	114.7	N10-C9-O8	121.7	C10-N9-O8	128.5
O11-C10-N9	123.1	O11-N10-C9	114.9	O11-C10-N9	128.3
H12-O8-H7	84.2	H12-O11-N10	103.9	H12-O11-C10	110.1
H13-C10-N9	112.5	H13-C9-O8	123.8	H13-C10-N9	117.1
H14-N9-C10	115.7	H14-N10-C9	116.5	H14-O8-N9	101.7
O4-N2-C1-O3	161.4	O4-N2-C1-O3	161.1	O4-N2-C1-O3	0.0
H5-O4-N2-C1	117.8	H5-O4-N2-C1	119.5	H5-O3-C1-N2	4.9
H6-C1-N2-O4	-21.9	H6-C1-N2-O4	-22.0	H6-C1-N2-O4	180.3

Contd.....

Parameters	1E-1E (3)	Parameters	1E-1E (4)	Parameters	2Z-2Z (1)
H7-N2-C1-O3	26.0	H7-N2-C1-O3	27.9	H7-O4-N2-C1	166.8
O8-H7-N2-O4	-161.8	O8-H7-N2-C1	-21.8	O8-H5-O3-C1	127.9
N9-O8-H7-N2	125.2	C9-O8-H7-N2	1.8	N9-O8-H5-O3	158.3
C10-N9-O8-H7	10.2	N10-C9-O8-H7	172.0	C10-N9-O8-H5	67.0
O11-C10-N9-O8	158.4	O11-N10-C9-O8	159.3	O11-C10-N9-O8	-0.0
H12-O8-H7-O3	0.5	H12-O11-N10-C9	115.2	H12-O11-C10-O4	25.5
H13-C10-N9-O8	-25.2	H13-C9-O8-O3	-9.2	H13-C10-N9-O8	179.7
H14-N9-C10-O11	27.8	H14-N10-C9-H13	203.0	H14-O8-N9-C10	193.2

* At MP2/6-31G

Table 2.17: Geometrical parameters of FHA dimers at MP2/6-31+G* [L3] theoretical level. All the bond distances are in angstrom (Å), angles and dihedrals are in degrees (°).

Parameters	2Z-2Z (2)	Parameters	2Z-2Z (3)	Parameters	2E-2E (1)	Parameters	2E-2E (2)
N2-C1	1.288	N2-C1	1.291	N2-C1	1.288	N2-C1	1.283
O3-C1	1.350	O3-C1	1.348	O3-C1	1.340	O3-C1	1.356
O4-N2	1.432	O4-N2	1.437	O4-N2	1.430	O4-N2	1.420
H5-O3	0.991	H5-O3	0.982	H5-O3	0.998	H5-O3	0.980
H6-C1	1.083	H6-C1	1.084	H6-C1	1.088	H6-C1	1.087
H7-O4	0.974	H7-O4	0.974	H7-O4	0.975	H7-O4	0.983
N8-H5	2.074	N8-H6	2.523	N8-H5	1.873	O8-H7	1.873
O9-N8	1.429	C9-N8	1.437	C9-N8	1.283	N9-O8	1.437
C10-N8	1.289	O10-C9	1.291	O10-C9	1.352	C10-N9	1.285
O11-C10	1.346	O11-N8	1.348	O11-N8	1.414	O11-C10	1.352
H12-O11	0.982	H12-O11	0.982	H12-O11	0.990	H12-O11	0.980
H13-C10	1.082	H13-O10	1.084	H13-O10	0.979	H13-C10	1.088
H14-O9	0.980	H14-C9	0.974	H14-C9	1.086	H14-O8	0.975
O3-C1-N2	126.7	O3-C1-N2	126.3	O3-C1-N2	122.1	O3-C1-N2	121.1

Contd......

Parameters	2Z-2Z (2)	Parameters	2Z-2Z (3)	Parameters	2E-2E (1)	Parameters	2E-2E (2)
O4-N2-C1	108.1	O4-N2-C1	108.0	O4-N2-C1	109.2	O4-N2-C1	109.8
H5-O3-C1	109.9	H5-O3-C1	107.3	H5-O3-C1	109.4	H5-O3-C1	107.8
H6-C1-N2	118.3	H6-C1-N2	118.2	H6-C1-N2	123.1	H6-C1-N2	125.0
H7-O4-N2	102.2	H7-O4-N2	102.4	H7-O4-N2	102.4	H7-O4-N2	102.2
N8-H5-O3	123.1	N8-H5-O3	137.7	N8-H5-O3	172.5	O8-H7-O4	173.5
O9-N8-H5	96.6	C9-N8-H5	148.0	C9-N8-H5	138.9	N9-O8-H7	116.9
C10-N8-H5	154.8	O10-C9-N8	103.9	O10-C9-N8	121.9	C10-N9-O8	109.1
O11-C10-N8	126.2	O11-N8-C9	126.3	O11-N8-C9	110.8	O11-C10-N9	120.1
H12-O11-C10	107.1	H12-O11-N8	107.3	H12-O11-N8	101.6	H12-O11-C10	108.0
H13-C10-N8	118.7	H13-C10-C9	118.2	H13-C10-C9	108.9	H13-C10-N9	125.4
H14-O9-N8	102.0	H14-C9-N8	102.4	H14-C9-N8	124.1	H14-O8-N9	101.8
O4-N2-C1-O3	0.2	O4-N2-C1-O3	-0.2	O4-N2-C1-O3	180.7	O4-N2-C1-O3	180.6
H5-O3-C1-N2	-2.0	H5-O3-C1-N2	-0.2	H5-O3-C1-N2	0.3	H5-O3-C1-N2	-0.5
H6-C1-N2-O4	179.7	H6-C1-N2-O4	179.9	H6-C1-N2-O4	0.6	H6-C1-N2-O4	0.5
H7-O4-N2-C1	179.4	H7-O4-N2-C1	179.9	H7-O4-N2-C1	184.8	H7-O4-N2-C1	186.1
N8-H5-O3-C1	202.2	N8-H5-O3-C1	-5.5	N8-H5-O3-C1	2.8	O8-H7-O4-N2	-14.8
O9-N8-H5-O3	6.8	C9-N8-H5-O3	178.4	C9-N8-H5-O3	177.6	N9-O8-H7-O4	-50.5
C10-N8-H5-O3	187.3	O10-C9-N8-H5	4.4	O10-C9-N8-H5	0.2	C10-N9-O8-H7	44.9
O11-C10-N8-H5	179.5	O11-N8-C9-C1	176.8	O11-N8-C9-O10	180.0	O11-C10-N9-O8	176.2
H12-O11-C10-N8	-0.2	H12-O11-C10-N8	0.2	H12-O11-N8-N2	0.2	H12-O11-C10-N9	0.1
H13-C10-N8-O9	180.0	H13-C10-N8-N2	-2.4	H13-O10-C9-N8	-0.0	H13-C10-N9-O4	27.8
H14-O9-N8-O3	1.6	H14-O9-N8-C10	181.0	H14-C9-N8-O11	-0.0	H14-O8-N9-C10	176.5

Table 2.18: Geometrical parameters of FHA dimers at MP2/6-31+G* [L3] theoretical level. All the bond distances are in angstrom (Å), angles and dihedrals are in degrees (°).

Parameters	2E-2E (3)	Parameters	2E-1E (1)	Parameters	2E-1E (2)	Parameters	2E-2Z (1)
N2-C1	1.285	N2-C1	1.366	N2-C1	1.386	N2-C1	1.290
O3-C1	1.355	O3-C1	1.236	O3-C1	1.225	O3-C1	1.336
O4-N2	1.433	O4-N2	1.418	O4-N2	1.420	O4-N2	1.432
H5-O3	0.980	H5-O4	0.977	H5-O4	0.989	H5-O3	0.999
H6-C1	1.087	H6-C1	1.099	H6-C1	1.102	H6-C1	1.088
H7-O4	0.975	H7-N2	1.029	H7-N2	1.020	H7-O4	0.975
H8-O4	2.261	H8-O3	1.857	H8-O4	1.932	N8-H5	1.858
C9-H8	1.087	O9-H8	0.989	O9-H8	0.987	C9-N8	1.290
N10-C9	1.285	N10-O9	1.417	C10-O9	1.345	O10-C9	1.343
O11-N10	1.433	C11-N10	1.284	N11-C10	1.288	O11-N8	1.418
O12-C9	1.355	O12-C11	1.355	H12-N11	1.426	H12-O11	0.995
H13-O12	0.980	H13-O12	0.979	H13-O12	0.975	H13-O10	0.983
H14-O11	0.975	H14-C11	1.086	H14-C10	1.087	H14-C9	1.083
O3-C1-N2	119.9	O3-C1-N2	122.8	O3-C1-N2	122.8	O3-C1-N2	122.6
O4-N2-C1	109.4	O4-N2-C1	114.8	O4-N2-C1	114.7	O4-N2-C1	109.0
H5-O3-C1	107.7	H5-O4-N2	103.8	H5-O4-N2	105.1	H5-O3-C1	109.8
H6-C1-N2	125.8	H6-C1-N2	113.8	H6-C1-N2	112.8	H6-C1-N2	122.6
H7-O4-N2	101.6	H7-N2-C1	116.6	H7-N2-C1	115.2	H7-O4-N2	102.5
N8-O4-N2	110.9	H8-O3-C1	118.1	H8-O4-N2	131.3	N8-H5-O3	169.5
C9-H8-O4	160.9	O9-H8-O3	169.0	O9-H8-O4	154.3	C9-N8-H5	138.9
N10-C9-H8	125.8	N10-O9-H8	102.6	C10-O9-H8	109.4	C10-C9-N8	125.5
O11-N10-C9	109.4	O11-C10-N9	110.2	N11-C10-O9	122.3	O11-N8-C9	109.4
H12-C9-N10	119.9	O12-C11-N10	121.5	O12-N11-C10	109.3	H12-O11-N8	101.7
H13-O12-C9	107.7	H13-O12-C11	108.3	H13-C12-N11	102.4	H13-O10-C9	106.7
H14-O11-N10	101.6	H14-C11-N10	124.4	H14-C10-N11	123.4	H14-C9-N8	119.3
O4-N2-C1-O3	180.1	O4-N2-C1-O3	161.8	O4-N2-C1-O3	158.3	O4-N2-C1-O3	177.4

Contd......

Parameters	2E-2E (3)	Parameters	2E-1E (1)	Parameters	2E-1E (2)	Parameters	2E-2Z (1)
H5-O3-C1-N2	-0.1	H5-O4-N2-C1	116.8	H5-O4-N2-C1	114.6	H5-O3-C1-N2	-1.5
H6-C1-N2-O4	-0.4	H6-C1-N2-O4	-21.4	H6-C1-N2-O4	-25.6	H6-C1-N2-O4	-2.3
H7-O4-N2-C1	182.7	H7-N2-C1-O3	27.4	H7-N2-C1-O3	29.0	H7-O4-N2-C1	161.3
H8-O4-N2-C1	48.1	H8-O3-C1-N2	-8.1	H8-O4-N2-C1	17.0	N8-H5-O3-C1	8.5
C9-H8-O4-N2	-20.0	O9-H8-O4-N2	-14.3	O9-H8-O4-N2	107.4	C9-N8-H5-O3	187.5
N10-C9-H8-O4	-56.8	N10-O9-H8-H7	-19.4	C10-O9-H8-O4	0.8	O10-C9-N8-H5	179.0
O11-N10-C9-H6	-31.0	C11-N10-O9-H8	191.9	N11-C10-O9-H5	0.7	O11-N8-C9-O10	0.0
H12-C9-N10-O11	180.1	O12-C11-N10-O9	181.1	O12-N11-C10-O9	179.3	H12-O11-N8-N2	-0.5
H13-C12-C9-N10	-0.0	H13-O12-C11-N10	-2.0	H13-O12-N11-C10	176.5	H13-O10-C9-N8	-0.0
H14-O11-N10-C9	182.6	H14-C11-N10-O9	0.9	H14-C10-N11-O12	-0.6	H14-C9-N8-O11	180.1

Table 2.19: Geometrical parameters of FHA dimers at MP2/6-31+G* [L3] theoretical level. All the bond distances are in angstrom (Å), angles and dihedrals are in degrees (°).

Parameters	2E-2Z (2)	Parameters	2Z-1E (1)	Parameters	1Z-2Z (1)	Parameters	1Z-2E (1)
N2-C1	1.287	N2-C1	1.361	N2-C1	1.359	N2-C1	1.353
O3-C1	1.343	O3-C1	1.239	O3-C1	1.233	O3-C1	1.237
O4-N2	1.427	O4-N2	1.415	O4-N2	1.400	O4-N2	1.395
H5-O3	0.991	H5-O4	0.977	H5-O4	0.989	H5-O4	0.998
H6-C1	1.088	H6-C1	1.099	H6-C1	1.012	H6-C1	1.101
H7-O4	0.975	H7-N2	1.030	H7-N2	1.015	H7-N2	1.015
O8-H5	1.832	H8-O3	1.805	H8-O3	1.809	H8-O3	1.768
N9-O8	1.448	O9-H8	0.992	O9-H8	0.995	O9-H8	0.998
C10-N9	1.291	N10-O9	1.421	C10-O9	1.332	C10-O9	1.331
O11-C10	1.337	C11-N10	1.291	N11-C10	1.293	N11-C10	1.291
H12-O11	0.998	O12-C11	1.345	O12-N11	1.451	O12-N11	1.436
H13-C10	1.085	H13-O12	0.984	H13-O12	0.976	H13-O12	0.975
H14-O8	0.976	H14-C11	1.083	H14-C10	1.085	H14-C10	1.088
O3-C1-N2	121.7	O3-C1-N2	123.2	O3-C1-N2	125.5	O3-C1-N2	125.6

Contd......

Parameters	2E-2Z (2)	Parameters	2Z-1E (1)	Parameters	1Z-2Z (1)	Parameters	1Z-2E (1)
O4-N2-C1	109.6	O4-N2-C1	115.0	O4-N2-C1	119.2	O4-N2-C1	119.9
H5-O3-C1	108.7	H5-O4-N2	103.9	H5-O4-N2	103.4	H5-O4-N2	104.5
H6-C1-N2	123.5	H6-C1-N2	113.8	H6-C1-N2	111.7	H6-C1-N2	112.0
H7-O4-N2	102.4	H7-N2-C1	117.6	H7-N2-C1	119.9	H7-N2-C1	120.3
O8-H5-O3	162.7	H8-O3-C1	120.0	H8-O3-C1	130.4	H8-O3-C1	132.1
N9-O8-H5	131.0	O9-H8-O3	169.1	O9-H8-O3	159.7	O9-H8-O3	168.5
C10-N9-O8	109.1	N10-O9-H8	102.9	C10-O9-H8	114.3	C10-O9-H8	111.5
O11-C10-N9	129.9	C11-N10-O9	108.7	N11-C10-O9	130.7	N11-C10-O9	123.2
H12-O11-C10	113.3	O12-C11-N10	125.6	O12-N11-C10	109.1	O12-N11-C10	108.9
H13-C10-N9	115.9	H13-O12-C11	106.4	H13-O12-N11	101.2	H13-O12-N11	102.3
H14-O8-N9	101.2	H14-C10-N10	119.2	H14-C10-N11	115.4	H14-C10-N11	122.2
O4-N2-C1-O3	181.0	O4-N2-C1-O3	162.7	O4-N2-C1-O3	-18.5	O4-N2-C1-O3	17.7
H5-O3-C1-N2	4.7	H5-O4-N2-C1	117.1	H5-O4-N2-C1	79.1	H5-O4-N2-C1	-78.5
H6-C1-N2-O4	1.1	H6-C1-N2-O4	-20.3	H6-C1-N2-O4	166.4	H6-C1-N2-O4	193.4
H7-O4-N2-C1	186.6	H7-N2-C1-O3	25.4	H7-N2-C1-O3	-161.4	H7-N2-C1-O3	-195.5
O8-H5-O3-C1	59.9	H8-O3-C1-N2	-11.5	H8-O3-C1-N2	-4.6	H8-O3-C1-N2	17.6
N9-O8-H5-O3	204.0	O9-H8-O3-C1	-16.9	O9-H8-O3-C1	137.6	O9-H8-O3-C1	126.2
C10-N9-O8-H5	70.4	N10-O9-H8-O7	-9.1	C10-O9-H8-O3	177.7	C10-O9-H8-O3	252.5
O11-C10-N9-O8	-0.4	C11-N10-O9-H8	187.7	N11-C10-O9-H5	-0.1	N11-C10-O9-H5	16.1
H12-O11-C10-N2	8.9	O12-C11-N10-O9	0.6	O12-N11-C10-O9	-35.4	O12-N11-C10-O9	182.4
H13-C10-N9-O8	179.7	H13-O12-C11-N10	1.1	H13-O12-N11-C10	196.4	H13-O12-N11-C10	205.1
H14-O8-N9-C10	196.3	H14-C11-N10-O9	180.6	H14-C10-N11-O12	180.7	H14-C10-N11-O12	1.5

Table 2.20: Geometrical parameters of FHA dimers at MP2/6-31+G* [L3] theoretical level. All the bond distances are in angstrom (Å), angles and dihedrals are in degrees (°).

Parameters	1Z-1E (1)	Parameters	1Z-1E (2)	Parameters	1Z-1E (3)
N2-C1	1.355	N2-C1	1.361	N2-C1	1.369
O3-C1	1.236	O3-C1	1.235	O3-C1	1.234
O4-N2	1.394	O4-N2	1.401	O4-N2	1.419
H5-O4	0.996	H5-O4	0.986	H5-O4	0.976
H6-C1	1.102	H6-C1	1.102	H6-C1	1.101
H7-N2	1.015	H7-N2	1.015	H7-N2	1.029
O8-H5	1.760	O8-H5	1.904	O8-H7	1.921
C9-O8	1.242	N9-O8	1.416	C9-O8	1.248
N10-C9	1.354	C10-N9	1.381	N10-C9	1.347
O11-N10	1.413	O11-C10	1.226	O11-N10	1.404
H12-O11	0.977	H12-O8	0.990	H12-O11	0.989
H13-C9	1.099	H13-C10	1.103	H13-C9	1.096
H14-N10	1.032	H14-N9	1.019	H14-N10	1.014
O3-C1-N2	125.7	O3-C1-N2	125.3	O3-C1-N2	123.2
O4-N2-C1	119.6	O4-N2-C1	119.8	O4-N2-C1	114.3
H5-O4-N2	105.0	H5-O4-N2	104.5	H5-O4-N2	103.7
H6-C1-N2	111.7	H6-C1-N2	111.7	H6-C1-N2	113.1
H7-N2-C1	120.0	H7-N2-C1	119.8	H7-N2-C1	117.3
O8-H5-O4	168.2	O8-H5-O4	145.5	O8-H7-N2	164.5
C9-O8-H5	118.4	N9-O8-H5	131.2	C9-O8-H7	111.6
N10-C9-O8	123.8	C10-N9-O8	115.1	N10-C9-O8	120.8
O11-N10-C9	114.9	O11-C10-N9	123.2	O11-N10-C9	117.6
H12-O11-N10	103.9	H12-O8-N9	104.9	H12-O11-N10	101.6
H13-C9-O8	122.6	H13-C10-N9	112.5	H13-C9-O8	123.6
H14-N10-C9	120.3	H14-N9-C10	116.1	H14-N10-C9	123.0
O4-N2-C1-O3	-17.9	O4-N2-C1-O3	-21.1	O4-N2-C1-O3	161.6
H5-O4-N2-C1	88.5	H5-O4-N2-C1	76.8	H5-O4-N2-C1	119.0
H6-C1-N2-O4	166.3	H6-C1-N2-O4	164.0	H6-C1-N2-O4	-21.5

Contd.....

Parameters	1Z-1E (1)	Parameters	1Z-1E (2)	Parameters	1Z-1E (3)
H7-N2-C1-O3	-164.3	H7-N2-C1-O3	-164.0	H7-N2-C1-O3	27.3
O8-H5-O4-N2	260.7	O8-H5-O4-N2	-66.0	O8-H7-N2-C1	-21.1
C9-O8-H5-O4	3.3	N9-O8-H5-O4	117.1	C9-O8-H7-N2	1.5
N10-C9-O8-H5	9.0	C10-N9-O8-H5	11.4	N10-C9-O8-H7	181.2
O11-N10-C9-O8	166.1	O11-C10-N9-O8	159.1	O11-N10-C9-O8	-10.7
H12-O11-N10-C9	118.2	H12-O8-N9-O3	19.7	H12-O11-N10-C9	5.6
H13-C9-O8-H5	191.9	H13-C10-N9-O8	-24.8	H13-C9-O8-O3	-7.4
H14-N10-C9-O3	-5.2	H14-N9-C10-O11	27.7	H14-N10-C9-H13	26.3

Out of twenty five minima, twelve are keto-keto dimers in which five are **1Z-1Z** and four are characterized as **1E-1E** dimers and three are **1Z-1E** dimers. Eight are enol-enol dimers out of which three are **2Z-2Z** dimer and three **2E-2E** dimer and two **2Z-2E** dimers. Five mixed keto-enol dimeric structures have also been optimized. The H-bond interaction depends on distance between the hydrogen and H-bond acceptor (H...B) and A-H...B angle, these parameters are important for locating H-bonds. Few such important parameters indicating presence of H-bonds are reported in Table 2.21 at **[L3]** theoretical level. The stabilization energies resulting from the dimer formation, evaluated as $E_{dimer}-(E_{monomer1}+ E_{monomer2})$ and corrected for basis set superposition error (BSSE) are also reported in the Table 2.21. The dimerization is accompanied by some distortion of the monomer geometry. The energy change for the distortion of the monomers to the geometry in the dimer is also reported in the Table 2.21. The values indicate that in few cases the distortion energy is quite substantial e.g. (3.85 kcal/mol in case of most stable **1Z-1Z (1)** dimer). The distortion energies are comparatively smaller in comparison to stabilization energies. The atomic charges on atoms involved in H-bonding are recorded in Table 2.21 while charges on all the atoms of FHA and its dimers are recorded in Table 2.22. The high values of positive and negative atomic charges on the hydrogen and H-bond acceptor B, lend support to the point that electrostatic interactions form an important component of H-bonding.

The stabilization energies indicate that **1Z-1Z (1)** is the strongest homodimer. It is a symmetrical homodimer where H-bond interactions result from hydroxyl group of one FHA acting as H-bond donor and oxygen of the carbonyl group of the second FHA molecule acting as H-bond acceptor and vice versa. Dimerization has effect on the intramolecular H-bonds of monomers, for instance the intramolecular H-bond in one of the FHA units of **1Z-1Z (2)** is strengthened (cooperativity) as the H5...O3 distance decreases from 2.046 Å to 1.969 Å and in **1Z-1Z (5)** the intramolecular H-bond in both FHA units is strengthened as is evident from decrease in the H5...O3 distance from 2.046 Å (MP2/6-31+G*) to 1.989 Å in both the units. The stabilization energies in the homodimeric forms involving **1Z** conformation are comparatively larger in comparison to other tautomeric forms. Since the relative abundance of **1Z** tautomer is reported to be largest of all the tautomeric forms, the dimerization energies may be another reason favoring the dominance of **1Z**.

Table 2.21: Important intermolecular H-bonding parameters (hydrogen bond distances[a], hydrogen bond angles[b]), charges on hydrogen bond acceptor and hydrogen along with stabilization energies S.E.[c] and distortion energies E_{Dis}[d] of various hydrogen bonded complexes of FHA dimers at MP2/6-31+G* **[L3]** theoretical level.

Species	Hydrogen bond distances[a]		Hydrogen bond angles[b]		Atomic charges		S.E.[c]	E_{Dis}[d]
1Z-1Z (1)	H5...O8	1.808	O4-H5...O8	153.4	$q_H(q_O)$	0.561(-0.752)	16.12	3.85
	H12...O3	1.808	O11-H12...O3	153.3	$q_H(q_O)$	0.561(-0.752)		
1Z-1Z (2)	H7...O8	1.872	N2-H7...O8	163.4	$q_H(q_O)$	0.477(-0.757)	12.93	1.88
	H12...O4	1.825	O11-H12...O4	160.4	$q_H(q_O)$	0.551(-0.659)		
1Z-1Z (3)	H5...O8	1.771	O4-H5...O8	168.1	$q_H(q_O)$	0.559(-0.807)	11.36	2.08
	H13...O3	2.217	C9-H13...O3	141.1	$q_H(q_O)$	0.228(-0.749)		
1Z-1Z (4)	H13...O4	2.467	C9-H13...O4	121.5	$q_H(q_O)$	0.206(-0.648)	8.47	0.61
	H7...O8	1.907	N2-H7...O8	154.7	$q_H(q_O)$	0.471(-0.792)		
1Z-1Z (5)	H7...O8	2.013	N2-H7...O8	144.9	$q_H(q_O)$	0.460(-0.651)	8.09	4.85
	H14...O4	2.013	N9-H14...O4	145.0	$q_H(q_O)$	0.460(-0.651)		
1E-1E (1)	H7...O8	1.903	N2-H7...O8	169.1	$q_H(q_O)$	0.471(-0.761)	12.44	1.48
	H14...O3	1.903	N10-H14...O3	169.1	$q_H(q_O)$	0.471(-0.761)		
1E-1E (2)	H7...O8	2.002	N2-H7...O8	164.0	$q_H(q_O)$	0.464(-0.766)	12.81	5.01
	H12...O3	1.961	O11-H12...O3	147.5	$q_H(q_O)$	0.541(-0.784)		
1E-1E (3)	H12...O3	1.886	O8-H12...O3	145.7	$q_H(q_O)$	0.558(-0.737)	7.91	0.55
	H7...O8	2.155	N2-H7...O8	132.9	$q_H(q_O)$	0.447(-0.678)		

Contd.....

Species	Hydrogen bond distances[a]		Hydrogen bond angles[b]		Atomic charges		S.E.[c]	E_{Dis}[d]
1E-1E (4)	H13…O3	2.405	C9-H13…O3	132.0	$q_H(q_O)$	0.207(-0.736)	7.67	0.61
	H7…O8	1.936	N2-H7…O8	164.9	$q_H(q_O)$	0.463(-0.741)		
2Z-2Z (1)	H5…O8	1.941	O3-H5…O8	140.2	$q_H(q_O)$	0.561(-0.720)	6.06	1.16
	H12…O4	1.941	O11-H12…O4	140.1	$q_H(q_O)$	0.561(-0.720)		
2Z-2Z (2)	H5…N8	2.074	O3-H5…N8	123.1	$q_H(q_N)$	0.573(-0.320)	6.03	0.39
	H14…O3	2.062	O9-H14…O3	135.3	$q_H(q_O)$	0.546(-0.828)		
2Z-2Z (3)	H6…N8	2.523	C1-H6…N8	137.7	$q_H(q_N)$	0.233(-0.311)	3.02	0.09
	H13…N2	2.520	C10-H13…N2	137.8	$q_H(q_N)$	0.233(-0.311)		
2E-2E (1)	H5…N8	1.873	O3-H5…N8	172.5	$q_H(q_N)$	0.548(-0.362)	11.38	0.84
	H12…N2	1.910	O11-H12…N2	162.8	$q_H(q_N)$	0.543(-0.369)		
2E-2E (2)	H7…O8	1.873	O4-H7…O8	173.5	$q_H(q_O)$	0.541(-0.678)	5.42	0.17
	H13…N2	2.470	C10-H13…N2	142.6	$q_H(q_O)$	0.226(-0.328)		
2E-2E (3)	H6…O11	2.265	C1-H6…O11	160.4	$q_H(q_O)$	0.230(-0.664)	2.64	-0.13
	H8…O4	2.261	C9-H8…O4	160.9	$q_H(q_O)$	0.230(-0.664)		
1Z-1E (1)	H5…O8	1.760	O4-H5…O8	168.2	$q_H(q_O)$	0.560(-0.771)	15.55	2.92
	H14…O3	1.867	N10-H14…O3	170.5	$q_H(q_O)$	0.477(-0.758)		
1Z-1E (2)	H5…O8	1.904	O4-H5…O8	145.5	$q_H(q_O)$	0.543(-0.680)	9.42	1.72
	H12…O3	1.849	O8-H12…O3	150.6	$q_H(q_O)$	0.565(-0.747)		
1Z-1E (3)	H7…O8	1.921	N2-H7…O8	164.5	$q_H(q_O)$	0.465(-0.793)	8.67	0.69

Contd……

Species	Hydrogen bond distances[a]		Hydrogen bond angles[b]		Atomic charges		S.E.[c]	E_{Dis}[d]
	H13...O3	2.323	C9-H13...O3	134.7	$q_H(q_O)$	0.219(-0.744)		
1Z-2Z (1)	H8...O3	1.800	O9-H8...O3	159.7	$q_H(q_O)$	0.575(-0.746)	11.68	3.28
	H5...O12	1.800	O4-H5...O12	158.2	$q_H(q_O)$	0.550(-0.720)		
1Z-2E (1)	H8...O3	1.768	O9-H8...O3	168.5	$q_H(q_O)$	0.569(-0.767)	15.26	3.05
	H5...N11	1.810	O4-H5...N11	167.6	$q_H(q_N)$	0.549(-0.395)		
2Z-1E (1)	H7...N10	2.032	N2-H7...N10	152.6	$q_H(q_N)$	0.461(-0.332)	12.28	1.01
	H8...O3	1.805	O9-H8...O3	169.1	$q_H(q_O)$	0.565(-0.752)		
2Z-2E (1)	H5...N8	1.858	O3-H5...N8	169.5	$q_H(q_N)$	0.555(-0.343)	13.47	1.33
	H12...N2	1.866	O11-H12...N2	161.9	$q_H(q_N)$	0.551(-0.380)		
2Z-2E (2)	H5...O8	1.880	O3-H5...O8	162.7	$q_H(q_O)$	0.552(-0.722)	9.98	1.84
	H12...N2	1.832	O11-H12...N2	157.1	$q_H(q_N)$	0.561(-0.364)		
2E-1E (1)	H8...O3	1.857	O9-H8...O3	169.0	$q_H(q_O)$	0.551(-0.741)	9.97	0.73
	H7...N10	2.046	N2-H7...N10	154.5	$q_H(q_N)$	0.454(-0.352)		
2E-1E (2)	H8...O4	1.932	O9-H8...O4	154.3	$q_H(q_O)$	0.545(-0.680)	7.15	0.47
	H5...N11	1.969	O4-H5...N11	137.2	$q_H(q_N)$	0.554(-0.368)		

a - in angstrom (Å)
b - in degrees (°)
c, d - in kcal/mol

Table 2.22: Atomic charges (NPA) on atoms of FHA monomers and dimers in gas phase at MP2/6-31+G* [L3] theoretical level.

Species	Unit	C1	N2	O3	O4	H5	H6	H7
1Z		0.636	-0.426	-0.732	-0.616	0.535	0.173	0.430
1Z-1Z (1)	1	0.671	-0.430	-0.752	-0.635	0.561	0.162	0.422
	2	0.671	-0.430	-0.752	-0.635	0.561	0.162	0.423
1Z-1Z (2)	1	0.640	-0.438	-0.750	-0.659	0.550	0.180	0.477
	2	0.670	-0.431	-0.757	-0.627	0.551	0.166	0.430
1Z-1Z (3)	1	0.659	-0.435	-0.749	-0.637	0.559	0.155	0.423
	2	0.635	-0.404	-0.807	-0.608	0.539	0.228	0.442
1Z-1Z (4)	1	0.636	-0.440	-0.754	-0.648	0.540	0.174	0.471
	2	0.639	-0.407	-0.792	-0.605	0.538	0.206	0.442
1Z-1Z (5)	1	0.638	-0.433	-0.739	-0.651	0.546	0.179	0.460
	2	0.638	-0.433	-0.739	-0.651	0.546	0.179	0.460
1E		0.661	-0.454	-0.685	-0.626	0.517	0.166	0.422
1E-1E (1)	1	0.680	-0.462	-0.761	-0.615	0.516	0.171	0.471
	2	0.680	-0.462	-0.761	-0.615	0.516	0.171	0.471
1E-1E (2)	1	0.684	-0.453	-0.784	-0.610	0.518	0.174	0.464
	2	0.672	-0.421	-0.766	-0.615	0.541	0.176	0.419
1E-1E (3)	1	0.676	-0.453	-0.737	-0.614	0.519	0.178	0.447
	2	0.660	-0.443	-0.693	-0.678	0.558	0.158	0.422
1E-1E (4)	1	0.670	-0.469	-0.736	-0.624	0.515	0.164	0.463
	2	0.669	-0.444	-0.741	-0.616	0.518	0.207	0.424
2Z		0.478	-0.282	-0.783	-0.703	0.547	0.214	0.528
2Z-2Z (1)	1	0.493	-0.285	-0.796	-0.720	0.561	0.213	0.533
	2	0.493	-0.285	-0.796	-0.720	0.561	0.213	0.533
2Z-2Z (2)	1	0.478	-0.278	-0.828	-0.689	0.573	0.212	0.525
	2	0.499	-0.320	-0.780	-0.703	0.549	0.215	0.546
2Z-2Z (3)	1	0.485	-0.311	-0.780	-0.700	0.546	0.233	0.527
	2	0.485	-0.311	-0.780	-0.700	0.546	0.233	0.526
2E		0.488	-0.312	-0.772	-0.650	0.524	0.206	0.516
2E-2E (1)	1	0.536	-0.369	-0.800	-0.641	0.548	0.205	0.514
	2	0.520	-0.361	-0.767	-0.661	0.524	0.210	0.543
2E-2E (2)	1	0.490	-0.328	-0.773	-0.671	0.521	0.205	0.541
	2	0.495	-0.315	-0.767	-0.677	0.525	0.226	0.527
2E-2E (3)	1	0.483	-0.320	-0.770	-0.664	0.522	0.230	0.519
	2	0.483	-0.320	-0.770	-0.664	0.522	0.230	0.519
1Z-1E (1)	1	0.669	-0.429	-0.758	-0.638	0.560	0.162	0.426
	2	0.682	-0.457	-0.771	-0.613	0.516	0.175	0.477
1Z-1E (2)	1	0.664	-0.440	-0.747	-0.615	0.543	0.170	0.432
	2	0.658	-0.440	-0.690	-0.680	0.565	0.154	0.426
1Z-1E (3)	1	0.672	-0.468	-0.744	-0.622	0.515	0.166	0.465
	2	0.634	-0.410	-0.793	-0.610	0.538	0.219	0.439
1Z-2Z (1)	1	0.671	-0.436	-0.746	-0.628	0.550	0.163	0.428
	2	0.511	-0.311	-0.795	-0.720	0.575	0.210	0.529
1Z-2E (1)	1	0.673	-0.429	-0.767	-0.636	0.549	0.165	0.429
2Z-1E (1)	1	0.682	-0.460	-0.752	-0.617	0.517	0.174	0.461

Contd.....

Species	Unit	C1	N2	O3	O4	H5	H6	H7
	2	0.497	-0.332	-0.783	-0.715	0.552	0.210	0.565
2E-2Z (1)	1	0.546	-0.380	-0.795	-0.641	0.555	0.205	0.512
	2	0.511	-0.343	-0.779	-0.712	0.551	0.215	0.556
2E-2Z (2)	1	0.534	-0.364	-0.791	-0.639	0.552	0.206	0.516
	2	0.504	-0.295	-0.806	-0.722	0.561	0.213	0.532
2E-1E (1)	1	0.680	-0.463	-0.741	-0.621	0.517	0.173	0.454
	2	0.501	-0.352	-0.771	-0.664	0.520	0.209	0.551
2E-1E (2)	1	0.659	-0.444	-0.683	-0.680	0.554	0.160	0.424
	2	0.526	-0.368	-0.637	-0.783	0.515	0.210	0.545

1E-1E (1) homodimeric form is symmetrical with NH of one unit acting as H-bond donor and carbonyl group of second unit acting as H-bond acceptor. These interactions are comparable to amide-amide interactions. A statistical study of amide-amide interactions in hundreds of high resolution crystal structures in the Cambridge structural database suggested mean value of 1.9 Å for H...O bond distance and 171° for N-H...O angle [24]. The bond distance in **1E-1E (1)** dimeric form is 1.903 Å with N-H...O angle 169.1° at MP2/6-31+G* level which is comparable to the corresponding values for amides. Leutwyler et al. calculated the stabilization of formamide dimer with two N-H...O bonds to be 14.9 kcal/mol at MP2+ΔCCSD(T) while for **1E-1E (1)** of FHA with two similar N-H...O bonds it is 12.44 kcal/mol at [L3] theoretical level [25].

The stabilization arising out of **2Z** and **2E** dimerization is comparatively smaller than that for **1Z** and **1E** tautomers. The angular property suggests that deviation from linearity is comparatively larger. Among the dimeric forms of **2E** tautomer, **2E-2E (1)** homodimer results with stabilization energy of 11.38 kcal/mol at [L3] theoretical level which is relatively much higher than its other dimeric structures. In this dimeric form, proximity of hydroxyl groups attached to carbon and nitrogen of the two different monomers with 'NOH' group of other monomer along with near to linear angle centering hydrogen suggest stronger interactions. All the dimers studied are doubly H-bonded. **1Z-1Z (3)**, **1Z-1Z (4)**, **1E-1E (4)**, **1Z-1E (3)** and **2E-2E (3)** contain one of the H-bonds as weak C-H...O H-bond ranging from 2.217 Å to 2.467 Å, characteristic of C-H...O H-bond. Among C-H...O interactions, the strongest C-H...O H-bonding in **1Z-1Z (3)** may be assigned to the synergic strengthening of the weak C-H...O interaction by simultaneous formation of strong O4-H5...O8 H-bond. Out of all the optimized dimers **1Z-1Z (6)**, **1E-1E (3)**, **1E-1E (5)**, **1Z-1E (2)**, **2Z-2Z**

(2) and **2E-1E (2)** are chiefly end on dimers. **1Z-1Z (1), 1Z-1Z (5), 1E-1E (1), 2Z-2Z (1), 2Z-2Z (3), 2E-2E (3)** are centrosymmetric dimers with the two A-H...B bonds geometrically and energetically equivalent as is evident from equivalent bond lengths, charges and associated angles of the two interacting FHA units.

A total of ten mixed dimeric structures are formed among **1Z, 1E, 2Z** and **2E** tautomeric forms. The stabilization energies for the mixed dimerization suggest strong H-bond interaction between the monomer units having comparable stabilization energies as in homodimers.

Effect of solvent on dimerization is also studied employing polarized continuum model (PCM). Table 2.23 records the interaction energy in gas phase $(S.E.)_g$, in aqueous phase $(S.E.)_{aq}$, free energy of solvation (ΔG_{sol}) and change in free energy of association $(\Delta\Delta G_{Assoc})$. The $(S.E.)_{aq}$ and $\Delta\Delta G_{Assoc}$ are defined by equations 1.5.4.1 and 1.5.4.2 respectively (chapter 1). It is found that interaction of solvent with water molecules present in the aggregates tends to weaken the dimers and it results in reduced interaction energy concluding that the dimerization phenomena becomes unfavorable in aqueous conditions. The intermolecular interactions of two FHA units compete with interactions of monomeric FHA with medium. Even in aqueous phase **1Z-1Z (1)** is the most stable dimeric form.

A comparison of the stabilization energy of water aggregates and the dimers of FHA (N-hydroxy formamide) at **[L3]** theoretical level depicts that for the keto **1Z** and **1E** form the amide-amide H-bonds are stronger than the amide-water interactions. The interaction energy of **1Z-1Z (1)** is 16.12 kcal/mol and of **1ZW1** is 10.88 kcal/mol at **[L3]** theoretical level. For **1E-1E (1)** it is 12.44 kcal mol^{-1} as compared to **1EW1** which is 8.38 kcal/mol.

Table 2.23: Stabilization energies in gas phase (S.E.)$_g$, in aqueous solution (S.E.)$_{aq}$ free energy of solvation (ΔG_{sol}), difference in free energy of solvation of the dimer and the free energy of solvation of isolated monomers ($\Delta\Delta G_{Assoc}$) in kcal/mol at MP2/6-31+G* [L3] theoretical level.

Sr. No	Species	Gas Phase	PCM (Water)		
		(S.E.)$_g^a$ (-ive)	ΔG_{sol}	($\Delta\Delta G_{Assoc}$)^b	(S.E.)$_{aq}^c$
1	1Z	-	11.33	-	-
2	1E	-	12.13	-	-
3	2Z	-	10.02	-	-
4	2E	-	11.90	-	-
5	1Z-1Z (1)	19.97	15.37	7.29	-12.68
6	1Z-1Z (2)	14.81	11.82	10.84	-3.97
7	1Z-1Z (3)	13.44	17.24	5.42	-8.02
8	1Z-1Z (4)	9.08	14.05	8.61	-0.47
9	1Z-1Z (5)	12.94	11.03	11.63	-1.31
10	1E-1E (1)	13.92	12.27	11.99	-1.93
11	1E-1E (2)	17.82	12.70	11.56	-6.26
12	1E-1E (3)	8.46	14.44	9.82	1.36
13	1E-1E (4)	8.28	15.44	8.82	0.54
14	2Z-2Z (1)	7.22	12.74	7.30	0.08
15	2Z-2Z (2)	6.42	10.94	9.10	2.68
16	2Z-2Z (3)	3.11	13.97	6.07	2.96
17	2E-2E (1)	12.22	12.49	11.31	-0.91
18	2E-2E (2)	5.59	16.27	7.53	1.94
19	2E-2E (3)	2.77	18.09	5.71	2.94
20	1Z-1E (1)	18.47	13.84	9.62	-8.85
21	1Z-1E (2)	11.14	14.41	9.05	-2.09
22	1Z-1E (3)	9.36	14.65	8.81	-0.55
23	1Z-2Z (1)	14.96	14.82	6.53	-8.43
24	1Z-2E (1)	18.31	13.48	9.75	-8.56
25	2Z-1E (1)	13.29	10.50	11.65	-1.64
26	2E-2Z (1)	14.80	9.42	12.5	-2.30
27	2E-2Z (2)	11.82	11.99	9.93	-1.89
28	2E-1E (1)	10.70	12.80	11.23	0.53
29	2E-1E (2)	7.62	12.73	11.30	3.68

(S.E.)$_g^a$ is stabilization energies including E$_{Dis}$
($\Delta\Delta G_{Assoc}$)^b = (ΔG_{sol})$_{dimer}$-[(ΔG_{sol})$_{monomer1}$+(ΔG_{sol})$_{monomer2}$]
(S.E.)$_{aq}^c$ = (S.E.)$_g^a$ + $\Delta\Delta G_{Assoc}$

2.2.6 NATURAL BOND ORBITAL (NBO) ANALYSIS:

NBO analysis is a reliable tool to derive information about the charge densities spreading over bonding and antibonding regions of the molecule. It has been found to explain various conjugative interactions in all types of molecules. Since the H-bonding involves change in H-bond donor and acceptor charge densities, NBO analysis can be utilized for understanding the component of H-bond formation that arises as the result of charge transfer. NBO theory describes the H-bond formation of A-H...B as the charge transfer from the lone pair, n (B), of the base B into the vacant antibonding orbital $\sigma^*_{(A-H)}$ of H-bond donor.

The second order delocalization energies arising from the charge transfer indicates the presence of strong conjugative interactions in various isomeric forms of FHA and only one $n_{O3} \rightarrow \sigma^*_{O4-H5}$ delocalization in **1Z** contributes toward H-bond formation with $E^{(2)}$ value of 5.22 kcal/mol at **[L4]** theoretical level. Since the lone pairs present on basic centers (N, O) are also involved in conjugative interactions with rest of the molecule itself, the occupancies of σ^*_{H-A} (accepting electron density from H-bond donor) are better indicator of H-bonding interactions. The electron delocalizations that are important for H-bond formation in aggregates of FHA with water and dimeric forms are recorded in Tables 2.24 and 2.25 respectively. The second order delocalization energies associated with these interactions are also included. The $E^{(2)}$ values associated with charge transfer between the H-bond acceptor and donor groups reflect the stabilization of the H-bond as the result of covalent character arising from the charge transfer. The σ^*_{A-H} occupancies characterize the extent of charge transfer.

Table 2.24: The second order delocalization energies $E^{(2)}$ (kcal/mol) and occupancies of acceptor orbitals associated with electron delocalizations responsible for H-bond formation in FHA-H$_2$O aggregates at MP2/Aug-cc-pVDZ **[L4]** theoretical level.

Species	Donor FHA	Acceptor H$_2$O	$E^{(2)}$	Donor H$_2$O	Acceptor FHA	$E^{(2)}$	Occupancies Acceptor H$_2$O	Acceptor FHA
1ZW1	n$_{O3(1)}$ →	σ*$_{O8-H9}$	5.09	n$_{O8(1)}$ →	σ*$_{O4-H5}$	0.53	σ*$_{O8-H9}$ 0.019	σ*$_{O4-H5}$ 0.036
	n$_{O3(2)}$ →	σ*$_{O8-H9}$	6.76	n$_{O8(2)}$ →	σ*$_{O4-H5}$	25.13		
1ZW2	n$_{O4(1)}$ →	σ*$_{O8-H9}$	0.71	n$_{O8(1)}$ →	σ*$_{N2-H7}$	0.27	σ*$_{O8-H9}$ 0.004	σ*$_{N2-H7}$ 0.022
	n$_{O4(2)}$ →	σ*$_{O8-H9}$	1.76	n$_{O8(2)}$ →	σ*$_{N2-H7}$	7.27		
1ZW3	n$_{O3(1)}$ →	σ*$_{O9-H8}$	2.00	n$_{O9(1)}$ →	σ*$_{C1-H6}$	0.37	σ*$_{O9-H8}$ 0.011	σ*$_{C1-H6}$ 0.039
	n$_{O3(2)}$ →	σ*$_{O9-H8}$	5.84	n$_{O9(2)}$ →	σ*$_{C1-H6}$	0.05		
1ZW4				n$_{O8(1)}$ →	σ*$_{N2-H7}$	0.32		σ*$_{N2-H7}$ 0.032
				n$_{O8(2)}$ →	σ*$_{N2-H7}$	17.93		
1EW1	n$_{O3(1)}$ →	σ*$_{O8-H9}$	2.80	n$_{O8(1)}$ →	σ*$_{N2-H7}$	0.26	σ*$_{O8-H9}$ 0.015	σ*$_{N2-H7}$ 0.027
	n$_{O3(2)}$ →	σ*$_{O8-H9}$	7.00	n$_{O8(2)}$ →	σ*$_{N2-H7}$	8.90		
1EW2	n$_{N2}$ →	σ*$_{O8-H10}$	0.97	n$_{O8(1)}$ →	σ*$_{O4-H5}$	0.29	σ*$_{O8-H10}$ 0.002	σ*$_{O4-H5}$ 0.019
				n$_{O8(2)}$ →	σ*$_{O4-H5}$	13.23		
1EW3	n$_{N2}$ →	σ*$_{O8-H10}$	0.84	n$_{O8(1)}$ →	σ*$_{O4-H5}$	0.29	σ*$_{O8-H10}$ 0.002	σ*$_{O4-H5}$ 0.019
				n$_{O8(2)}$ →	σ*$_{O4-H5}$	13.58		
1EW4	n$_{O3(1)}$ →	σ*$_{O9-H8}$	1.96	n$_{O9(1)}$ →	σ*$_{C1-H6}$	0.30	σ*$_{O9-H8}$ 0.014	σ*$_{C1-H6}$ 0.044

Contd......

Species	Donor FHA	Acceptor H$_2$O	E$^{(2)}$	Donor H$_2$O	Acceptor FHA	E$^{(2)}$	Occupancies Acceptor H$_2$O		Acceptor FHA	
1EW5	n$_{O3(2)}$ →	σ*$_{O9\text{-}H8}$	6.88							
	n$_{O4(1)}$ →	σ*$_{O9\text{-}H8}$	2.21	n$_{O9(1)}$ →	σ*$_{C1\text{-}H6}$	0.38	σ*$_{O9\text{-}H8}$	0.009	σ*$_{C1\text{-}H6}$	0.046
	n$_{O4(2)}$ →	σ*$_{O9\text{-}H8}$	4.40	n$_{O9(2)}$ →	σ*$_{C1\text{-}H6}$	0.99				
2ZW1	n$_{N2}$ →	σ*$_{O8\text{-}H10}$	3.70	n$_{O8(1)}$ →	σ*$_{O4\text{-}H7}$	0.31	σ*$_{O8\text{-}H10}$	0.006	σ*$_{O4\text{-}H7}$	0.019
				n$_{O8(2)}$ →	σ*$_{O4\text{-}H7}$	12.63				
2ZW2	n$_{O4(1)}$ →	σ*$_{O8\text{-}H9}$	4.09	n$_{O8(1)}$ →	σ*$_{O3\text{-}H5}$	0.52	σ*$_{O8\text{-}H9}$	0.008	σ*$_{O3\text{-}H5}$	0.037
	n$_{O4(2)}$ →	σ*$_{O8\text{-}H9}$	0.49	n$_{O8(2)}$ →	σ*$_{O3\text{-}H5}$	22.96				
2ZW3	n$_{O3(1)}$ →	σ*$_{O9\text{-}H8}$	2.32	n$_{O9(1)}$ →	σ*$_{C1\text{-}H6}$	0.76	σ*$_{O9\text{-}H8}$	0.004	σ*$_{C1\text{-}H6}$	0.016
2EW1	n$_{N2}$ →	σ*$_{O9\text{-}H8}$	9.10	n$_{O9(1)}$ →	σ*$_{O3\text{-}H5}$	0.49	σ*$_{O9\text{-}H8}$	0.016	σ*$_{O3\text{-}H5}$	0.034
				n$_{O9(2)}$ →	σ*$_{O3\text{-}H5}$	21.14				
2EW2	n$_{N2}$ →	σ*$_{O8\text{-}H10}$	4.22	n$_{O8(1)}$ →	σ*$_{O4\text{-}H7}$	0.26	σ*$_{O8\text{-}H10}$	0.007	σ*$_{O4\text{-}H7}$	0.015
				n$_{O8(2)}$ →	σ*$_{O4\text{-}H7}$	10.01				
2EW3	n$_{O3(1)}$ →	σ*$_{O9\text{-}H8}$	2.44	n$_{O9(1)}$ →	σ*$_{C1\text{-}H6}$	0.68	σ*$_{O9\text{-}H8}$	0.004	σ*$_{C1\text{-}H6}$	0.025
				n$_{O9(2)}$ →	σ*$_{C1\text{-}H6}$	0.30				

The $E^{(2)}$ values for FHA-H_2O aggregates suggest strong intermolecular H-bond interactions resulting from charge transfer. The $E^{(2)}$ values and σ^*_{A-H} occupancies of FHA aggregates suggest that FHA has better electron acceptor ability than water. Both the lone pair of electrons present on each oxygen of carbonyl of FHA and water molecule are delocalized but to different extent. The water molecule has better electron donor ability than FHA. The sum of the two $E^{(2)}$ values for the delocalization of lone pairs present on oxygen is highest in **1ZW1**. This stabilization also favors the existence of **1Z** conformation of FHA in aqueous medium. The strong H-bond interaction results where hydroxyl group of FHA acts as H-bond donor to oxygen of water.

The $E^{(2)}$ values for various delocalizations responsible for increasing H-bonding interactions in dimers are given in Table 2.25. Relatively higher second order delocalization energies and occupancies of acceptor antibonding orbitals are observed for the dimers in comparison to the values for FHA-H_2O aggregates and suggest that H-bonding in dimers has significant contribution from the charge transfer component. As a result of intermolecular $n_{(B)} \to \sigma^*_{A-H}$ charge transfer, it is observed that some internal delocalizations are enhanced. The conjugative interactions are affected on aggregation with water and also during self aggregation. These are recorded in Tables 2.26 and 2.27 respectively. Upon coordination with water, the adduct **1ZW2** shows an enhancement of the intramolecular H-bonding delocalization $n_{O3} \to \sigma^*_{O4-H5}$ from an $E^{(2)}$ of 5.22 kcal/mol in **1Z** to $E^{(2)}$ of 6.22 kcal/mol which is also indicated by decrease in O3…H5 distance (cooperativity). As a result of synergic enhancement, the intramolecular H-bond in **1Z-1Z (5)** and in **1Z-1Z (2)** gets strengthened on the formation of intermolecular H-bonds between two FHA units upon dimerization. The delocalization representing intramolecular H-bonding interaction $n_{O3} \to \sigma^*_{O4-H5}$ has $E^{(2)}$ value of 7.05 kcal/mol in FHA$^{(1)}$ and $n_{O11} \to \sigma^*_{O8-H12}$ in FHA$^{(2)}$ in **1Z-1Z (5)** as compared to $E^{(2)}$ value of 5.36 kcal/mol in its **1Z** monomer. Similarly in **1Z-1Z (2)** $n_{O3} \to \sigma^*_{O4-H5}$ has $E^{(2)}$ value of 7.88 kcal/mol which is a clear case of synergic enhancement of intramolecular H-bonding upon the formation of intermolecular H-bond (Table 2.27). This conclusion is also supported by the Wiberg Bond Index (WBI) which for intramolecular H5…O3 H-bond is 0.019 in **1Z** conformation and in the dimer **1Z-1Z (5)** it is 0.025 and 0.026 in the constituent FHA$^{(1)}$ and FHA$^{(2)}$. For **1Z-1Z (2)** dimer Wiberg bond order in FHA$^{(1)}$ for intramolecular H-bond increases from 0.019 in **1Z** to 0.028. This increase of WBI is also supportive of increased strength of intramolecular H5…O3 hydrogen bond upon dimerization. It is seen that the magnitude of the conjugative interactions is same in both the units of the symmetrical homodimers. When intramolecular H-bond is intact e.g. in **1Z-1Z (5)**, the $n_{N2} \rightarrow \pi^*_{C1-O3}$ shows small increase but when it ruptures e.g. in **1Z-1Z (1)**, there is huge increase in this delocalization. The $E^{(2)}$ values for the orbital interactions involving lone pair of electrons present at N in dimers of **1Z** and **1E** are enhanced suggesting that conjugative interactions are more favorable in the dimers in comparison to monomers. Similar increase in orbital interactions is also reflected for aggregates of **1Z** isomer with water.

Table 2.25: The second order delocalization energies $E^{(2)}$ in kcal/mol and occupancies of acceptor orbitals associated with electron delocalizations responsible for H-bond formation in dimeric forms of FHA in gas phase at MP2/6-31+G* [L3] theoretical level.

Species	Donor FHA$^{(1)}$	Acceptor FHA$^{(2)}$	$E^{(2)}$	Donor FHA$^{(2)}$	Acceptor FHA$^{(1)}$	$E^{(2)}$	Occupancies Acceptor FHA$^{(2)}$		Acceptor FHA$^{(1)}$	
1Z-1Z (1)	$n_{O3(1)} \to \sigma^*_{O11-H12}$		8.79	$n_{O8(1)} \to \sigma^*_{O4-H5}$		8.79	$\sigma^*_{O11-H12}$	0.032	σ^*_{O4-H5}	0.032
	$n_{O3(2)} \to \sigma^*_{O11-H12}$		10.83	$n_{O8(2)} \to \sigma^*_{O4-H5}$		10.84				
1Z-1Z (2)	$n_{O4(1)} \to \sigma^*_{O11-H12}$		1.04	$n_{O8(1)} \to \sigma^*_{N2-H7}$		10.44	$\sigma^*_{O11-H12}$	0.030	σ^*_{N2-H7}	0.037
	$n_{O4(2)} \to \sigma^*_{O11-H12}$		20.40	$n_{O8(2)} \to \sigma^*_{N2-H7}$		9.92				
1Z-1Z (3)	$n_{O3(1)} \to \sigma^*_{C9-H13}$		3.18	$n_{O8(1)} \to \sigma^*_{O4-H5}$		9.27	σ^*_{C9-H13}	0.038	σ^*_{O4-H5}	0.036
	$n_{O3(2)} \to \sigma^*_{C9-H13}$		1.69	$n_{O8(2)} \to \sigma^*_{O4-H5}$		17.59				
1Z-1Z (4)	$n_{O4(1)} \to \sigma^*_{C9-H13}$		1.28	$n_{O8(1)} \to \sigma^*_{N2-H7}$		7.65	σ^*_{C9-H13}	0.038	σ^*_{N2-H7}	0.033
	$n_{O4(2)} \to \sigma^*_{C9-H13}$		0.69	$n_{O8(2)} \to \sigma^*_{N2-H7}$		10.60				
1Z-1Z (5)	$n_{O4(1)} \to \sigma^*_{N9-H14}$		1.14	$n_{O8(1)} \to \sigma^*_{N2-H7}$		1.14	σ^*_{N9-H14}	0.026	σ^*_{N2-H7}	0.026
	$n_{O4(2)} \to \sigma^*_{N9-H14}$		11.28	$n_{O8(2)} \to \sigma^*_{N2-H7}$		11.28				
1E-1E (1)	$n_{O3(1)} \to \sigma^*_{N10-H14}$		9.22	$n_{O8(1)} \to \sigma^*_{N2-H7}$		9.22	$\sigma^*_{N10-H14}$	0.040	σ^*_{N2-H7}	0.040
	$n_{O3(2)} \to \sigma^*_{N10-H14}$		11.65	$n_{O8(2)} \to \sigma^*_{N2-H7}$		11.64				
1E-1E (2)	$n_{O3(1)} \to \sigma^*_{O11-H12}$		8.62	$n_{O8(1)} \to \sigma^*_{N2-H7}$		5.12	$\sigma^*_{O11-H12}$	0.017	σ^*_{N2-H7}	0.034
	$n_{O3(2)} \to \sigma^*_{O11-H12}$		0.83	$n_{O8(2)} \to \sigma^*_{N2-H7}$		0.73				
1E-1E (3)	$n_{O3(1)} \to \sigma^*_{O8-H12}$		6.05	$n_{O8(1)} \to \sigma^*_{N2-H7}$		2.64	σ^*_{O8-H12}	0.026	σ^*_{N2-H7}	0.024
	$n_{O3(2)} \to \sigma^*_{O8-H12}$		10.98	$n_{O8(2)} \to \sigma^*_{N2-H7}$		3.38				
1E-1E (4)	$n_{O3(1)} \to \sigma^*_{C9-H13}$		1.54	$n_{O8(1)} \to \sigma^*_{N2-H7}$		6.56	σ^*_{C9-H13}	0.042	σ^*_{N2-H7}	0.036
	$n_{O3(2)} \to \sigma^*_{C9-H13}$		1.58	$n_{O8(2)} \to \sigma^*_{N2-H7}$		11.56				
2Z-2Z (1)	$n_{O4(1)} \to \sigma^*_{O11-H12}$		4.66	$n_{O8(1)} \to \sigma^*_{O3-H5}$		4.66	$\sigma^*_{O11-H12}$	0.027	σ^*_{O3-H5}	0.027
	$n_{O4(2)} \to \sigma^*_{O11-H12}$		8.12	$n_{O8(2)} \to \sigma^*_{O3-H5}$		8.14				
2Z-2Z (2)	$n_{O3(1)} \to \sigma^*_{O9-H14}$		4.54	$n_{N8} \to \sigma^*_{O3-H5}$		10.61	σ^*_{O9-H14}	0.010	σ^*_{O3-H5}	0.028
	$n_{O3(2)} \to \sigma^*_{O9-H14}$		1.31							
2Z-2Z (3)	$n_{N2} \to \sigma^*_{C10-H13}$		2.94	$n_{N8} \to \sigma^*_{C1-H6}$		2.91	$\sigma^*_{C10-H13}$	0.020	σ^*_{C1-H6}	0.020

Contd......

Species	Donor FHA(1)	Acceptor FHA(2)	E(2)	Donor FHA(2)	Acceptor FHA(1)	E(2)	Occupancies			
							Acceptor FHA(2)		Acceptor FHA(1)	
2E-2E (1)	$n_{N2} \rightarrow$	$\sigma^*_{O11\text{-}H10}$	23.01	$n_{N8} \rightarrow$	$\sigma^*_{O3\text{-}H5}$	28.94	$\sigma^*_{O11\text{-}H10}$	0.037	$\sigma^*_{O3\text{-}H5}$	0.051
2E-2E (2)	$n_{N2} \rightarrow$	$\sigma^*_{C10\text{-}H13}$	4.10	$n_{O8(1)} \rightarrow$	$\sigma^*_{O4\text{-}H7}$	5.36	$\sigma^*_{C10\text{-}H13}$	0.029	$\sigma^*_{O4\text{-}H7}$	0.024
				$n_{O8(2)} \rightarrow$	$\sigma^*_{O4\text{-}H7}$	13.46				
2E-2E (3)	$n_{O4(1)} \rightarrow$	$\sigma^*_{C9\text{-}H8}$	3.24	$n_{O11(1)} \rightarrow$	$\sigma^*_{C1\text{-}H6}$	3.24	$\sigma^*_{C9\text{-}H8}$	0.030	$\sigma^*_{C1\text{-}H6}$	0.030
	$n_{O4(2)} \rightarrow$	$\sigma^*_{C9\text{-}H8}$	2.98	$n_{O11(2)} \rightarrow$	$\sigma^*_{C1\text{-}H6}$	2.89				
1Z-1E (1)	$n_{O3(1)} \rightarrow$	$\sigma^*_{N10\text{-}H14}$	10.74	$n_{O8(1)} \rightarrow$	$\sigma^*_{O4\text{-}H5}$	9.90	$\sigma^*_{N10\text{-}H14}$	0.043	$\sigma^*_{O4\text{-}H5}$	0.040
	$n_{O3(2)} \rightarrow$	$\sigma^*_{N10\text{-}H14}$	11.33	$n_{O8(2)} \rightarrow$	$\sigma^*_{O4\text{-}H5}$	19.99				
1Z-1E (2)	$n_{O3(1)} \rightarrow$	$\sigma^*_{O8\text{-}H12}$	7.82	$n_{O8(1)} \rightarrow$	$\sigma^*_{O4\text{-}H5}$	2.09	$\sigma^*_{O8\text{-}H12}$	0.028	$\sigma^*_{O4\text{-}H5}$	0.023
	$n_{O3(2)} \rightarrow$	$\sigma^*_{O8\text{-}H12}$	8.90	$n_{O8(2)} \rightarrow$	$\sigma^*_{O4\text{-}H5}$	12.58				
1Z-1E (3)	$n_{O3(1)} \rightarrow$	$\sigma^*_{C9\text{-}H13}$	2.02	$n_{O8(1)} \rightarrow$	$\sigma^*_{N2\text{-}H7}$	7.62	$\sigma^*_{C9\text{-}H13}$	0.039	$\sigma^*_{N2\text{-}H7}$	0.037
	$n_{O3(2)} \rightarrow$	$\sigma^*_{C9\text{-}H13}$	2.09	$n_{O8(2)} \rightarrow$	$\sigma^*_{N2\text{-}H7}$	10.86				
1Z-2Z (1)	$n_{O3(1)} \rightarrow$	$\sigma^*_{O9\text{-}H8}$	10.90	$n_{O12(1)} \rightarrow$	$\sigma^*_{O4\text{-}H5}$	3.16	$\sigma^*_{O9\text{-}H8}$	0.038	$\sigma^*_{O4\text{-}H5}$	0.032
	$n_{O3(2)} \rightarrow$	$\sigma^*_{O9\text{-}H8}$	13.15	$n_{O12(2)} \rightarrow$	$\sigma^*_{O4\text{-}H5}$	20.39				
1Z-2E (1)	$n_{O3(1)} \rightarrow$	$\sigma^*_{O9\text{-}H8}$	12.77	$n_{N11} \rightarrow$	$\sigma^*_{O4\text{-}H5}$	31.08	$\sigma^*_{O9\text{-}H8}$	0.042	$\sigma^*_{O4\text{-}H5}$	0.053
	$n_{O3(2)} \rightarrow$	$\sigma^*_{O9\text{-}H8}$	14.51							
2Z-1E (1)	$n_{O3(1)} \rightarrow$	$\sigma^*_{O9\text{-}H8}$	8.99	$n_{N10} \rightarrow$	$\sigma^*_{N2\text{-}H7}$	17.71	$\sigma^*_{O9\text{-}H8}$	0.034	$\sigma^*_{N2\text{-}H7}$	0.040
	$n_{O3(2)} \rightarrow$	$\sigma^*_{O9\text{-}H8}$	16.23							
2Z-2E (1)	$n_{N2} \rightarrow$	$\sigma^*_{O11\text{-}H12}$	27.84	$n_{N8} \rightarrow$	$\sigma^*_{O3\text{-}H5}$	29.12	$\sigma^*_{O11\text{-}H12}$	0.045	$\sigma^*_{O3\text{-}H5}$	0.050
2Z-2E (2)	$n_{N2} \rightarrow$	$\sigma^*_{O11\text{-}H12}$	26.87	$n_{O8(1)} \rightarrow$	$\sigma^*_{O3\text{-}H5}$	3.76	$\sigma^*_{O11\text{-}H12}$	0.050	$\sigma^*_{O3\text{-}H5}$	0.035
2E-1E (1)	$n_{O3(1)} \rightarrow$	$\sigma^*_{O9\text{-}H8}$	7.46	$n_{N10} \rightarrow$	$\sigma^*_{N2\text{-}H7}$	17.19	$\sigma^*_{O9\text{-}H8}$	0.029	$\sigma^*_{N2\text{-}H7}$	0.040
	$n_{O3(2)} \rightarrow$	$\sigma^*_{O9\text{-}H8}$	13.40							
2E-1E (2)	$n_{O4(1)} \rightarrow$	$\sigma^*_{O9\text{-}H8}$	1.89	$n_{N11} \rightarrow$	$\sigma^*_{O4\text{-}H5}$	17.05	$\sigma^*_{O9\text{-}H8}$	0.027	$\sigma^*_{O4\text{-}H5}$	0.031
	$n_{O4(2)} \rightarrow$	$\sigma^*_{O9\text{-}H8}$	11.75							

Table 2.26: Second order delocalization energies $E^{(2)}$ in kcal/mol associated with orbital interactions obtained by NBO analysis of isolated monomers and aggregates of FHA with water at MP2/Aug-cc-pVDZ [L4] theoretical level.

Species	$n_{N2} \rightarrow \sigma^*_{C1-O3}$	$n_{N2} \rightarrow \pi^*_{C1-O3}$	$n_{O3} \rightarrow \sigma^*_{C1-N2}$	$n_{O3} \rightarrow \pi^*_{C1-N2}$	$n_{O4} \rightarrow \pi^*_{C1-N2}$	$n_{O3} \rightarrow \sigma^*_{O4-H5}$	$n_{O4} \rightarrow \sigma^*_{O3-H5}$
1Z	15.37	29.99	29.97	-	-	0.87+5.22	-
1ZW1	-	93.40	28.44	-	-	-	-
1ZW2	13.33	32.58	29.15	-	-	1.06+6.22	-
1ZW3	-	50.23	27.90	-	-	0.99+4.43	-
1ZW4	-	58.83	28.58	-	-	0.88+5.80	-
1E	9.92	23.50	32.87	-	-	-	-
1EW1	-	38.56	28.40	-	-	-	-
1EW2	10.05	22.89	32.50	-	-	-	-
1EW3	10.06	24.08	32.38	-	-	-	-
1EW4	-	33.42	30.25	-	-	-	-
1EW5	-	21.66	33.59	-	-	-	-
2Z	13.55	-	-	54.30	13.33	-	3.78
2ZW1	13.06	-	-	55.03	14.83	-	4.36
2ZW2	15.29	-	59.62	11.24	-	-	1.17
2ZW3	13.79	-	-	50.50	13.33	-	4.18
2E	-	-	-	47.81	13.76	-	-
2EW1	-	-	-	50.12	10.92	-	-
2EW2	-	-	-	47.73	15.23	-	-
2EW3	-	-	-	44.03	14.15	-	-

Table 2.27: Second order delocalization energies $E^{(2)}$ in kcal/mol associated with orbital interactions obtained by NBO analysis of isolated monomers and dimeric aggregates of FHA at MP2/6-31+G* **[L3]** theoretical level.

Species	$n_{N2} \to \sigma^*_{C1-O3}$	$n_{N2} \to \pi^*_{C1-O3}$	$n_{O3} \to \sigma^*_{O4-H5}$	$n_{O3} \to \sigma^*_{C1-N2}$	$n_{O3} \to \pi^*_{C1-N2}$	$n_{O4} \to \sigma^*_{O3-H5}$	$n_{O4} \to \pi^*_{C1-N2}$
1Z	14.98	32.68	1.24+5.36	29.25	-	-	-
1Z-1Z (1)	-	83.99	-	26.48	-	-	-
	-	83.90	-	26.48	-	-	-
1Z-1Z (2)	-	44.53	1.76+7.88	27.87	-	-	-
	-	80.81	-	26.68	-	-	-
1Z-1Z (3)	-	62.05	-	30.02	-	-	-
	-	66.39	2.20+3.96	26.33	-	-	-
1Z-1Z (4)	-	48.13	1.43+6.49	28.00	-	-	-
	-	67.82	1.73+4.24	27.30	-	-	-
1Z-1Z (5)	12.06	37.63	1.62+7.05	28.53	-	-	-
	-	37.77	1.62+7.05	28.52	-	-	-
1E	-	20.90	-	32.25	-	-	-
1E-1E (1)	12.11	41.35	-	25.19	-	-	-
	-	41.29	-	25.19	-	-	-
1E-1E (2)	14.85	36.05	-	27.72	-	-	-
	-	73.72	-	27.96	-	-	-
1E-1E (3)	12.30	31.36	-	26.80	-	-	-
	12.69	23.53	-	31.56	-	-	-
1E-1E (4)	10.96	32.15	-	29.18	-	-	-
	11.08	29.02	-	29.42	-	-	-
2Z	13.50	-	-	8.63	54.42	5.06	13.22
2Z-2Z (1)	13.90	-	-	15.82	21.85	3.61	-
	13.90	-	-	15.81	21.86	3.60	-
2Z-2Z (2)	13.90	-	-	-	42.77	5.12	11.28
	13.08	-	-	-	39.59	5.13	11.34
2Z-2Z (3)	13.12	-	-	-	55.12	5.49	12.51
	13.11	-	-	-	55.37	5.29	12.56

Contd.....

Species	$n_{N2} \to \sigma^*_{C1-O3}$	$n_{N2} \to \pi^*_{C1-O3}$	$n_{O3} \to \sigma^*_{O4-H5}$	$n_{O3} \to \sigma^*_{C1-N2}$	$n_{O3} \to \pi^*_{C1-N2}$	$n_{O4} \to \sigma^*_{O3-H5}$	$n_{O4} \to \pi^*_{C1-N2}$
2E	3.82	-	-	6.12	48.37	-	13.80
2E-2E (1)	-	-	-	-	51.67	-	11.09
2E-2E (2)	-	-	-	-	50.23	-	15.60
2E-2E(3)	-	-	-	48.90	43.62	-	14.22
							8.16
1Z-1E (1)	-	65.95	-	26.22	48.24	-	12.27
	-	52.63	-	22.79	48.31	-	12.30
1Z-1E (2)	-	84.58	-	27.44	-	-	-
	12.28	24.94	-	31.85	-	-	-
1Z-1E (3)	10.75	34.13	-	28.75	-	-	-
	-	58.85	-	27.36	-	4.48	-
2Z-2E (1)	-	-	-	17.57	20.95	5.84	15.34
	12.81	-	-	-	57.12	-	-
2Z-2E (2)	-	-	-	-	42.44	-	-
	14.29	-	-	13.73	27.07	2.61	-
2Z-1E (1)	10.12	39.98	-	24.43	-	-	12.68
	13.14	-	-	-	48.96	6.24	-
2E-1E (1)	10.07	36.16	-	25.74	-	-	11.14
	-	-	-	-	29.84	-	-
2E-1E (2)	12.37	22.18	-	-	50.11	-	11.75
	-	-	-	-	-	-	-
1Z-2Z (1)	15.96	35.58	-	26.25	-	2.20	-
	14.29	-	-	18.77	2.70	-	-
1Z-2E (1)	-	58.14	-	24.75	-	-	-
	-	-	-	55.56	-	-	-

2.3 CONCLUSIONS:

The present study has been carried out to understand the intramolecular and intermolecular H-bonding interactions involving FHA molecule in different isomeric forms. The studies indicate that in spite of the presence of multiple potential sites for H-bonding in FHA; the intra molecular H-bond interactions are present mainly in **1Z** and **2Z** isomers. Though many distances between H-bond donor and acceptor atoms in FHA are less than the sum of their respective van der Waals radii, the A-H...B bond angle does not allow the H-bond stabilization.

The aggregation of the four isomeric forms **1Z, 1E, 2Z** and **2E** of FHA with single water molecule at MP2/Aug-cc-pVDZ level leads to stabilization energies that are recorded to fall in the range 3.33 to 11.30 kcal/mol. The **1ZW1** FHA-H_2O adduct has maximum stabilization energy of 11.30 kcal/mol. In this aggregate the H5...O3 intramolecular H-bond is replaced by intermolecular H-bonds; between O3 and 'H' of water molecule and between O-H of FHA and 'O' of water. The stabilization energy clearly indicates that the intermolecular H-bond is stronger than the intramolecular H-bond. The dimerization of various isomeric forms at MP2/6-31+G* level resulted in stabilization energies of the order 2.64 to 16.12 kcal/mol. **1Z-1Z (1)** homodimer is most stable dimer and the optimized structure indicates two symmetrical strong H-bond interactions between carbonyl oxygen of one FHA unit with the hydroxyl hydrogen of the second unit.

The carbonyl oxygen is the strongest H-bond acceptor site and O-H is strongest H-bond donor site in FHA as is suggested by the dimerization and FHA-H_2O adduct formation studies. Cooperativity and synergic enhancement of intramolecular H-bonding is also observed in some of the optimized structures. The atomic charges and NBO analysis indicated that electrostatic interactions and charge transfer stabilization are the important components responsible for intra and intermolecular H-bonding in FHA. The intra and intermolecular H-bond interactions favor the existence of **1Z** conformer of FHA. In aqueous medium, the specific interactions per monomeric unit are stronger than the same for dimeric structures. Thus the FHA molecule is expected to exist as monomers in aqueous medium.

2.4 REFERENCES:

1. V. Makarov, B.M. Pettitt, M. Feig, Acc. Chem. Res. 35 (2002) 376.
2. T. Wyttenbach, B. Paizs, P. Barran, L. Breci, D. Liu, S. Suhai, V.H. Wysocki, M.T. Bowers, J. Am. Chem. Soc. 125 (2003) 13768.
3. M. Schmitt, M. Böhm, C. Ratzer, C. Vu, I. Kalkman, W.L. Meerts, J. Am. Chem. Soc. 127 (2005) 10356.
4. M.F. Jarrold, Acc. Chem. Res. 32 (1999) 360.
5. V.V. Gromak, J. Mol. Struct. (Theochem) 726 (2005) 213.
6. A. Nowroozi, H. Raissi, F. Farzad, J. Mol. Struct. (Theochem) 730 (2005) 161.
7. F.T.T. Huque, J.A. Platts, Org. Biomol. Chem. 1 (2003) 1419.
8. G.R. Desiraju, T. Steiner, The Weak Hydrogen Bond In Structural Chemistry and Biology; Oxford University Press, 1999.
9. M. Saldyka, Z. Mielke, J. Phys. Chem. A 106 (2002) 3714.
10. L. Bauer, O. Exner, Angew. Chem. Int. Ed. 6 (1974) 376.
11. H.R. Bravo, W. Lazo, J. Agric. Food Chem. 44 (1996) 1569.
12. A.E. Fazary, M.M. Khalil, A. Fahmy, T.A. Tantawy, Medical Journal of Islamic Academy of Sciences 14 (2001) 107.
13. J.B. Neilands, Science 156 (1967) 1443.
14. M. Barbarić, S. Uršic, V. Pilepić, B. Zorc, A.H. Brundić, A. Nagl, M. Grdiša, K. Pavelic, R. Snoeck, G. Andrei, J. Balzarini, E.D. Clercq, M. Mintas, J. Med. Chem. 48 (2005) 884.
15. S. Pepeljnjak, B. Zorc, I. Butula, Acta. Pharm. 55 (2005) 401.
16. M. Saldyka, Z. Mielke, Chem. Phys. 308 (2005) 59.
17. D.M. Stinchcomb, J. Pranta, J. Mol. Struct. (Theochem) 370 (1996) 25.
18. R. Kakkar, R. Grover, P. Chaddha, Org. Biomol. Chem. 1 (2003) 2200.
19. D.-H., Wu, J.-J., Ho, J. Phys. Chem. A 102 (1998) 3582.
20. M. Saldyka, Z. Mielke, Chem. Phys. Lett. 371 (2003) 713.
21. B. Lindberg, A. Berndtsson, R. Nilsson, R. Nyholm, O. Exner, Acta. Chem. Scand. A32 (1978) 353.
22. D.A. Brown, W.K. Glass, R. Mangeswaran, B. Girmay, Magn. Reson. Chem. 26 (1988) 970.

23. B. García, S. Ibeas, J.M. Leal, F. Secco, M. Venturini, M.L. Senent, A. Niño, C. Muñoz, Inorg. Chem. 44 (2005) 2908.
24. O. Gálvez, P.C. Gómez, L.F. Pacios, J. Chem. Phys. 118 (2003) 4878.
25. J.A. Frey, S. Leutwyler, J. Phys. Chem. A 110 (2006) 12512.

INTRA- AND INTERMOLECULAR HYDROGEN BONDING IN THIOFORMOHYDROXAMIC ACID

3.1 INTRODUCTION:

The nature of hydrogen bonds (H-bonds) formed by sulfur is quite different from that formed by oxygen atom and is known further from much familiar examples of H_2O and H_2S. According to electrostatics, sulfur atoms are usually positive and oxygen is negative, so the fundamental nature of H-bonds is anticipated to be different in them [1]. Though the electronegativity difference between H and S is just 0.38 on the Pauling scale, yet there are reports that the H-bond acceptor ability of thioamide sulfur could surprisingly be equal to or exceed that of amide oxygen [2-4]. There is scarce information available on H-bonding involving sulfur atoms.

The living systems contain a number of sulfur containing molecules including thiopeptides, amino acids e.g. cysteine and methionine etc. Hence understanding biomolecular interactions during protein folding by studying S...H bonding carries prime importance in biochemical research [5]. N-H...S H-bonds are suggested to be present at the active site of various cytochrome P450s and these H-bonds play crucial role in stabilizing the Fe (III) state and protecting the complex against decomposition [6,7]. The influence of N-H...S interactions in regulation of redox potential of the metal-sulfur proteins and other complexes is well known [8,9]. Since the fundamental building block in thiopurines and thiopyrimidines is S=C-NH- and these are H-bonded in the nucleic acids (DNA and RNA), thus thioformohydroxamic acid (TFHA) can serve as a model for studying H-bonds in these systems [10].

The presence of N-methylthioformohydroxamic acid has been recognised in bacterial sources and is commonly termed Thioformin [11]. The antibiotic activity of cupric and ferric complexes of N-methylthioformohydroxamic acid is also reported [12,13]. Thiohydroxamic acids are important in biological and analytical chemistry and are utilized in detection and quantitative determination of metals [14]. Using 'S' and 'O' atoms thiohydroxamic acids coordinate with metal ions like Fe^{+3}, Ni^{+2}, Cu^{+2} forming colored metal complexes and hence find applications in detection and quantitative estimation of metals [15,16]. A number of thiohydroxamic acids found applications as biocidal compounds for bacteria, fungi, insects, mites and weeds etc. They are effective as antiperspirant and antihypertensive agents, as enzyme inhibitors, and as drugs for treatment of leukaemia. Thiohydroxamic acids have also been used to counteract the effect of war toxins and to alleviate paralysis [17]. Their other important applications include gravimetric and spectrophotometric determination of metals. Cyclic thiohydroxamic acids such as 1-hydroxy-2(1H)-pyridinethione and 3-hydroxy-4-methyl-2(3H)-thiazolethione find diverse applications as fungicides and alkoxy-radical precursors in synthetic procedures and mechanistic studies [18]. It is used in personal care formulation such as hand lotions, emollient creams or in shampoos etc. Zinc chloride, acetate or oxide complex of this compound gives enhanced microbiological activity [19].

Inspite of numerous applications of thiohydroxamic acid, very few studies have been carried out to understand the chemistry of these molecules. Formohydroxamic acid (FHA) has undergone structural analysis by a number of research groups due to its vast number of applications [20,21]. The intra and intermolecular H-bonding ability of FHA has already been analyzed by us (previous chapter) and reported [22]. Thiohydroxamic acid is expected to undergo tautomerism as observed in FHA. The earlier experimental reports based on IR spectral studies suggested that thiohydroxamic acid exists in thione form in solid state [23,24]. However the natural products separated from plants (e.g. mustard) contain sulfated S-glucosyl thiohydroxamates that are derivatives of thiol form [17]. The intra and intermolecular H-bonding interactions play key role in the stability of tautomers and in the process of tautomerism if the strength of interactions is magnificent. The understanding of C=S...H and S-H...X (X=O, N, S) H-bonds in these molecules is important for inferring the nature and strength of H-bonds in biomolecules.

The interactions with water are particularly important because of water being universal solvent and its abundance in the biological systems. The present study explores the H-bonding ability of thiohydroxamic acid in its various tautomeric forms and in 1:1 adduct with H_2O and the comparison with the FHA analogues has been made.

3.2 RESULTS AND DISCUSSION:
3.2.1 CONFORMATIONAL STABILITY AND TAUTOMERISM:

Keto↔enol tautomerism in FHA has been studied by a number of research groups. The replacement of carbonyl oxygen by sulfur can alter the conformational stability, H-bonding ability and the overall reactivity of the molecule. The present study is aimed at understanding the characteristics of thiohydroxamic acids in comparison to formohydroxamic acids. There are two reports on the gas phase intramolecular tautomerism in thioformohydroxamic acid (TFHA) [9,17]. Both the studies concentrate on similar pathways for proton transfer in tautomerization as adopted by FHA [25]. It has been recognized that H-bonding interactions play important role in stability of tautomers and proton transfer in case of FHA. In thiohydroxamic acid, where sulfur has replaced the oxygen of carbonyl group of FHA, H-bonding ability of the molecule can be anticipated to decrease by a significant amount.

To understand the role of H-bonding interactions on conformational stability, seven minima including one zwitter ionic structure have been located employing MP2/Aug-cc-pVDZ **[L4]** theoretical method. All the seven structures and three transition states for intramolecular proton transfer among isomeric forms of TFHA are shown in Figure 3.1. The barriers calculated with respect to the more stable ground state conformation are also shown on arrow heads in the same figure.

The relative energy differences among these forms at **[L4]** theoretical level are given in Table 3.1. The gas phase order of stabilities in TFHA is **2Z > 2Z2 > 1Z > 2E > 2E2 >1E > 3** at **[L4]** theoretical level. This order of relative energies for TFHA isomers is the same as determined by R. Kakkar et al. and J. J. Ho et al [17,9]. The gas phase order of stabilities is **1Z > 2Z > 1E > 2E > 2Z2 > 2E2 > 3** in FHA at **[L4]** theoretical level [22]. The **2Z** conformation is the most stable conformation in TFHA while in FHA it was the conformation similar to **1Z**.

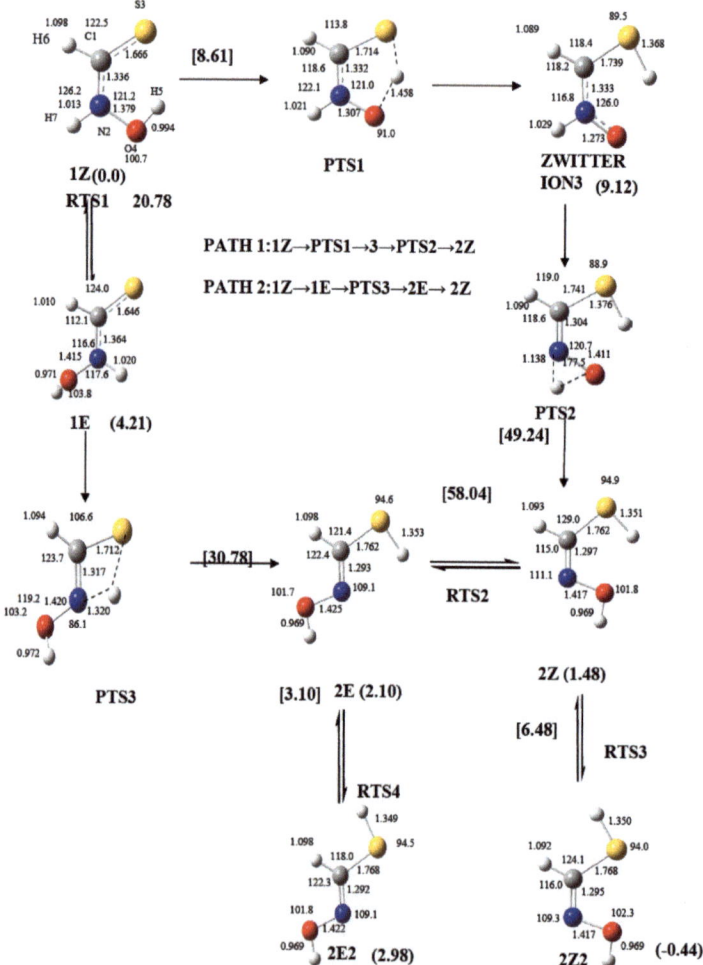

Figure 3.1: Intramolecular proton transfer among isomeric forms of TFHA. (Relative energies) and the [activation barriers] in kcal/mol calculated with respect to most stable form at [L4] theoretical level. All distances in angstrom and angles in degrees.

Table 3.1: Relative energies (kcal/mol) and second order electron delocalizations $E^{(2)}$ (kcal/mol) from NBO analysis of isomeric forms of TFHA (FHA) at **[L4]** theoretical level.

Species	Electron Delocalization	$E^{(2)}$	Relative Energy
1Z	$n_{N2(1)} \rightarrow \pi^*_{C1-X3}$	143.97 (29.99)	0.00 (0.00)
	$n_{N2(1)} \rightarrow \sigma^*_{C1-X3}$	-(15.37)	
	$n_{X3(2)} \rightarrow \sigma^*_{C1-N2}$	15.82 (29.97)	
	$n_{X3(2)} \rightarrow \sigma^*_{C1-H6}$	13.78 (25.94)	
	$n_{X3(2)} \rightarrow \sigma^*_{O4-H5}$	15.80 (5.22)	
1E	$n_{N2(1)} \rightarrow \pi^*_{C1-X3}$	62.73 (23.50)	4.21 (1.22)
	$n_{N2(1)} \rightarrow \sigma^*_{C1-X3}$	1.84 (9.92)	
	$n_{X3(2)} \rightarrow \sigma^*_{C1-N2}$	18.99 (32.87)	
	$n_{X3(2)} \rightarrow \sigma^*_{C1-H6}$	13.56 (26.64)	
2Z	$n_{N2(1)} \rightarrow \sigma^*_{C1-X3}$	14.65 (13.55)	-1.48 (1.14)
	$n_{X3(2)} \rightarrow \pi^*_{C1-N2}$	38.60 (54.30)	
	$n_{O4(2)} \rightarrow \pi^*_{C1-N2}$	18.46 (13.33)	
	$n_{O4(1)} \rightarrow \sigma^*_{X3-H5}$	2.45 (3.78)	
	$n_{X3(1)} \rightarrow \sigma^*_{C1-N2}$	6.50 (-)	
2E	$n_{N2(1)} \rightarrow \sigma^*_{C1-H6}$	10.26 (10.93)	2.10 (5.16)
	$n_{X3(2)} \rightarrow \pi^*_{C1-N2}$	34.13 (47.81)	
	$n_{O4(2)} \rightarrow \pi^*_{C1-N2}$	16.76 (13.76)	
	$n_{N2(1)} \rightarrow \sigma^*_{C1-X3}$	0.96 (3.66)	
	$n_{X3(1)} \rightarrow \sigma^*_{C1-N2}$	3.96 (5.87)	
2Z2	$n_{N2(1)} \rightarrow \sigma^*_{C1-X3}$	13.68 (14.80)	-0.44 (7.13)
	$n_{X3(2)} \rightarrow \pi^*_{C1-N2}$	36.51 (47.90)	
	$n_{O4(2)} \rightarrow \pi^*_{C1-N2}$	18.99 (17.72)	
	$n_{O4(1)} \rightarrow \sigma^*_{X3-H5}$	1.25 (-)	
2E2	$n_{N2(1)} \rightarrow \sigma^*_{C1-H6}$	10.21 (10.54)	2.98 (8.88)
	$n_{X3(2)} \rightarrow \pi^*_{C1-N2}$	23.03 (45.06)	
	$n_{O4(2)} \rightarrow \pi^*_{C1-N2}$	13.81 (14.22)	
3	$n_{X3(1)} \rightarrow \sigma^*_{C1-N2}$	3.93 (5.61)	9.12 (12.76)
	$n_{X3(2)} \rightarrow \pi^*_{C1-N2}$	44.17 (62.82)	
	$n_{O4(1)} \rightarrow \sigma^*_{C1-N2}$	7.25 (5.49)	
	$n_{O4(2)} \rightarrow \sigma^*_{C1-N2}$	9.21 (8.09)	
	$n_{O4(1)} \rightarrow \sigma^*_{X3-H5}$	1.74 (2.13)	
	$n_{O4(2)} \rightarrow \sigma^*_{X3-H5}$	8.41 (13.73)	
	$n_{O4(3)} \rightarrow \pi^*_{C1-N2}$	81.37 (51.88)	

The comparison of important geometrical parameters of **1Z** conformation of TFHA and FHA (Table 3.2-3.3) shows that C-N and O-N bond distances are shorter for the thio analogs (C-N and O-N bond distances are 1.336 Å and 1.379 Å respectively in TFHA as compared to 1.368 Å and 1.406 Å in FHA respectively), thereby suggesting that conjugative interactions in -C(=S)-N-O- unit of TFHA are stronger than the corresponding interactions in -C(=O)-N-O- unit of FHA. Inspite of enhanced conjugative interactions in **1Z** conformer of TFHA, it is the **2Z** conformation which is more stable than the **1Z**. The comparatively weaker strength of C=S π bond in comparison to C=N π bond is the rationale behind it. The C-N bond distance is shorter in keto (**1Z, 1E**) for TFHA while it is shorter in enol forms (**2Z, 2E, 2Z2, 2E2**) for FHA. Only **1E** conformer shows non planarity in case of TFHA while both **1Z** and **1E** conformers of FHA are nonplanar. As thioformamide has planar arrangement of atoms around N, certainly the non planarity in **1E** of TFHA is the result of presence of hydroxyl group. The repulsive interactions between lone pairs present on oxygen and nitrogen distort the orientation of lone pairs present on nitrogen away from oxygen that favors conjugation with C=S in **1Z** while disfavors in **1E** which is apparent from NBO analysis also (Table 3.1). The second order stabilization energy $E^{(2)}$ for $n_{N2} \to \pi^*_{C1-S3}$ electron delocalization is 143.97 kcal/mol in **1Z** relative to 62.73 kcal/mol in **1E** of TFHA. The C-N and N-O bond distances in **2Z** and **2E** conformations are comparable for TFHA and FHA. The electron delocalization from lone pair present on N is reduced considerably in **2Z** and **2E** due to presence of C=N double bond as is evident from $E^{(2)}$ values recorded in the Table 3.1 for $n_{N2} \to \pi^*_{C1-S3}$ and $n_{N2} \to \sigma^*_{C1-S3}$ delocalization. **2Z2** and **2E2** conformers are rotamers of C-S bond of **2Z** and **2E** forms respectively and corresponding transition states are shown in Figure 3.2. These conformers (**2Z2** and **2E2**) differ from their stable forms **2Z** and **2E** only by 1.04 kcal/mol and 0.88 kcal/mol respectively while in case of FHA, similar conformations had relative energies of 5.99 and 3.72 kcal/mol from respective **2Z** and **2E** conformers. Zwitterionic structure (**3**) being least stable with relative energy of 9.12 kcal/mol in comparison to **1Z**.

Table 3.2: Geometrical parameters of **1Z**, **1E** and dipolar form (**3**) of TFHA (FHA) at MP2/6-31+G* [**L3**] and MP2/Aug-cc-pVDZ [**L4**] theoretical levels. All the bond distances are in angstrom (Å), angles and dihedrals are in degrees (°).

Parameters	1Z		1E		Parameters	3	
	L3	L4	L3	L4		L3	L4
N2-C1	1.333	1.336	1.366	1.364	N2-C1	1.326	1.333
	(1.362)	(1.368)	(1.388)	(1.393)		(1.311)	(1.317)
X3-C1	1.652	1.666	1.631	1.646	X3-C1	1.727	1.739
	(1.236)	(1.235)	(1.225)	(1.224)		(1.333)	(1.335)
O4-N2	1.386	1.379	1.420	1.415	O4-N2	1.280	1.273
	(1.410)	(1.406)	(1.423)	(1.421)		(1.303)	(1.297)
H5-O4	0.997	0.994	0.977	0.971	H5-X3	1.350	1.368
	(0.988)	(0.983)	(0.976)	(0.969)		(0.999)	(0.998)
H6-C1	1.091	1.098	1.092	1.099	H6-C1	1.082	1.089
	(1.099)	(1.106)	(1.101)	(1.109)		(1.080)	(1.086)
H7-N2	1.013	1.013	1.020	1.020	H7-N2	1.029	1.029
	(1.016)	(1.017)	(1.020)	(1.020)		(1.024)	(1.024)
X3-C1-N2	123.7	122.5	124.2	124.0	X3-C1-N2	120.2	118.4
	(122.0)	(121.9)	(122.7)	(122.8)		(116.8)	(115.9)
O4-N2-C1	121.6	121.2	115.5	116.6	O4-N2-C1	126.3	126.0
	(115.9)	(115.3)	(113.5)	(113.7)		(121.1)	(120.5)
H5-O4-N2	101.6	100.7	103.8	103.8	H5-X3-C1	91.7	89.5
	(101.8)	(101.2)	(103.6)	(103.4)		(103.1)	(101.5)
H6-C1-N2	112.4	112.6	112.1	112.1	H6-C1-N2	118.2	118.2
	(113.5)	(113.0)	(112.6)	(112.1)		(123.4)	(123.0)
H7-N2-C1	126.6	126.2	117.0	117.6	H7-N2-C1	117.3	116.8
	(120.6)	(119.5)	(114.7)	(114.4)		(120.6)	(120.0)
O4-N2-C1-X3	0.0	0.0	158.2	160.1	O4-N2-C1-X3	-0.0	-0.0
	(-13.2)	(-12.0)	(157.7)	(157.7)		(0.0)	(0.0)
H5-O4-N2-C1	-0.0	-0.0	117.8	115.2	H5-X3-C1-N2	0.0	0.0
	(7.7)	(4.5)	(119.9)	(119.0)		(-0.0)	(-0.0)

Contd......

119

Parameters	1Z		1E		Parameters	3	
	L3	L4	L3	L4		L3	L4
H6-C1-N2-O4	180.0 (170.9)	180.0 (171.4)	-25.4 (-25.9)	-22.7 (-25.9)	H6-C1-N2-O4	180.0 (180.0)	180.0 (180.0)
H7-N2-C1-X3	-180.0 (-150.0)	-180.0 (-148.1)	25.3 (30.1)	23.1 (29.7)	H7-N2-C1-X3	-180.0 (180.0)	180.0 (180.0)

Table 3.3: Geometrical parameters of **2Z**, **2E**, **2Z2** and **2E2** of TFHA (FHA) at MP2/6-31+G* **[L3]** and MP2/Aug-cc-pVDZ **[L4]** theoretical levels. All the bond distances are in angstrom (Å), angles and dihedrals are in degrees (°).

Parameters	2Z		2E		2Z2		2E2	
	L3	L4	L3	L4	L3	L4	L3	L4
N2-C1	1.293 (1.290)	1.297 (1.294)	1.288 (1.283)	1.293 (1.287)	1.290 (1.284)	1.295 (1.289)	1.287 (1.281)	1.292 (1.286)
X3-C1	1.748 (1.348)	1.762 (1.352)	1.748 (1.355)	1.762 (1.359)	1.754 (1.359)	1.768 (1.363)	1.755 (1.361)	1.768 (1.365)
O4-N2	1.420 (1.436)	1.417 (1.435)	1.425 (1.427)	1.425 (1.427)	1.417 (1.417)	1.417 (1.418)	1.422 (1.430)	1.422 (1.429)
H5-X3	1.341 (0.982)	1.351 (0.975)	1.343 (0.980)	1.353 (0.973)	1.341 (0.973)	1.350 (0.967)	1.341 (0.973)	1.349 (0.966)
H6-C1	1.085 (1.082)	1.093 (1.089)	1.091 (1.087)	1.098 (1.093)	1.085 (1.086)	1.092 (1.092)	1.091 (1.091)	1.098 (1.097)
H7-O4	0.975 (0.974)	0.969 (0.967)	0.976 (0.974)	0.969 (0.967)	0.975 (0.974)	0.969 (0.968)	0.976 (0.974)	0.969 (0.967)
X3-C1-N2	129.5 (126.7)	129.0 (126.5)	121.8 (120.8)	121.4 (120.7)	124.6 (124.2)	124.1 (124.0)	118.3 (117.9)	118.0 (117.8)
O4-N2-C1	110.9 (107.7)	111.1 (107.7)	109.2 (109.3)	109.1 (109.2)	109.4 (110.6)	109.5 (110.3)	109.1 (108.1)	109.1 (108.1)

Contd......

Parameters	2Z		2E		2Z2		2E2	
	L3	L4	L3	L4	L3	L4	L3	L4
H5-X3-C1	95.6 (107.2)	94.9 (106.3)	94.9 (107.7)	94.6 (107.1)	93.9 (109.1)	94.0 (108.4)	94.7 (108.9)	94.5 (108.2)
H6-C1-N2	115.0 (118.4)	115.0 (118.2)	122.2 (125.2)	122.4 (124.9)	115.8 (116.7)	116.0 (116.6)	121.9 (123.2)	122.3 (123.1)
H7-O4-N2	102.1 (102.3)	101.8 (102.0)	102.0 (101.8)	101.7 (101.6)	102.6 (101.5)	102.3 (101.3)	102.1 (101.5)	101.8 (101.4)
O4-N2-C1-X3	0.0 (0.0)	-0.0 (0.0)	180.0 (180.0)	180.0 (180.0)	-0.0 (-0.0)	-0.0 (0.0)	175.2 (180.0)	178.2 (180.0)
H5-X3-C1-N2	0.1 (-0.0)	0.0 (0.0)	0.0 (-0.0)	-0.0 (-0.0)	180.0 (180.0)	180.0 (180.0)	158.1 (180.0)	172.2 (180.0)
H6-C1-N2-O4	180.0 (180.0)	180.0 (180.0)	-0.0 (0.0)	-0.0 (0.0)	180.0 (180.0)	180.0 (180.0)	-1.5 (-0.0)	-0.5 (0.0)
H7-O4-N2-C1	180.1 (179.9)	180.0 (180.0)	180.0 (180.0)	180.0 (180.0)	180.0 (180.0)	180.0 (180.0)	177.4 (180.0)	179.1 (180.0)

The structural parameters of all rotational transition states (RTS) and proton transfer transition states (PTS) is listed in Table 3.4 and 3.5. For conversion of **1Z** conformer to **1E** in TFHA, rotational barrier of 20.78 kcal/mol (evaluated at [L4] theoretical level) has to be crossed that involves rotation around C-N bond through rotataional transition state (**RTS1**). The C-N rotation in formamide and thioformamide requires energy of 16.12 and 18.64 kcal/mol respectively at [L4] theoretical level. The higher relative activation barrier for TFHA from the thioamide can be rationalized in terms of intramolecular H-bonding interactions that are to be ruptured during rotation in addition to the breaking of C-N π bond and steric repulsive interactions due to the presence of lone pairs of electrons present on hydroxyl oxygen. The conjugative interactions $n_{N2} \rightarrow \pi^*_{C1-S3}$ in keto forms are significantly higher in TFHA than FHA, while the conjugative interaction $n_{X3} \rightarrow \sigma^*_{C1-N2}$ is higher in FHA relative to TFHA. The higher interconversion barrier for **1Z** → **1E** in the TFHA in comparison to that in FHA (20.78 kcal/mol vs. 15.86 kcal/mol), is the result of relatively higher conjugative interactions in TFHA. The higher conjugative interactions are favored by larger size and charge holding capacity of sulfur in comparison to oxygen and the weaker C=S π bond that also favors the formation of C=N π bond. The transition state for interconversion of **2Z** → **2E** (**RTS2**) involves the breaking of C=N π bond, hence the activation barriers are significantly large being 58.04 kcal/mol in TFHA and 54.77 kcal/mol in FHA. The tautomerism between **1Z** and **2Z** forms is indicated to involve high energy barriers for both the pathways shown in Figure 3.1. Path 2 is thermodynamically favorable out of the two as in case of FHA.

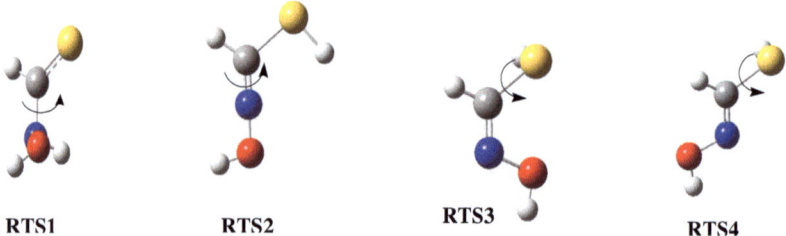

RTS1   RTS2   RTS3   RTS4

Figure 3.2: The rotational transition states

Table 3.4: Geometrical parameters of rotational transition states (RTS) of TFHA (FHA) at MP2/Aug-cc-pVDZ [**L4**] theoretical level. All the bond distances are in angstrom (Å), angles and dihedrals are in degrees (°).

Parameters	RTS1	Parameters	RTS2	RTS3	RTS4
N2-C1	1.457 (1.469)	N2-C1	1.239 (1.239)	1.295 (1.290)	1.293 (1.287)
X3-C1	1.629 (1.214)	X3-C1	1.849 (1.409)	1.795 (1.375)	1.786 (1.377)
O4-N2	1.463 (1.461)	O4-N2	1.333 (1.343)	1.399 (1.409)	1.415 (1.419)
H5-O4	0.970 (0.969)	H5-X3	1.351 (0.971)	1.353 (0.969)	1.352 (0.968)
H6-C1	1.099 (1.108)	H6-C1	1.112 (1.103)	1.093 (1.093)	1.098 (1.098)
H7-N2	1.027 (1.028)	H7-O4	0.977 (0.974)	0.970 (0.968)	0.969 (0.968)
X3-C1-N2	125.9 (124.5)	X3-C1-N2	126.2 (127.1)	126.7 (125.4)	118.4 (118.6)
O4-N2-C1	106.6 (104.8)	O4-N2-C1	177.4 (179.3)	113.0 (111.4)	109.4 (108.6)
H5-O4-N2	100.9 (101.0)	H5-X3-C1	94.8 (107.0)	94.7 (109.6)	95.6 (109.1)
H6-C1-N2	112.7 (112.5)	H6-C1-N2	125.9 (126.3)	114.1 (116.0)	121.6 (123.1)
H7-N2-C1	106.8 (105.9)	H7-O4-N2	106.1 (106.0)	101.7 (101.4)	101.9 (101.7)
O4-N2-C1-X3	53.2 (54.3)	O4-N2-C1-X3	-106.8 (-133.4)	-3.4 (-4.9)	179.4 (176.7)
H5-O4-N2-C1	122.2 (125.4)	H5-X3-C1-N2	1.2 (0.2)	94.5 (99.6)	80.9 (93.2)
H6-C1-N2-O4	-126.1 (-125.5)	H6-C1-N2-O4	74.2 (46.9)	180.3 (180.1)	1.9 (1.0)
H7-N2-C1-X3	-55.1 (-53.4)	H7-O4-N2-C1	-73.9 (133.4)	179.5 (179.5)	180.8 (181.1)

Table 3.5: Geometrical parameters of proton transition states (PTS) for tautomerization in TFHA (FHA) at MP2/Aug-cc-pVDZ **[L4]** theoretical level. All the bond distances are in angstrom (Å), angles and dihedrals are in degrees (°).

Parameters	PTS1	Parameters	PTS2	Parameters	PTS3
N2-C1	1.332 (1.324)	N2-C1	1.304 (1.303)	N2-C1	1.317 (1.319)
X3-C1	1.714 (1.300)	X3-C1	1.741 (1.324)	X3-C1	1.712 (1.296)
O4-N2	1.307 (1.341)	O4-N2	1.411 (1.453)	O4-N2	1.420 (1.410)
H5-O4	1.497 (1.154)	H5-X3	1.376 (1.008)	H5-O4	0.972 (0.971)
H6-C1	1.090 (1.089)	H6-C1	1.090 (1.088)	H6-C1	1.094 (1.093)
H7-N2	1.021 (1.015)	H7-N2	1.138 (1.120)	H7-N2	1.320 (1.317)
X3-C1-N2	113.8 (112.1)	X3-C1-N2	119.0 (117.7)	X3-C1-N2	106.6 (106.2)
O4-N2-C1	121.0 (113.1)	O4-N2-C1	120.7 (112.3)	O4-N2-C1	119.2 (122.6)
H5-X3-C1	78.7 (92.3)	H5-X3-C1	88.9 (100.8)	H5-O4-N2	103.2 (103.5)
H6-C1-N2	118.6 (123.2)	H6-C1-N2	118.6 (122.2)	H6-C1-N2	123.7 (128.0)
H7-N2-C1	122.1 (127.0)	H7-N2-C1	177.5 (152.6)	H7-N2-C1	86.1 (76.6)
O4-N2-C1-X3	0.0 (-0.0)	O4-N2-C1-X3	-0.5 (6.2)	O4-N2-C1-X3	-169.9 (165.9)
H5-X3-C1-N2	-0.0 (0.0)	H5-X3-C1-N2	0.5 (-6.1)	H5-O4-N2-C1	-142.3 (128.2)
H6-C1-N2-O4	180.0 (180.0)	H6-C1-N2-O4	179.8 (182.4)	H6-C1-N2-O4	9.4 (-15.0)
H7-N2-C1-X3	-180.0 (-180.0)	H7-N2-C1-O4	-134.2 (70.7)	H7-N2-C1-X3	-2.2 (3.3)

3.2.2 INTRAMOLECULAR HYDROGEN BONDING:

The presence of multiple atoms that contain lone pair of electrons enable them to act as H-bond acceptors and many hydrogen donor groups A-H (A=O, N) provides opportunity for intramolecular H-bond formation if allowed by the relative distances and orientations. The atomic separations between different probable H-bond donors and acceptors have been scrutinized and the distances that are less than the sum of van der Waals radii along with the A-H...B angles are recorded in Table 3.6. The shorter the distance from the sum of van der Waals radii and closer the angle to 180°;

stronger is the H-bond interaction. The results suggest that though in many cases, H...B distance is far less than the sum of their van der Waals radii (Δr) but H-bonding interactions are still expected to be negligible because of acute angle, the bridging hydrogen makes.

Table 3.6: Important hydrogen bond distances in angstrom (Å) and intramolecular hydrogen bonding angles in degrees (°) in isomeric forms of TFHA.

Species	Atomic separations H...B (r)		Δr	Angles at bridging hydrogen A-H...B	
1Z	H6...S3	2.460	0.590	C1-H6...S3	33.6
	H5...S3	2.248	0.802	O4-H5...S3	126.8
	H7...O4	2.000	0.600	N2-H7...O4	39.5
	H6...N2	2.029	0.711	C1-H6...N2	37.4
1E	H7...S3	2.830	0.220	N2-H7...S3	70.1
	H6...O4	2.450	0.150	C1-H6...O4	71.9
	H5...N2	1.890	0.850	O4-H5...N2	46.4
	H6...S3	2.430	0.620	C1-H6...S3	34.0
	H6...N2	2.049	0.691	C1-H6...N2	38.1
2Z	H5...O4	2.180	0.420	S3-H5...O4	109.2
	H6...S3	2.440	0.610	C1-H6...S3	40.4
	H7...N2	1.870	0.870	O4-H7...N2	47.8
	H6...N2	2.019	0.721	C1-H6...N2	35.6
2E	H6...O4	2.383	0.217	C1-H6...O4	67.7
	H6...S3	2.450	0.600	C1-H6...S3	40.1
	H7...N2	1.870	0.870	O4-H7...N2	48.0
	H5...N2	2.550	0.190	S3-H5...N2	80.0
	H6...N2	2.097	0.643	C1-H6...N2	31.4
2Z2	H7...N2	1.879	0.861	O4-H7...N2	47.4
	H6...N2	2.020	0.720	C1-H6...N2	35.0
	H6...S3	2.490	0.560	C1-H6...S3	37.8
2E2	H7...N2	1.877	0.863	O4-H7...N2	47.9
	H6...O4	2.380	0.220	C1-H6...O4	67.7
	H6...N2	2.090	0.650	C1-H6...N2	31.4
	H6...S3	2.500	0.550	C1-H6...S3	37.9

Difference from sum of van der Waal radii, $\Delta r = r_{vw} - r$

The **1Z** conformer has H5...S3 distance of 2.248 Å and angle O4-H5...S3 of 126.8° that suggests the presence of H-bond interactions. The **2Z** conformer on the other hand has S-H as H-bond donor and oxygen of hydroxyl group as H-bond acceptor with S3-H5...O4 contact distance being 2.180 Å and 109.2° angle which indicates it to be very weaker bond. Platts et al. in their study on directionality of H-bond to sulfur and oxygen concluded that both experimental and theoretical results suggest the presence of more perpendicular approach of acidic hydrogen towards the donor sulfur [1]. They rationalize the fact in terms of dipole and quadrupole of sulfur interacting with the hydrogen. Desiraju et al. also describe the H-bond interactions between sulfur and H-bond donor with the X-H...S in the range 90° to 180° [26]. The presence of ring critical point (using AIM method) in **1Z** and **2Z** also supports the same [27-35]. In the several previous reports on H-bonding, the factors contributing to the H-bonding interactions have been listed as (1) electrostatic (2) polarization (3) van der Waals (4) charge transfer [26]. Electrostatic interactions play the major role. The charges on atoms involved in H-bonding are important for the coulombic interactions.

The analysis of natural atomic charges obtained using NBO method at **[L4]** theoretical level (Table 3.7) indicates that polarization of C-X (X=O, S) bond is reduced considerably in TFHA in comparison to that in FHA as expected from the small electronegativity difference between carbon and sulfur. Thus, the electrostatic component of H-bond involving sulfur as H-bond acceptor is anticipated to be weak, also supported by the hard and soft acid base (HSAB) principle, there is poorer match between the hard proton and soft sulfur. The hydrogen atom H5 attached to O4 carries positive charge of 0.524 units in **1Z** conformation while the magnitude is reduced to 0.175 units in **2Z** conformation of TFHA. The H5 (Figure 3.1) interacts with sulfur in **1Z** that has charge of -0.278 units while in **2Z** it interacts with hydroxylic oxygen having negative charge of 0.653 units.

Table 3.7: Atomic charges from NBO analysis of the isomeric forms of TFHA (FHA) (X=S, O respectively) at MP2/Aug-cc-pVDZ **[L4]** theoretical level.

Species	1Z	1E	2Z	2E	2Z2	2E2	3
C1	0.041	0.048	-0.037	-0.026	-0.043	-0.035	-0.079
	(0.696)	(0.719)	(0.535)	(0.541)	(0.495)	(0.508)	(0.552)
N2	-0.338	-0.401	-0.213	-0.231	-0.206	-0.220	-0.104
	(-0.423)	(-0.449)	(-0.300)	(-0.328)	(-0.275)	(-0.276)	(-0.205)
X3	-0.278	-0.181	0.004	0.026	0.045	0.043	0.015
	(-0.750)	(-0.703)	(-0.782)	(-0.771)	(-0.761)	(-0.760)	(-0.782)
O4	-0.596	-0.592	-0.653	-0.632	-0.645	-0.631	-0.672
	(-0.603)	(-0.605)	(-0.683)	(-0.630)	(-0.628)	(-0.638)	(-0.725)
H5	0.524	0.499	0.175	0.151	0.124	0.129	0.211
	(0.522)	(0.497)	(0.534)	(0.511)	(0.502)	(0.501)	(0.558)
H6	0.217	0.212	0.219	0.214	0.222	0.215	0.227
	(0.148)	(0.138)	(0.187)	(0.181)	(0.174)	(0.168)	(0.196)
H7	0.430	0.415	0.505	0.497	0.503	0.498	0.401
	(0.410)	(0.403)	(0.509)	(0.497)	(0.493)	(0.498)	(0.407)

Electrostatic interaction energy is calculated using Coulomb's formula considering net atomic charges on the atoms interacting to form H-bond to be point charges obtained from NBO analysis. In **1Z** of FHA, it is 64.1 kcal/mol and in **1Z** of TFHA, it is 21.5 kcal/mol while for **2Z** form of FHA, it is 59.7 kcal/mol in FHA and found to be 16.8 kcal/mol in **2Z** of TFHA. This is a crude approximation and is suggested to be around 10-fold greater than experimentally determined values [36]. These values conclude that electrostatic interactions are considerably weaker for TFHA than that for FHA and therefore electrostatic component of intramolecular H-bonding is higher in FHA in comparison to TFHA. NBO analysis (Table 3.1) indicates that the charge transfer interactions strengthening the H-bond are higher in **1Z** conformation in comparison to **2Z** of TFHA. The $E^{(2)}$ for $n_{S3} \to \sigma^*_{O4-H5}$ in **1Z** is 15.80 kcal/mol and for $n_{O4} \to \sigma^*_{S3-H5}$ electron delocalization in **2Z** is 2.45 kcal/mol in TFHA. The similar delocalization in FHA have $E^{(2)}$ values 5.22 kcal/mol and 3.78 kcal/mol ($n_{O3} \to \sigma^*_{O4-H5}$ and $n_{O4} \to \sigma^*_{O3-H5}$ respectively). These values suggest that contribution to intramolecular H-bonding from charge transfer is much larger for **1Z** of TFHA.

The activation barriers for **2Z → 2Z2 (RTS3)** and **2E → 2E2 (RTS4)** are 6.48 kcal/mol and 3.10 kcal/mol respectively in case of TFHA. Both the transformations involve C-S single bond rotation. The difference in activation barriers (3.38 kcal/mol) for the two transformations is reflective of intramolecular H-bond interactions present in **2Z**. Barriers for similar rotation in FHA i.e. **2Z** → **2Z2** is 9.47 kcal/mol and for **2E** → **2E2** is 6.16 kcal/mol.

3.2.3 INTERMOLEULAR HYDROGEN BONDING WITH WATER:

The characteristic properties of sulfur like polarizability and electronegativity result in dipole moment of TFHA being 3.905 debye in comparison to 3.283 debye in case of FHA. The low polarity associated with bonds involving C-S or S-H bonds anticipate that the H-bonding with these bonds will be weak. To explore further, the aggregates of single water molecule with each of the six conformations (**1Z, 1E, 2Z, 2E, 2Z2, 2E2**) have been optimized for different relative positions of H_2O and TFHA conformers at B3LYP/6-31+G* [**L1**], MP2/6-31+G* [**L3**] and MP2/Aug-cc-pVDZ [**L4**] theoretical level and eighteen aggregates (Figure 3.3) are observed to be minima on the potential energy surface. The Tables 3.8-3.12 include the geometrical parameters of adducts of TFHA with water at MP2/Aug-cc-pVDZ [**L4**] level. In these aggregates, TFHA acts both as an H-bond acceptor as well as an H-bond donor, as is evident from H-bond distances (r) between hydrogen and hydrogen bond acceptors and the angles (θ) which the hydrogen atom makes with the two atoms being bridged.

The Table 3.13 displays the important H-bonded distances between the atoms of TFHA-H_2O aggregates which are less than sum of their van der Waals radii and the angles associated with H-bonding interactions. Stabilization energies (S.E.) arising as a result of intermolecular H-bond formation are also tabulated (Table 3.13).

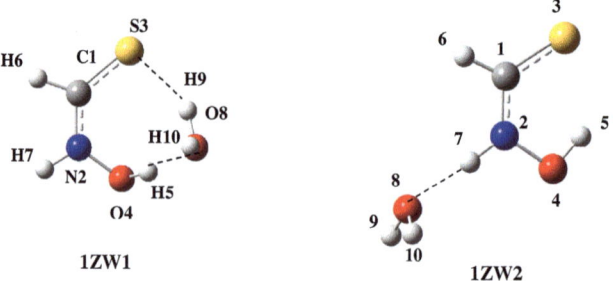

Contd.....

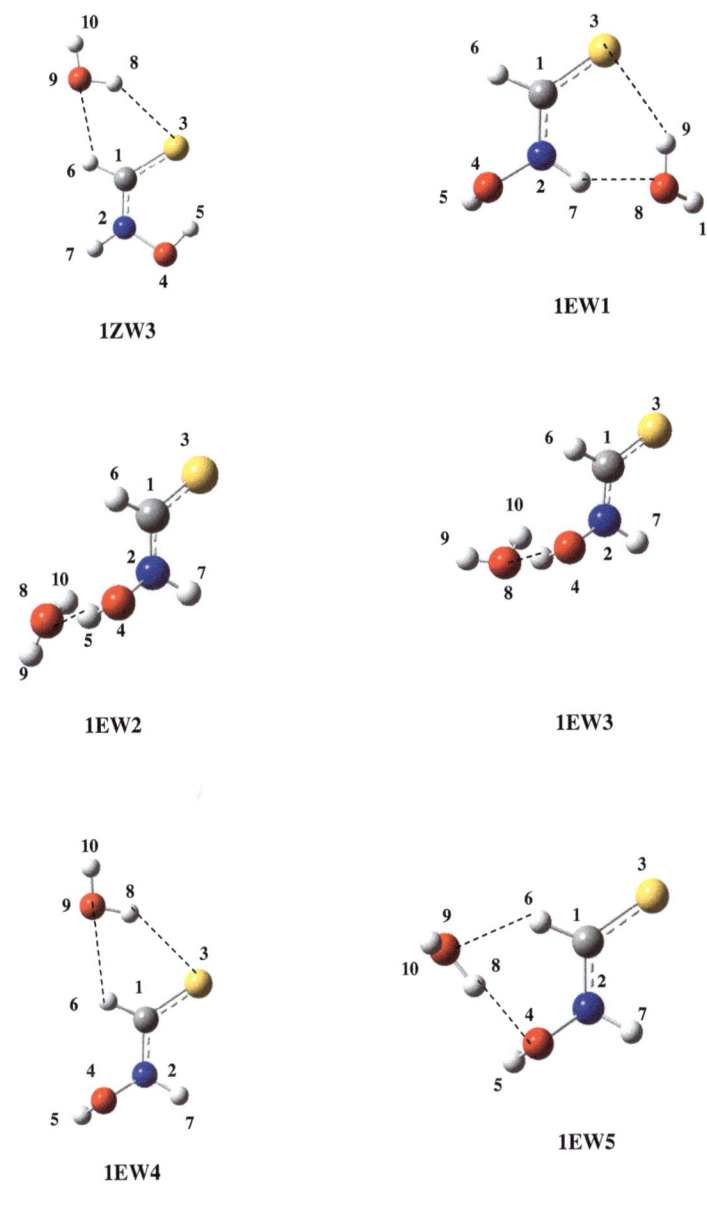

1ZW3

1EW1

1EW2

1EW3

1EW4

1EW5

Contd…..

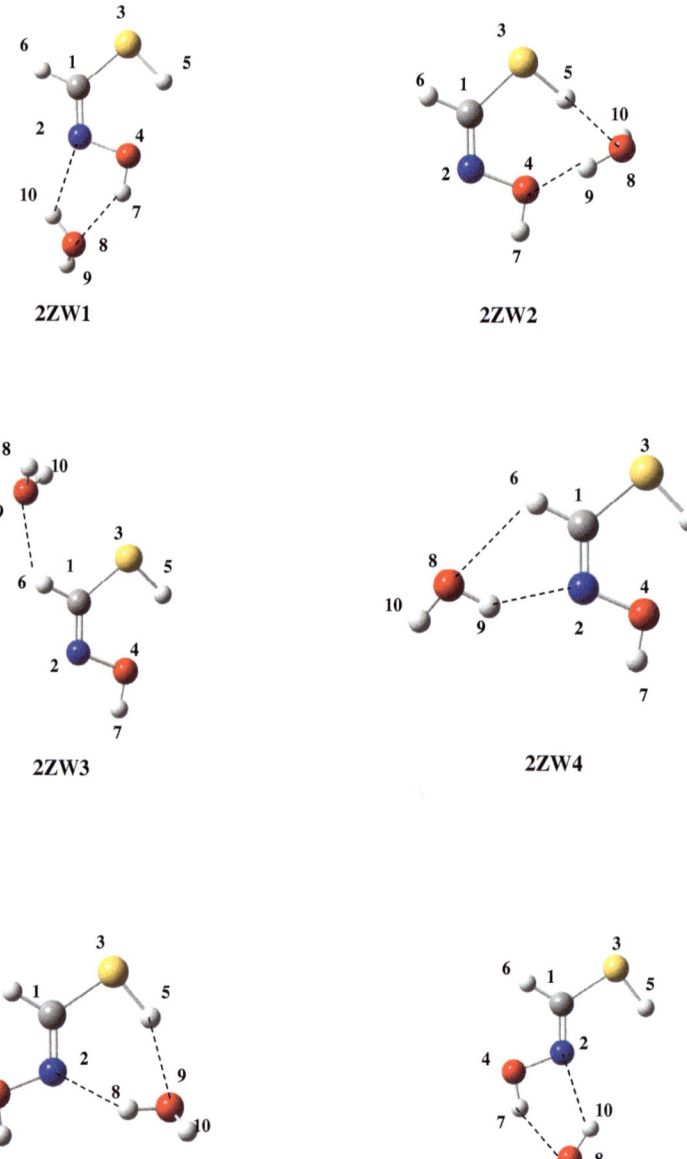

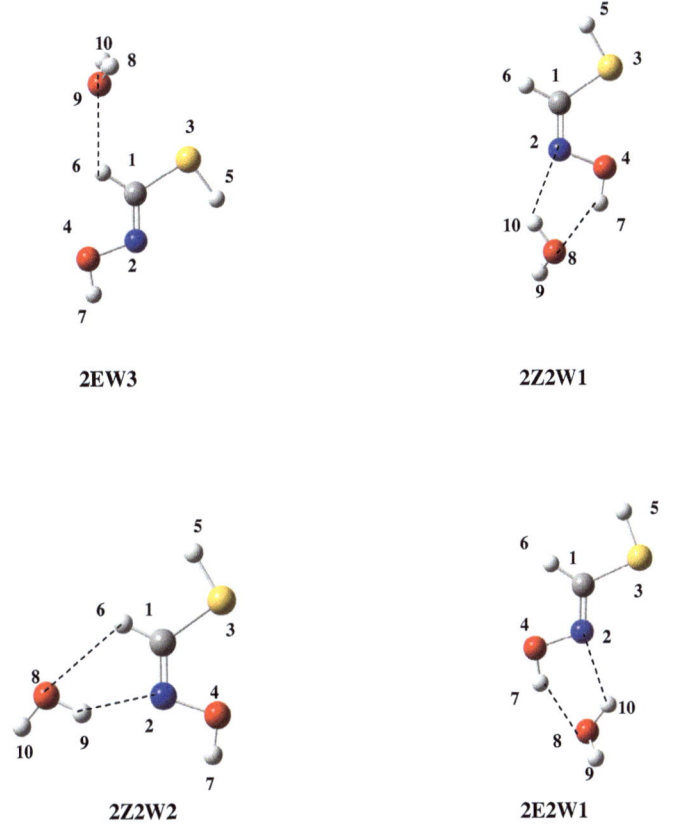

Figure 3.3: The optimized structures of hydrogen bonded TFHA-H$_2$O (thione and thiol) aggregates

Table 3.8: Geometrical parameters of adducts of TFHA with water at MP2/Aug-cc-pVDZ [L4] theoretical level. All the bond distances are in angstrom (Å), angles and dihedrals are in degrees (°).

Parameters	1ZW1	Parameters	1ZW2	Parameters	1ZW3	Parameters	1EW1
N2-C1	1.339	N2-C1	1.331	N2-C1	1.331	N2-C1	1.342
S3-C1	1.664	S3-C1	1.673	S3-C1	1.673	S3-C1	1.663
O4-N2	1.378	O4-N2	1.380	O4-N2	1.379	O4-N2	1.406
H5-O4	0.995	H5-O4	0.996	H5-O4	0.992	H5-O4	0.971
H6-C1	1.101	H6-C1	1.098	H6-C1	1.098	H6-C1	1.099
H7-N2	1.015	H7-N2	1.026	H7-N2	1.014	H7-N2	1.030
O8-H5	1.695	O8-H7	1.838	H8-S3	2.517	O8-H7	1.880
H9-O8	0.981	H9-O8	0.967	O9-H8	0.973	H9-O8	0.979
H10-O8	0.967	H10-O8	0.967	H10-O9	0.965	H10-O8	0.966
S3-C1-N2	128.5	S3-C1-N2	122.7	S3-C1-N2	122.3	S3-C1-N2	125.1
O4-N2-C1	125.0	O4-N2-C1	120.3	O4-N2-C1	121.8	O4-N2-C1	118.2
H5-O4-N2	105.7	H5-O4-N2	100.9	H5-O4-N2	101.2	H5-O4-N2	104.3
H6-C1-N2	109.9	H6-C1-N2	112.5	H6-C1-N2	113.4	H6-C1-N2	112.5
H7-N2-C1	121.5	H7-N2-C1	126.7	H7-N2-C1	125.7	H7-N2-C1	121.7
O8-H5-O4	167.8	O8-H7-N2	179.8	H8-S3-C1	77.0	O8-H7-N2	151.1
H9-O8-H5	89.5	H9-O8-H7	117.8	O9-H8-S3	140.9	H9-O8-H7	86.8
H10-O8-H9	104.8	H10-O8-H7	118.0	H10-O9-H8	104.5	H10-O8-H9	105.0
O4-N2-C1-S3	-8.0	O4-N2-C1-S3	-0.0	O4-N2-C1-S3	-0.3	O4-N2-C1-S3	168.7
H5-O4-N2-C1	60.6	H5-O4-N2-C1	0.0	H5-O4-N2-C1	0.0	H5-O4-N2-C1	109.0
H6-C1-N2-O4	173.2	H6-C1-N2-O4	180.0	H6-C1-N2-O4	179.8	H6-C1-N2-O4	-12.8
H7-N2-C1-S3	-176.7	H7-N2-C1-S3	-179.9	H7-N2-C1-S3	-179.4	H7-N2-C1-S3	13.8
O8-H5-O4-N2	-89.0	O8-H7-N2-C1	110.3	H8-S3-C1-N2	-181.0	O8-H7-N2-O4	199.3
H9-O8-H5-S3	-7.2	H9-O8-H7-N2	6.4	O9-H8-S3-C1	0.0	H9-O8-H7-S3	-3.6
H10-O8-H9-S3	254.9	H10-O8-H7-N2	133.4	H10-O9-H8-S3	177.7	H10-O8-H9-S3	139.4

Table 3.9: Geometrical parameters of adducts of TFHA with water at MP2/Aug-cc-pVDZ [L4] theoretical level. All the bond distances are in angstrom (Å), angles and dihedrals are in degrees (°).

Parameters	1EW2	Parameters	1EW3	Parameters	1EW4	Parameters	1EW5
N2-C1	1.354	N2-C1	1.353	N2-C1	1.354	N2-C1	1.370
S3-C1	1.652	S3-C1	1.650	S3-C1	1.654	S3-C1	1.645
O4-N2	1.403	O4-N2	1.400	O4-N2	1.408	O4-N2	1.421
H5-O4	0.981	H5-O4	0.980	H5-O4	0.971	H5-O4	0.971
H6-C1	1.100	H6-C1	1.090	H6-C1	1.099	H6-C1	1.099
H7-N2	1.019	H7-N2	1.010	H7-N2	1.019	H7-N2	1.020
O8-H5	1.829	H8-S3	1.830	H8-S3	2.478	H8-O4	2.078
H9-O8	0.966	O9-H8	0.960	O9-H8	0.973	O9-H8	0.969
H10-O8	0.967	H10-O9	0.960	H10-O9	0.965	H10-O9	0.965
S3-C1-N2	124.2	S3-C1-N2	124.4	S3-C1-N2	123.6	S3-C1-N2	122.9
O4-N2-C1	119.0	O4-N2-C1	119.2	O4-N2-C1	117.9	O4-N2-C1	116.8
H5-O4-N2	102.4	H5-O4-N2	102.3	H5-O4-N2	104.1	H5-O4-N2	103.7
H6-C1-N2	112.0	H6-C1-N2	111.9	H6-C1-N2	113.1	H6-C1-N2	112.9
H7-N2-C1	119.3	H7-N2-C1	119.7	H7-N2-C1	119.5	H7-N2-C1	117.2
O8-H5-O4	158.4	H8-S3-C1	158.0	H8-S3-C1	78.0	H8-O4-N2	114.5
H9-O8-H5	128.9	O9-H8-S3	126.0	O9-H8-S3	145.1	O9-H8-O4	145.6
H10-O8-H5	108.5	H10-O9-H8	107.4	H10-O9-H8	104.4	H10-O9-H8	104.9
O4-N2-C1-S3	166.4	O4-N2-C1-S3	167.4	O4-N2-C1-S3	163.5	O4-N2-C1-S3	159.1
H5-O4-N2-C1	105.9	H5-O4-N2-C1	106.4	H5-O4-N2-C1	109.9	H5-O4-N2-C1	112.6
H6-C1-N2-O4	-16.2	H6-C1-N2-O4	-15.1	H6-C1-N2-O4	-19.4	H6-C1-N2-O4	-24.2
H7-N2-C1-S3	19.0	H7-N2-C1-S3	18.2	H7-N2-C1-S3	19.6	H7-N2-C1-S3	23.6
O8-H5-O4-N2	-14.6	H8-S3-C1-N2	-2.7	H8-S3-C1-N2	-189.6	H8-O4-N2-C1	-1.5
H9-O8-H5-O4	143.6	O9-H8-S3-C1	-136.5	O9-H8-S3-C1	-2.7	O9-H8-O4-H6	45.7
H10-O8-H5-O4	15.3	H10-O9-H8-S3	-12.5	H10-O9-H8-S3	164.1	H10-O9-H8-O4	-176.0

Table 3.10: Geometrical parameters of adducts of TFHA with water at MP2/Aug-cc-pVDZ [L4] theoretical level. All the bond distances are in angstrom (Å), angles and dihedrals are in degrees (°).

Parameters	2ZW1	Parameters	2ZW2	Parameters	2ZW3	Parameters	2ZW4
N2-C1	1.298	N2-C1	1.297	N2-C1	1.296	N2-C1	1.297
S3-C1	1.760	S3-C1	1.750	S3-C1	1.765	S3-C1	1.758
O4-N2	1.406	O4-N2	1.420	O4-N2	1.420	O4-N2	1.414
H5-S3	1.352	H5-S3	1.350	H5-S3	1.351	H5-S3	1.351
H6-C1	1.093	H6-C1	1.090	H6-C1	1.092	H6-C1	1.093
H7-O4	0.980	H7-O4	0.960	H7-O4	0.969	H7-O4	0.969
O8-H7	1.898	O8-H5	2.250	H8-S3	3.794	O8-H6	2.560
H9-O8	0.966	H9-O8	0.970	O9-H8	0.966	H9-O8	0.973
H10-O8	0.974	H10-O8	0.960	H10-O9	0.967	H10-O8	0.965
S3-C1-N2	128.3	S3-C1-N2	131.3	S3-C1-N2	128.7	S3-C1-N2	128.8
O4-N2-C1	112.1	O4-N2-C1	111.5	O4-N2-C1	111.3	O4-N2-C1	111.8
H5-S3-C1	94.5	H5-S3-C1	97.7	H5-S3-C1	95.2	H5-S3-C1	95.2
H6-C1-N2	115.6	H6-C1-N2	113.9	H6-C1-N2	116.4	H6-C1-N2	114.8
H7-O4-N2	101.0	H7-O4-N2	101.3	H7-O4-N2	101.7	H7-O4-N2	102.3
O8-H7-O4	145.3	O8-H5-S3	156.3	H8-S3-C1	79.2	O8-H6-C1	113.5
H9-O8-H7	123.9	H9-O8-H5	68.2	O9-H8-S3	52.2	H9-O8-H6	67.6
H10-O8-H7	81.1	H10-O8-H5	117.0	H10-O9-H8	104.0	H10-O8-H9	104.9
O4-N2-C1-S3	-0.1	O4-N2-C1-S3	0.1	O4-N2-C1-S3	-0.0	O4-N2-C1-S3	-0.1
H5-S3-C1-N2	-0.3	H5-S3-C1-N2	5.4	H5-S3-C1-N2	1.5	H5-S3-C1-N2	0.0
H6-C1-N2-O4	180.0	H6-C1-N2-O4	-179.6	H6-C1-N2-O4	180.0	H6-C1-N2-O4	180.0
H7-O4-N2-C1	179.4	H7-O4-N2-C1	174.1	H7-O4-N2-C1	180.2	H7-O4-N2-C1	180.0
O8-H7-O4-N2	-3.8	O8-H5-S3-C1	86.9	H8-S3-C1-N2	183.3	O8-H6-C1-N2	0.7
H9-O8-H7-O4	108.2	H9-O8-H5-O4	17.3	O9-H8-S3-C1	-51.7	H9-O8-H6-N2	-0.6
H10-O8-H7-N2	3.5	H10-O8-H5-S3	14.9	H10-O9-H8-S3	285.8	H10-O8-H9-H6	176.8

Table 3.11: Geometrical parameters of adducts of TFHA with water at MP2/Aug-cc-pVDZ [L4] theoretical level. All the bond distances are in angstrom (Å), angles and dihedrals are in degrees (°).

Parameters	2EW1	Parameters	2EW2	Parameters	2EW3
N2-C1	1.295	N2-C1	1.293	N2-C1	1.293
S3-C1	1.757	S3-C1	1.762	S3-C1	1.764
O4-N2	1.425	O4-N2	1.414	O4-N2	1.428
H5-S3	1.361	H5-S3	1.352	H5-S3	1.353
H6-C1	1.098	H6-C1	1.098	H6-C1	1.096
H7-O4	0.969	H7-O4	0.978	H7-O4	0.968
H8-N2	1.988	O8-H7	1.939	H8-S3	3.595
O9-H8	0.977	H9-O8	0.966	O9-H8	0.966
H10-O9	0.966	H10-O8	0.973	H10-O9	0.966
S3-C1-N2	123.0	S3-C1-N2	121.9	S3-C1-N2	120.9
O4-N2-C1	109.4	O4-N2-C1	110.2	O4-N2-C1	109.0
H5-S3-C1	95.4	H5-S3-C1	94.9	H5-S3-C1	94.5
H6-C1-N2	121.0	H6-C1-N2	122.0	H6-C1-N2	123.7
H7-O4-N2	102.3	H7-O4-N2	101.0	H7-O4-N2	101.6
H8-N2-C1	116.3	O8-H7-O4	144.5	H8-S3-C1	81.1
O9-H8-N2	152.5	H9-O8-H7	123.2	O9-H8-S3	62.9
H10-O9-H8	104.9	H10-O8-H7	79.3	H10-O9-H8	104.0
O4-N2-C1-S3	179.9	O4-N2-C1-S3	180.1	O4-N2-C1-S3	180.0
H5-S3-C1-N2	-0.0	H5-S3-C1-N2	0.9	H5-S3-C1-N2	-0.0
H6-C1-N2-O4	-0.3	H6-C1-N2-O4	0.0	H6-C1-N2-O4	0.0
H7-O4-N2-C1	176.9	H7-O4-N2-C1	179.3	H7-O4-N2-C1	180.0
H8-N2-C1-S3	1.1	O8-H7-O4-N2	-2.6	H8-S3-C1-N2	192.4
O9-H8-N2-H5	-8.7	H9-O8-H7-O4	106.9	O9-H8-S3-C1	-50.3
H10-O9-H8-S3	123.0	H10-O8-H7-N2	4.2	H10-O9-H8-S3	252.8

Table 3.12: Geometrical parameters of adducts of TFHA with water at MP2/Aug-cc-pVDZ [L4] theoretical level. All the bond distances are in angstrom (Å), angles and dihedrals are in degrees (°).

Parameters	2Z2W1	Parameters	2Z2W2	Parameters	2E2W1
N2-C1	1.295	N2-C1	1.295	N2-C1	1.292
S3-C1	1.767	S3-C1	1.764	S3-C1	1.768
O4-N2	1.405	O4-N2	1.413	O4-N2	1.410
H5-S3	1.350	H5-S3	1.350	H5-S3	1.349
H6-C1	1.091	H6-C1	1.092	H6-C1	1.097
H7-O4	0.979	H7-O4	0.969	H7-O4	0.979
O8-H7	1.919	O8-H6	2.562	O8-H7	1.930
H9-O8	0.966	H9-O8	0.973	H9-O8	0.966
H10-O8	0.973	H10-O8	0.965	H10-O8	0.973
S3-C1-N2	123.4	S3-C1-N2	123.8	S3-C1-N2	118.0
O4-N2-C1	110.5	O4-N2-C1	110.2	O4-N2-C1	110.4
H5-S3-C1	94.0	H5-S3-C1	93.9	H5-S3-C1	94.8
H6-C1-N2	116.5	H6-C1-N2	115.7	H6-C1-N2	122.1
H7-O4-N2	101.4	H7-O4-N2	102.8	H7-O4-N2	101.0
O8-H7-O4	144.1	O8-H6-C1	112.8	O8-H7-O4	144.4
H9-O8-H7	123.2	H9-O8-H6	68.0	H9-O8-H7	124.1
H10-O8-H7	80.5	H10-O8-H9	104.2	H10-O8-H7	80.4
O4-N2-C1-S3	-0.0	O4-N2-C1-S3	-0.0	O4-N2-C1-S3	176.4
H5-S3-C1-N2	180.0	H5-S3-C1-N2	180.0	H5-S3-C1-N2	164.2
H6-C1-N2-O4	179.4	H6-C1-N2-O4	180.0	H6-C1-N2-O4	-1.0
H7-O4-N2-C1	180.0	H7-O4-N2-C1	179.9	H7-O4-N2-C1	177.6
O8-H6-C1-N2	-3.4	O8-H6-C1-N2	0.7	O8-H7-O4-N2	-3.5
H9-O8-H6-N2	107.5	H9-O8-H6-N2	-0.6	H9-O8-H7-O4	108.0
H10-O8-H7-N2	3.5	H10-O8-H9-H6	176.7	H10-O8-H7-N2	4.0

Table 3.13: Important intermolecular hydrogen bonding parameters (hydrogen bond distances[a], hydrogen bond angles[b]), charges on hydrogen bond acceptor and hydrogen at MP2/Aug-cc-pVDZ [L4] theoretical level along with stabilization energies S.E.[c] and distortion energies E_{Dis}^{d} of various hydrogen bonded complexes of TFHA-H$_2$O.

Species	Hydrogen bond distances[a]	Hydrogen bond Angles[b]	Atomic Charges		S.E.[c]	E_{Dis}^{d}
1ZW1	H9...S3	O8-H9...S3	148.5	q$_H$(q$_S$)	0.520(-0.290)	10.51	4.78
	H5...O8	O4-H5...O8	167.8	q$_H$(q$_O$)	0.551(-1.008)		
1ZW2	H7...O8	N2-H7...O8	179.8	q$_H$(q$_O$)	0.469(-0.982)	7.53	0.16
1ZW3	H8...S3	O9-H8...S3	140.9	q$_H$(q$_S$)	0.504(-0.323)	5.05	0.11
	H6...O9	C1-H6...O9	121.6	q$_H$(q$_O$)	0.241(-1.003)		
1EW1	H7...O8	N2-H7...O8	151.1	q$_H$(q$_O$)	0.469(-1.012)	8.82	0.77
	H9...S3	O8-H9...S3	144.4	q$_H$(q$_S$)	0.515(-0.274)		
1EW2	H5...O8	O4-H5...O8	158.4	q$_H$(q$_O$)	0.538(-0.990)	7.44	0.34
1EW3	H5...O8	O4-H5...O8	158.0	q$_H$(q$_O$)	0.538(-0.988)	7.61	0.39
1EW4	H8...S3	O9-H8...S3	145.1	q$_H$(q$_S$)	0.503(-0.238)	4.82	0.17
	H6...O9	C1-H6...O9	120.7	q$_H$(q$_O$)	0.234(-1.002)		
1EW5	H8...O4	O9-H8...O4	145.6	q$_H$(q$_O$)	0.502(-0.615)	3.34	0.11
	H6...O9	C1-H6...O9	139.6	q$_H$(q$_O$)	0.230(-0.993)		
2ZW1	H10...N2	O8-H10...N2	122.7	q$_H$(q$_N$)	0.516(-0.250)	7.53	0.19
	H7...O8	O4-H7...O8	145.3	q$_H$(q$_O$)	0.541(-0.999)		
2ZW2	H5...O8	S3-H5...O8	156.3	q$_H$(q$_O$)	0.187(-0.992)	3.91	0.48

Contd.....

Species	Hydrogen bond distances[a]	Hydrogen bond Angles[b]	Atomic Charges		S.E.[c]	E_{Dis}^{d}
	H9...O4	1.998	O8-H9...O4	148.0	$q_H(q_O)$	0.507(-0.673)		
2ZW3	H6...O9	2.434	C1-H6...O9	125.4	$q_H(q_O)$	0.240(-0.978)	2.69	0.04
2ZW4	H6...O8	2.560	C1-H6...O8	113.5	$q_H(q_O)$	0.239(-1.000)	4.37	0.08
	H9...N2	2.060	O8-H9...N2	143.7	$q_H(q_N)$	0.508(-0.262)		
2EW1	H5...O9	2.098	S3-H5...O9	153.6	$q_H(q_O)$	0.189(-1.006)	6.21	0.24
	H8...N2	1.988	O9-H8...N2	152.5	$q_H(q_N)$	0.518(-0.290)		
2EW2	H10...N2	2.167	O8-H10...N2	125.1	$q_H(q_N)$	0.514(-0.267)	6.77	0.16
	H7...O8	1.939	O4-H7...O8	144.5	$q_H(q_O)$	0.532(-0.997)		
2EW3	H6...O9	2.430	C1-H6...O9	124.9	$q_H(q_O)$	0.233(-0.979)	2.60	0.03
2Z2W1	H10...N2	2.166	O8-H10...N2	123.8	$q_H(q_N)$	0.516(-0.244)	7.28	0.20
	H7...O8	1.919	O4-H7...O8	144.1	$q_H(q_O)$	0.538(-0.998)		
2Z2W2	H9...N2	2.064	O8-H9...N2	143.4	$q_H(q_N)$	0.508(-0.256)	4.61	0.08
	H6...O8	2.562	C1-H6...O8	112.8	$q_H(q_O)$	0.241(-1.002)		
2E2W1	H10...N2	2.171	O8-H10...N2	123.7	$q_H(q_N)$	0.516(-0.251)	7.23	0.21
	H7...O8	1.930	O4-H7...O8	144.4	$q_H(q_O)$	0.533(-0.999)		

a- in angstrom (Å)
b- in degrees (°)
c, d - in kcal/mol

Most of the aggregates of TFHA with water suggest that single water forms two H-bonds with the molecule. It is acting as H-bond donor to N, O or S and H-bond acceptor to A-H where A = N, O, C or S. Out of all the aggregates of **1Z** conformer and water in 1:1 ratio, the aggregate **1ZW1** is the most stabilized orientation of the aggregation. The aggregates **1ZW2, 1EW2, 1EW3, 2ZW3, 2EW3** contain single intermolecular H-bond between H_2O and TFHA. There are four aggregates (**1ZW1, 1ZW3, 1EW1, 1EW4**) where S acts as H-bond acceptor and two aggregates (**2ZW2, 2EW1**) where S-H bond is H-bond donor. In **1ZW2** aggregate there is single H-bond between water and TFHA while similar orientation of water and FHA resulted in a pair of H-bonds [22].

In **1ZW1**, water is involved in bidentate mode of binding forming two H-bonds, one as a donor and other as an acceptor. The water adduct **1ZW1** involves the insertion of water molecule into the intramolecular H-bond of **1Z**. As a consequence of that, the intramolecular H-bond distance H5...S3 is increased from 2.248 Å in **1Z** to 3.026 Å in **1ZW1** where no intramolecular H-bond interactions can be anticipated. It also distorts the dihedral H5-O4-N2-C1 from 0° in TFHA (**1Z**) to 60.6°(**1ZW1**). In this orientation the OH group is in more appropriate position for the intermolecular H-bonding. The Δr (difference from sum of van der Waals radii) for H9...S3 is 0.72 Å while the value for H9...O3 in similar conformation of FHA is 0.74 Å. The hydroxyl group attached to nitrogen acts as strong H-bond donor toward H_2O as suggested by close contact of H and O (H_2O) i.e. 1.695 Å (Δr =0.91 Å) with O4-H5...O8 angle of 167.8°. The similar interaction in FHA is weaker as suggested by H5...O8 distance being 1.743 Å and O4-H5...O8 angle of 159.8° [22]. The highest stabilization of the **1Z** conformation i.e. thione form suggests that it is the preferred conformation in aqueous solution.

The distortion of **1Z** conformation to **1ZW1** requires energy of 4.78 kcal/mol that involves breaking of intramolecular H-bonding. Inspite of large distortion energy requirement, the stabilization resulting from aggregation amounts to 10.51 kcal/mol at MP2/Aug-cc-pVDZ level. The analog of **1ZW1** in FHA has stabilization energy of 11.30 kcal/mol for aggregation with water while the distortion energy is 2.89 kcal/mol. The addition of distortion energy to stabilization energy (S.E.) indicates that H-bonding interactions are comparable in TFHA and FHA.

1ZW2, 1ZW3, 2ZW2 and **2ZW3** are the aggregates where intramolecular H-bonding interactions are also present as indicated by the geometrical parameters. The second order delocalization energies for intermolecular H-bonding between TFHA and water at **[L4]** theoretical level are listed in Table 3.14. Some of the internal delocalizations get affected upon the onset of intermolecular H-bonding interactions when TFHA forms aggregate with water. For instance, complete elimination of the delocalization responsible for intramolecular H-bond occurs in water adduct **1ZW1** while on the other hand the intramolecular H-bond in **1ZW2** shows synergic enhancement since $E^{(2)}$ for $n_{S3} \rightarrow \sigma^*_{O4-H5}$ increases by 1.4 kcal/mol (from 15.8 kcal/mol in **1Z** to 17.2 kcal/mol in **1ZW2**) supported by shortening of H-bonded distance by 0.021Å. The occupancies of the acceptor antibonding orbitals of water upon adduct formation are fairly high and range from 0.006 to 0.035 (Table 3.14). In the donor-acceptor interactions, the electron donor ability of water is more than that of TFHA as can be seen from $E^{(2)}$ values. Water acts as electron donor towards various sites of TFHA and it is found that its donor ability decreases towards various sites following the order O-H > N-H > S-H > C-H (Also reflected by their decreasing electronegativity values in the same order).

The aggregates **1ZW3, 1EW4, 1EW5, 2ZW3, 2ZW4, 2EW3, 2Z2W2** contain at least one of the two H-bond interactions as unconventional C-H...O interaction. C-H...O H-bonds in these water adducts range from 2.340 to 2.562 Å, characteristic of weak H-bond [26]. The stabilization energy of aggregation in these differs only slightly from the similar aggregates in case of FHA. The least stable water adduct of TFHA is **2EW3** with a stabilization energy of mere 2.6 kcal/mol. Adduct **1EW3** differs from **1EW2** only in orientation of water with respect to TFHA and is just 0.2 kcal/mol more stabilized over it (Table 3.13).

The sulfur of thiocarbonyl group acts as H-bond acceptor in **1ZW3** and **1EW4** along with C-H acting as H-bond donor to oxygen of water resulting in stabilization of 5.05 kcal/mol and 4.82 kcal/mol respectively. The stabilization energies show a difference of less than 1.00 kcal/mol from the values for similar interactions in FHA where C=O group acts as H-bond acceptor. This amazing strength of 'S' may result from more polarizable 'S' atom interacting with polar O-H bond of water in dipole induced dipole manner and due to greater basicity of lone pairs of sulfur.

The intermolecular H-bonding in **2ZW2** involves S-H group acting as H-bond donor to oxygen of water and oxygen of OH attached to nitrogen of TFHA acting as H-bond acceptor to water with interaction energy of 3.91 kcal/mol. The similar FHA aggregate has O-H group in place of S-H and the stabilization energy is nearly double (7.48 kcal/mol). The significant difference indicates that S-H is weaker H-bond donor than O-H group while H-bond acceptor abilities of sulfur and oxygen are comparable. This also explains why thione **1Z** form is preferred in aqueous solutions and solid state where intermolecular interactions are operative.

Table 3.15 records the lone pair occupancies of aggregates of TFHA with water along with those of FHA with water. As can be seen from the table, the lone pair of electrons present at N is the most delocalized lone pair in both the hydroxamic acids as well as their aggregates with water. The aggregation disrupts the intramolecular H-bond in most cases but the occupancy of the lone pair is not enhanced. Thus the conjugative interactions present in the molecule are not disturbed. The lone pair occupancy values also indicate that the conjugative interactions are relatively stronger in TFHA and its aggregates in comparison to FHA and its aggregates.

Table 3.14: The second order delocalization energies $E^{(2)}$ in kcal/mol and occupancies of acceptor antibonding orbitals associated with electron delocalizations important for intermolecular hydrogen bonding between TFHA and H_2O from NBO analysis at MP2/Aug-cc-pVDZ [L4] theoretical level.

Species	Donor TFHA	Acceptor H_2O	$E^{(2)}$	Donor H_2O	Acceptor TFHA	$E^{(2)}$	Occupancies Acceptor TFHA		Acceptor H_2O	
1ZW1	$n_{S3(1)} \to$	σ^*_{O8-H9}	2.39	$n_{O8(1)} \to$	σ^*_{O4-H5}	0.66	σ^*_{O8-H9}	0.028	σ^*_{O4-H5}	0.043
	$n_{S3(2)} \to$	σ^*_{O8-H9}	9.74	$n_{O8(2)} \to$	σ^*_{O4-H5}	30.82				
1ZW2			-	$n_{O8(1)} \to$	σ^*_{N2-H7}	0.34	-	-	σ^*_{N2-H7}	0.038
			-	$n_{O8(2)} \to$	σ^*_{N2-H7}	21.55				
1ZW3	$n_{S3(1)} \to$	σ^*_{O8-H9}	0.46	$n_{O9(2)} \to$	σ^*_{C1-H6}	1.25	σ^*_{O8-H9}	0.010	σ^*_{C1-H6}	0.030
	$n_{S3(2)} \to$	σ^*_{O8-H9}	4.98							
1EW1	$n_{S3(1)} \to$	σ^*_{O8-H9}	1.18	$n_{O8(2)} \to$	σ^*_{N2-H7}	0.45	σ^*_{O8-H9}	0.023	σ^*_{N2-H7}	0.041
	$n_{S3(2)} \to$	σ^*_{O8-H9}	10.16	$n_{O8(2)} \to$	σ^*_{N2-H7}	17.27				
1EW2			-	$n_{O8(1)} \to$	σ^*_{O4-H5}	0.23	-	-	σ^*_{O4-H5}	0.023
			-	$n_{O8(2)} \to$	σ^*_{O4-H5}	16.61				
1EW3			-	$n_{O8(1)} \to$	σ^*_{O4-H5}	0.27	-	-	σ^*_{O4-H5}	0.023
			-	$n_{O8(2)} \to$	σ^*_{O4-H5}	16.85				
1EW4	$n_{S3(1)} \to$	σ^*_{O9-H8}	0.30	$n_{O9(1)} \to$	σ^*_{C1-H6}	0.25	σ^*_{O9-H8}	0.013	σ^*_{C1-H6}	0.034
				$n_{O9(2)} \to$	σ^*_{C1-H6}	0.96				
1EW5	$n_{O4(1)} \to$	σ^*_{O8-H9}	1.66	$n_{O9(1)} \to$	σ^*_{C1-H6}	0.20	σ^*_{O8-H9}	0.035	σ^*_{C1-H6}	0.007
	$n_{O4(2)} \to$	σ^*_{O8-H9}	3.39	$n_{O9(2)} \to$	σ^*_{C1-H6}	1.65				
2ZW1	$n_{N2(1)} \to$	σ^*_{O8-H10}	3.43	$n_{O8(1)} \to$	σ^*_{O4-H7}	0.31	σ^*_{O8-H10}	0.006	σ^*_{O4-H7}	0.019
				$n_{O8(2)} \to$	σ^*_{O4-H7}	12.54				
2ZW2	$n_{O4(1)} \to$	σ^*_{O8-H9}	3.41	$n_{O8(1)} \to$	σ^*_{S3-H5}	0.14	σ^*_{O8-H9}	0.010	σ^*_{S3-H5}	0.015
	$n_{O4(2)} \to$	σ^*_{O8-H9}	3.05	$n_{O8(2)} \to$	σ^*_{S3-H5}	4.86				
2ZW3			-	$n_{O9(1)} \to$	σ^*_{C1-H6}	0.51	-	-	σ^*_{C1-H6}	0.015

Contd......

2Z2W4	-	-	1.38	$n_{O9(2)} \to \sigma^*_{C1-H6}$	-	-		
2EW1	$n_{N2(1)} \to \sigma^*_{O8-H9}$	5.88	0.38	$n_{O8(1)} \to \sigma^*_{C1-H6}$	σ^*_{O8-H9}	0.011	σ^*_{C1-H6}	0.014
	$n_{N2(1)} \to \sigma^*_{O8-H9}$	10.84	0.28	$n_{O9(1)} \to \sigma^*_{S3-H5}$	σ^*_{O8-H9}	0.018	σ^*_{S3-H5}	0.020
			9.12	$n_{O9(2)} \to \sigma^*_{S3-H5}$				
2EW2	$n_{N2(1)} \to \sigma^*_{O8-H10}$	3.75	0.27	$n_{O8(1)} \to \sigma^*_{O4-H7}$	σ^*_{O8-H10}	0.006	σ^*_{O4-H7}	0.016
			10.56	$n_{O8(2)} \to \sigma^*_{O4-H7}$				
2EW3	-	-	0.80	$n_{O9(1)} \to \sigma^*_{C1-H6}$	-	-	σ^*_{C1-H6}	0.024
	-	-	1.34	$n_{O9(2)} \to \sigma^*_{C1-H6}$				
2Z2W1	$n_{N2(1)} \to \sigma^*_{O8-H10}$	3.61	0.29	$n_{O8(1)} \to \sigma^*_{O4-H7}$	σ^*_{O8-H10}	0.006	σ^*_{O4-H7}	0.017
			11.50	$n_{O8(2)} \to \sigma^*_{O4-H7}$				
2Z2W2	$n_{N2(1)} \to \sigma^*_{O8-H9}$	5.79	0.39	$n_{O8(1)} \to \sigma^*_{C1-H6}$	σ^*_{O8-H9}	0.011	σ^*_{C1-H6}	0.014
2E2W1	$n_{N2(1)} \to \sigma^*_{O8-H10}$	3.55	0.28	$n_{O8(1)} \to \sigma^*_{O4-H7}$	σ^*_{O8-H10}	0.006	σ^*_{O4-H7}	0.016
			10.98	$n_{O8(2)} \to \sigma^*_{O4-H7}$				

Table 3.15: Lone pair occupancies of TFHA (FHA) and water in monomeric and in aggregate form with water at MP2/Aug-cc-pVDZ **[L4]** theoretical level.

Species	LP(1)N2	LP(1)X3	LP(2)X3	LP(1)O4	LP(2)O4	LP(1)O$_W$	LP(2)O$_W$
1Z	1.65	1.99	1.89	1.99	1.97	-	-
	(1.80)	(1.98)	(1.89)	(1.99)	(1.98)	(-)	(-)
1ZW1	1.67	1.98	1.91	1.99	1.97	2.00	1.96
	(1.75)	(1.98)	(1.90)	(1.99)	(1.98)	(2.00)	(1.97)
1ZW2	1.62	1.99	1.89	1.99	1.97	2.00	1.97
[1ZW4]	(1.80)	(1.98)	(1.90)	(1.99)	(1.98)	(2.00)	(1.99)
1ZW3	1.65	1.99	1.89	1.99	1.97	2.00	2.00
	(1.77)	(1.98)	(1.90)	(1.99)	(1.98)	(2.00)	(2.00)
1E	1.79	1.99	1.91	1.99	1.98	-	-
	(1.86)	(1.99)	(1.89)	(2.00)	(1.98)	(-)	(-)
1EW1	1.71	1.99	1.90	1.99	1.98	1.99	1.98
	(1.82)	(1.98)	(1.90)	(1.99)	(1.98)	(2.00)	(1.99)
1EW2	1.75	1.99	1.91	1.99	1.98	2.00	1.98
	(1.86)	(1.99)	(1.90)	(2.00)	(1.98)	(2.00)	(1.98)
1EW3	1.75	1.99	1.91	1.99	1.98	2.00	1.98
	(1.86)	(1.99)	(1.90)	(2.00)	(1.98)	(2.00)	(1.98)
1EW4	1.76	1.99	1.91	1.99	1.98	2.00	2.00
	(1.85)	(1.98)	(1.90)	(2.00)	(1.98)	(2.00)	(2.00)
1EW5	1.81	1.99	1.91	1.99	1.98	-	-
	(1.87)	(1.99)	(1.90)	(1.99)	(1.98)	(2.00)	(2.00)
2Z	1.96	1.99	1.89	1.99	1.95	-	-
	(1.97)	(1.98)	(1.89)	(1.99)	(1.97)	(-)	(-)
2ZW1	1.95	1.99	1.89	1.99	1.95	2.00	1.98
	(1.96)	(1.98)	(1.88)	(1.99)	(1.96)	(2.00)	(1.98)
2ZW2	1.96	1.99	1.88	1.99	1.96	2.00	1.99
	(1.96)	(1.98)	(1.87)	(1.99)	(1.97)	(2.00)	(1.97)
2ZW3	1.96	1.99	1.89	1.99	1.96	2.00	1.99
	(1.97)	(1.98)	(1.90)	(1.99)	(1.97)	(2.00)	(2.00)
2E	1.97	1.99	1.90	2.00	1.96	-	-
	(1.97)	(1.98)	(1.90)	(2.00)	(1.97)	(-)	(-)
2EW1	1.96	1.99	1.89	2.00	1.96	2.00	1.98
	(1.96)	(1.98)	(1.88)	(2.00)	(1.97)	(2.00)	(1.97)
2EW2	1.96	1.99	1.91	2.00	1.95	2.00	1.98
	(1.96)	(1.98)	(1.90)	(2.00)	(1.96)	(2.00)	(1.99)
2EW3	1.97	1.99	1.91	2.00	1.96	2.00	1.99
	(1.97)	(1.98)	(1.91)	(2.00)	(1.97)	(2.00)	(2.00)

3.3 CONCLUSIONS:

1. The present ab initio calculations on TFHA conclude that both **1Z** and **2Z** conformer possess intramolecular H-bonding interactions but **2Z** conformer is the most stable conformer in gas phase and is only 1.48 kcal/mol more stable than **1Z** conformer at MP2/Aug-cc-pVDZ theoretical level. In FHA, **1Z** conformer was observed to be most stable. The relative higher strength of C=N π bond in comparison to C=S π bond favor the **2Z** conformer.

2. The gas phase order of stabilities is **2Z > 2Z2 > 1Z > 2E > 2E2 > 1E > 3** in TFHA while the gas phase order of stabilities in FHA is **1Z > 2Z > 1E > 2E > 2Z2 > 2E2 > 3** at MP2/Aug-cc-pVDZ level.

3. The geometrical parameters and NBO analysis suggest that conjugative interactions in –C (=S)-NO- unit are stronger in comparison to –C (=O)-NO- in formohydroxamic acid.

4. The C-N rotational barriers for **1Z → 1E** and **2Z → 2E** are higher in TFHA than FHA, whereas the C-X barriers for **2Z → 2Z2** and **2E→2E2** are larger in FHA. All the activation barriers for tautomerizations are larger in FHA relative to TFHA.

5. The aggregation with water molecule indicates that stabilization as the result of intermolecular H-bonding favors the thione form of TFHA in gas phase and the most stable aggregate **1ZW1** has stabilization energy of 10.51 kcal/mol at MP2/Aug-cc-pVDZ level which is only 0.79 kcal/mol lower than that of FHA. The high stabilization energy associated with aggregations makes the **1Z** conformer most stable in the aggregate.

6. The electrostatic component of H-bonding interactions is weaker in TFHA in comparison to FHA while charge transfer component is higher in TFHA in comparison to FHA. Thus the H-bonds to oxygen are driven by charge-charge interaction while in case of sulfur; the stabilization principally results from the interaction of charge on the acidic hydrogen with the dipoles and quadrupoles of sulfur.

7. Though H-bond donor ability of 'SH' in TFHA is less than 'OH' in FHA, yet the H-bond acceptor abilities of sulfur and oxygen of thiocarbonyl and carbonyl group in TFHA and FHA respectively are comparable.

3.4 REFERENCES:

1. J.A. Platts, S.T. Howard, B.R.F. Bracke, J. Am. Chem. Soc. 118 (1996) 2726.
2. B.K. Min, H.J. Lee, Y.S. Choi, J. Park, C.J. Yoon, J.A. Yu, J. Mol. Struct. 471 (1998) 283.
3. R.A. Shaw, E. Kollat, M. Hollosi, H.H. Mantsch, Spectrochim. Acta 51A (1995) 1399.
4. D.B. Sherman, A.F. Spatola, J. Am. Chem. Soc. 112 (1990) 433.
5. N. Ueyama, Y. Yamada, T. Okamura, S. Kimura, A. Nakamura, Inorg. Chem. 35 (1996) 6473.
6. T.L. Poulos, B.C. Finzel, A.J. Howard, J. Mol. Biol. 195 (1987) 687.
7. T. Ueno, N. Nishikawa, S. Moriyama, S. Adachi, K. Lee, T. Okamura, N. Ueyama, A. Nakamura, Inorg. Chem. 38 (1999) 1199.
8. N. Ueyama, T. Terekawa, M. Nakata, A. Nakamura, J. Am. Chem. Soc. 105 (1983) 7098.
9. S.J. Yen, J.J. Ho, J. Phys. Chem. A 104 (2000) 8551.
10. Y. Orita, A. Ando, H. Abe, S. Yamabe, H. Berthod, A. Pullman, Theoret. Chim. Acta. (Ber) 54 (1979) 83.
11. G. Winkelmann, V.D. Helm, J.B. Neilands, Iron Transport in microbes, Plants and Animals, VCH, Weinheim, New York, 1987.
12. Y. Egawa, K. Umino, Y. Ito, T. Okuda, J. Antibiot. 24 (1971) 124.
13. S. Itoh, K. Inuzuka, T. Suzuki, J. Antibiot. 23 (1970) 542.
14. W. Walter, E. Schaumann, Synthesis 3 (1971) 111.
15. A. Mitchell, K.S. Murray, P.J. Newman, P.E. Clark, Aust. J. Chem. 30 (1977) 2439.
16. A. Chimiak, W. Przychodzen, J. Rachon, Heteroat. Chem. 13 (2002) 169.
17. R. Kakkar, A. Dua, S. Zaidi, Org. Biomol. Chem. 5 (2007) 547.
18. A.D. Bond, W. Jones, J. Phys. Org. Chem. 13 (2000) 395.
19. P.W. Austin, M. Singer, United States patent (1996) 5500217.
20. A. Kaczor, L.M. Proneiewicz, J. Mol. Struct. (Theochem) 640 (2003) 133.
21. A. Nino, C.M. Caro, M.L. Senent, J. Mol. Struct. (Theochem) 530 (2000) 291.
22. D. Kaur, R. Kohli, Int. J. Quantum Chem. 108 (2008) 119.
23. K.A. Jensen, O. Buchardt, C. Christophersen, Acta. Chem. Scand. 21 (1967) 1936.
24. K. Nagata, S. Mizukami, Chem. Pharm. Bull. 14 (1966) 1255.

25. M. Saldyka, Z. Mielke, Chem. Phys. 308 (2005) 59.
26. G.R. Desiraju, T. Steiner, The Weak hydrogen Bond in Structural Chemistry and Biology, Oxford University Press: Oxford, 1999.
27. R.F.W. Bader, Atoms in Molecules: A Quantum Theory, Oxford University Press: Oxford, 1990.
28. J. Cioslowski, A. Nanayakkara, M. Challacombe, Chem. Phys. Lett. 203 (1993) 137.
29. J. Cioslowski, P.R. Surjan, J. Mol. Struc. 255 (1992) 9.
30. J. Cioslowski, B.B. Stefanov, Mol. Phys. 84 (1995) 707.
31. B.B. Stefanov, J.R. Cioslowski, J. Comp. Chem. 16 (1995) 1394.
32. J. Cioslowski, Int. J. Quant. Chem. Quant. Chem. Symp. 24 (1990) 15.
33. J. Cioslowski, S.T. Mixon, J. Am. Chem. Soc. 113 (1991) 4142.
34. J. Cioslowski, Chem. Phys. Lett. 194 (1992) 73.
35. J. Cioslowski, Chem. Phys. Lett. 219 (1994) 151.

THE ROLE OF ISOMERISM AND EFFECT OF MEDIUM ON STABILITY OF ANIONS OF FORMO- AND THIOFORMO- HYDROXAMIC ACIDS

4.1 INTRODUCTION:

The knowledge of favored ionization site of hydroxamic acids (HAs) is important in understanding the role played by them in biological processes and in drug design, as well as in metal ion complexation [1]. Literature studies show that structural chemistry of HAs contain unresolved problems like determining the preferred site of deprotonation (NH or OH), the structure of corresponding anions and the reason for their high acidity [2-4]. The –NHOH functionality of HAs is responsible for the acidity. Both N-H and O-H protons are labile, however different titration experiments with strong bases have shown that only one proton can be removed, even at high basicity levels. The pka values are of the order of 9 pk units i.e. approximately 6 units more acidic than amides [5]. HAs are stabilized with respect to simpler reference molecules of amines or N-alkylhydroxylamines respectively. However the anions are stabilized still more and are responsible for acidity. In a series of carboxylic acids, amides and HAs, the resonance stabilization decreases in the sequence carboxylic acids > amides > HAs. According to Exner et al. resonance is responsible for about one quarter of enhanced activity of carboxylic acids, one third in case of amides and one half in case of HAs [6].

Deprotonation of HAs is challenging due to their extreme sensitivity to structure and solvent. A number of experimental and theoretical studies have been carried out to understand the dominant site of deprotonation in HAs. NMR relaxation rate and Nuclear Overhauser Effect (NOE) measurements in aqueous solution indicated that acetohydroxamic acid (AHA) in water is predominantly an oxygen acid whereas benzohydroxamic acid (BHA) is mainly nitrogen acid in methanol [5]. The gas phase studies by Fourier transform cyclotron resonance together with collision induced dissociation spectra of corresponding deprotonated species concluded that AHA behaves essentially as N-acid in gas phase [7]. Few experimental studies concluded that AHA exists in N-anionic and O-anionic forms in equilibrium and the direction of equilibrium is decided by the reaction conditions [8]. The dominant coordination mode in metal hydroxamic acid complexes is the O, O bidentate chelate in which the ligand is either singly or doubly deprotonated. The O, O bidentate coordination mode to metal perhaps forces researchers to suggest that both anionic forms may be in equilibrium or N-anion transforms to O-anion in polar medium assisted by buffer [9].

Explaining the role of HAs as matrix metalloprotease inhibitors, Yazal and Pang suggested that anionic forms of AHA have higher affinity for zinc relative to the HA molecules, thus probable binding involves anions [10]. Hydroxamate bind to zinc of hydrolytic enzymes with O, O coordination, thus inhibiting them **(Scheme 4.1)**.

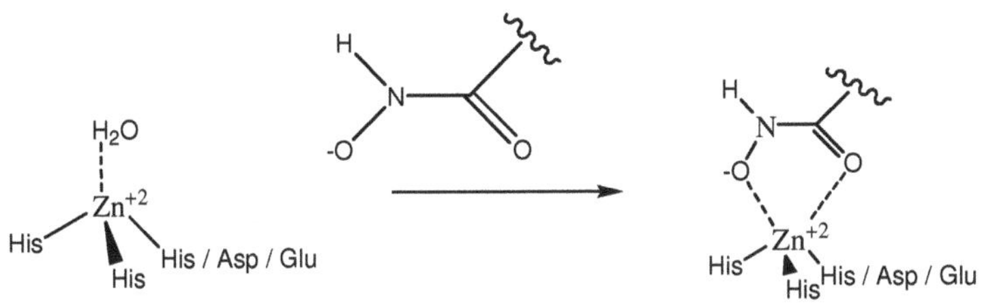

Scheme 4.1 Active sites and inhibition of hydrolytic enzymes

Hydroxamate anions are regarded as α-effect nucleophiles whose reactivity is higher than predicted by relationship between nucleophilicity and basicity [11,12]. Hydroxamate anions are known to be highly nucleophilic toward phenyl esters and attempts to use it as a catalyst for hydrolysis of phenyl esters have been carried out by Bender and coworkers in small molecule systems [13-16]. Hydroxamate anions are effective deacylating and dephosphorylating agents [17,18]. The anions of hydroxamic acid chelate with Fe (III), Ni (II) and Cu (II) and resultant complexes are highly colored and thus useful in colorimetric determination of metal ions [19].

In view of numerous applications of hydroxamic acid anions, the nature, stability and isomerism in anionic species of HAs is of utmost importance to understand their reactivity. The present chapter deals with the theoretical study of all the probable deprotonation pathways, the isomerism amongst the anionic forms, anion interconversions and free energy changes of deprotonation of formohydroxamic acid (FHA) and thioformohydroxamic acid (TFHA) in gas phase and in medium (H_2O) by applying both implicit and explicit solvation models on aggregates of neutral and anionic HAs with water.

4.2 RESULTS AND DISCUSSION:

4.2.1 DEPROTONATION ENTHALPIES OF FHA/TFHA IN GAS AND AQUEOUS PHASE:

As HAs are suggested to exist as equilibrium mixture of several tautomeric forms, the deprotonation of these forms can result in anions that are geometric or tautomeric isomers of one another. Considering the probable deprotonation sites (N-H, O-H in NHOH functionality and enolic X-H; X=O, S for FHA and TFHA respectively), eight of the anionic forms of FHA and TFHA each, generated upon several possible deprotonations have been optimized and presented in Figure 4.1. **1ZN⁻** and **1EN⁻** anionic forms are generated by the N-H deprotonation of **1Z** and **1E** neutral isomeric forms of HAs while the other six anionic forms are the O-H and X-H deprotonated forms. **2ZO4⁻** could be optimized only at [L1] theoretical level for FHA and at HF/6-31+G* for TFHA. **1EN⁻** and **2EX3⁻** (X=O, S) are descriptors of each other as they are resonating forms of each other **(Scheme 4.2)**.

The enthalpies associated with the N-H and O-H deprotonations of all the tautomeric forms of FHA and TFHA have been evaluated at B3LYP/6-31+G* [L1], B3LYP/Aug-cc-pVDZ [L2], MP2/6-31+G* [L3] and MP2/Aug-cc-pVDZ [L4] levels and are recorded in Table 4.1.

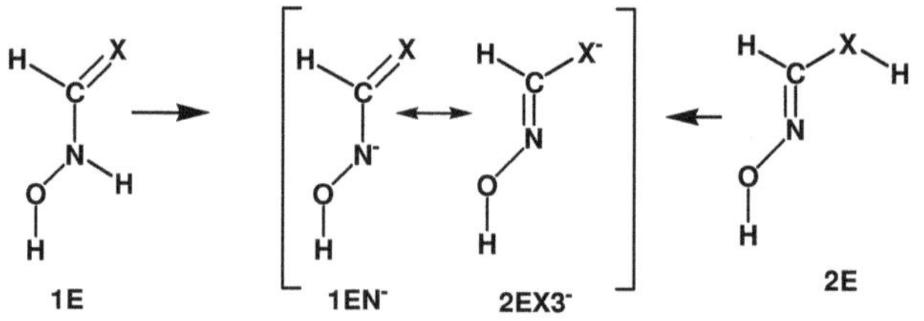

Scheme 4.2

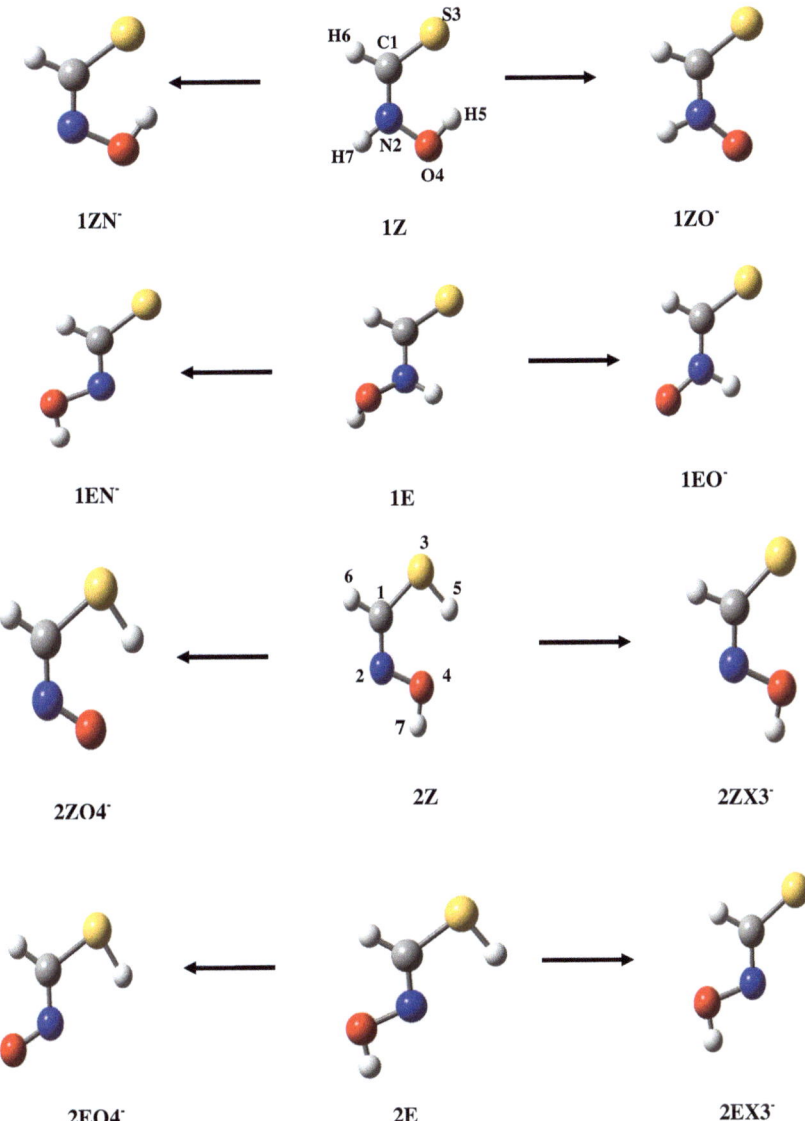

Figure 4.1: Various anions derived from deprotonation of TFHA at MP2/Aug-cc-pVDZ.

Table 4.1: Deprotonation enthalpies (kcal/mol) of FHA (TFHA) at B3LYP/6-31+G* [L1], B3LYP/Aug-cc-pVDZ [L2], MP2/6-31+G* [L3] and MP2/Aug-cc-pVDZ [L4] theoretical levels.

Species	L1	L2	L3	L4
1ZN⁻	340.35	342.53	335.34	338.48
	(326.00)	(328.15)	(320.97)	(323.00)
1ZO⁻	350.33	352.84	350.02	353.25
	(339.29)	(342.50)	(337.56)	(341.04)
1EN⁻	347.20	348.53	344.66	341.91
	(331.46)	(332.89)	(328.65)	(329.74)
1EO⁻	346.10	349.24	345.80	350.23
	(329.79)	(332.97)	(329.33)	(331.98)
2ZX3⁻	344.06	346.89	342.43	345.79
	(335.96)	(337.07)	(334.34)	(335.17)
2EO4⁻	359.29	362.27	357.72	362.05
	(351.99)	(354.57)	(352.21)	(354.77)
2EX3⁻	340.88	344.19	339.24	343.06
	(332.24)	(333.81)	(330.72)	(332.01)

The deprotonation enthalpies (gas phase acidities) for N-H and O-H heterolytic dissociations in several tautomeric forms range between 338.48-362.05 kcal/mol for FHA while for TFHA the range lies in 323.00-354.77 kcal/mol at [L4] theoretical level. Relatively lower energies are required for the deprotonation of the bonds in TFHA in comparison to respective values in FHA. The gas phase acidity values suggest that N-deprotonation from **1Z** conformation is preferred for both thio- and formo- hydroxamic acid over the O-deprotonation. The O-H deprotonation of **1Z** conformer of FHA and TFHA requires energy which is 14.77 and 18.04 kcal/mol respectively higher than that of the N-H deprotonation in the respective molecules. The intramolecular hydrogen bonding (H-bonding) present in **1Z** conformer involves OH group, making the deprotonation of OH comparatively difficult. The N-H deprotonation enthalpy of **1Z** is important as most of the studies show that **1Z** is the predominant form existing in neutral FHA and TFHA molecules in solution phase.

Table 4.2 lists the N- and O-deprotonation enthalpies of some related molecules. The N-H deprotonation enthalpy in case of **1Z** form of FHA and TFHA is lower than the enthalpy required for heterolytically dissociating N-H bond in case of formamide (359.39 kcal/mol at [L4] theoretical level) but is comparable to O-H deprotonation of formic acid (340.20 kcal/mol at [L4] theoretical level).

Table 4.2: Gas phase deprotonation enthalpies (acidities in kcal/mol) at MP2/Aug-cc-pVDZ [L4] theoretical level.

Species	Deprotonation site	Acidities
HCOOH	O	340.20
NH_2OH	O	382.22
NH_2OH	N	395.99
CH_3NH_2	N	398.49
$HCONH_2$	N	359.39

The comparison of deprotonation enthalpies of these molecules with that of FHA indicates formic acid to be more acidic than the O-deprotonated FHA. The N-H deprotonation of $HCONH_2$ requires 39.1 kcal/mol lower energy than the gas phase N-H deprotonation of CH_3NH_2 highlighting the importance of formyl substituent group. The FHA is formyl substituted derivative of hydroxylamine. The N-H deprotonation enthalpy of FHA is 57.51 kcal/mol lower than that of hydroxylamine. Thus, the larger change in N-H deprotonation enthalpy in hydroxylamine is due to the presence of formyl group substituent in comparison to the similar derivative of methylamine indicates the importance of –OH substituent.

4.2.2 RELATIVE STABILTY OF FHA/TFHA ANIONS:

Relative energies of the isomeric forms of anions have been evaluated at [L1], [L2], [L3] and [L4] theoretical levels are included in Table 4.3. The **1ZN⁻** is the most stable anionic structure for both FHA and TFHA which is in agreement with earlier reports by Remko et al. and Kakkar et al [20,21]. The relative energies in anionic isomeric forms of TFHA are higher in comparison to the respective values in case of FHA. The **2EO4⁻** is the least stable anionic form amongst all the isomeric forms in both FHA and TFHA. It is recalled that the order of relative energies among gas phase isomeric forms of FHA is **1Z > 2Z > 1E > 2E** while the order of relative stabilities of neutral forms of TFHA is **2Z > 1Z > 2E > 1E**. The relative stability of the isomeric anionic forms decreases in the order **1ZN⁻ > 2ZX3⁻ > 1EN⁻ = 2EX3⁻ > 1EO⁻ > 1ZO⁻ > 2EO4⁻** for both FHA and TFHA. The energy difference in the anionic forms ranges between 8.72-28.86 kcal/mol and 10.83-34.29 kcal/mol for FHA and TFHA respectively. The relative energy differences are found to be higher in the anions than neutral forms for both FHA and TFHA. For comparison of the relative energies of anions with neutral FHA and TFHA at different theoretical levels, refer to Table 2.2 and 3.1.

Table 4.3: Relative energies of isomeric anions of FHA (TFHA) in kcal/mol at B3LYP/6-31+G* **[L1]**, B3LYP/Aug-cc-pVDZ **[L2]**, MP2/6-31+G* **[L3]** and MP2/Aug-cc-pVDZ **[L4]** levels.

Species	L1	L2	L3	L4
1ZN⁻	0.00	0.00	0.00	0.00
	(0.00)	(0.00)	(0.00)	(0.00)
1ZO⁻	10.10	10.38	14.83	14.85
	(13.32)	(14.38)	(16.80)	(18.07)
1EN⁻	7.84	7.96	9.67	10.00
	(8.69)	(9.02)	(10.18)	(11.32)
1EO⁻	6.41	8.30	10.58	13.07
	(6.70)	(8.81)	(10.50)	(13.25)
2ZX3⁻	7.58	7.36	8.79	8.72
	(9.43)	(9.43)	(10.19)	(10.83)
2EO4⁻	26.03	25.77	28.03	28.86
	(28.64)	(30.05)	(31.51)	(34.29)
2EX3⁻	7.84	7.96	9.67	10.00
	(8.70)	(9.03)	(10.17)	(11.32)

4.2.3 VARIATION IN GEOMETRICAL PARAMETERS:

Geometrical parameters of the anions are recorded in the Tables 4.4 and 4.5. The analysis of the parameters indicate that the N-H deprotonation of **1Z** form keeps intramolecular H-bond between X3 (X=O, S) and O4-H5 intact and the anion **1ZN⁻** is planar (see Figure 4.1 for numbering). The O4-H5...O3 angle of 126.5° and non covalent distance H5...O3 of 1.860 Å in FHA suggest strengthening of intramolecular H-bond in the anion **1ZN⁻** in comparison to that in **1Z** form where the respective angle is 119.3° and the H-bond distance is 2.019 Å at **[L4]** theoretical level. For TFHA, the O4-H5...S3 angle of 131.5° and H-bond distance H5...S3 of 2.159 Å also indicates the strengthening of intramolecular H-bond relative to its **1Z** form where these values are 126.8° and 2.248 Å respectively.

Table 4.4: Geometrical parameters of **1Z** and **1E** forms of FHA (TFHA) and their N-H and O-H deprotonated anions at MP2/Aug-cc-pVDZ **[L4]** theoretical level. All the bond distances are in angstrom (Å), angles and dihedrals are in degrees (°).

Parameters	1Z	1ZN⁻	1ZO⁻	1E	1EN⁻	1EO⁻
C1-N2	1.368	1.332	1.351	1.393	1.338	1.337
	(1.336)	(1.320)	(1.345)	(1.364)	(1.311)	(1.328)
N2-O4	1.406	1.457	1.329	1.421	1.503	1.356
	(1.379)	(1.419)	(1.294)	(1.415)	(1.489)	(1.312)
O4-H5	0.983	0.991	-	0.969	0.967	-
	(0.994)	(0.998)	(-)	(0.971)	(0.967)	(-)
C1-H6	1.106	1.111	1.121	1.109	1.123	1.111
	(1.098)	(1.099)	(1.103)	(1.010)	(1.107)	(1.098)
C1-X3	1.235	1.292	1.259	1.224	1.273	1.274
	(1.666)	(1.743)	(1.706)	(1.646)	(1.728)	(1.726)
N2-H7	1.017	-	1.026	1.020	-	1.024
	(1.013)	(-)	(1.031)	(1.020)	(-)	(1.030)
C1-X3...H5	2.019	1.860				
	(2.248)	(2.159)				
X3-C1-N2	121.9	127.1	129.2	122.8	125.8	125.8
	(122.5)	(127.3)	(128.3)	(124.0)	(124.9)	(124.2)
O4-N2-C1	115.3	107.6	129.5	113.7	106.8	127.2
	(121.2)	(112.4)	(130.3)	(116.6)	(107.6)	(128.7)
H5-O4-N2	101.2	99.9	-	103.4	99.5	-
	(100.7)	(102.7)	(-)	(103.8)	(99.1)	(-)
H6-C1-N2	113.0	111.9	108.3	112.1	114.2	111.2
	(112.6)	(112.1)	(109.5)	(112.1)	(116.1)	(112.6)
H7-N2-C1	119.5	-	113.2	114.4	-	115.3
	(126.2)	(-)	(113.5)	(117.6)	(-)	(114.1)
C1-X3...H5-O4	119.3	126.5				
	(126.8)	(131.5)				
O4-N2-C1-X3	-12.0	0.0	-0.0	157.7	177.7	180.0
	(0.0)	(-0.1)	(0.0)	(160.1)	(180.0)	(180.0)
H5-O4-N2-C1	4.5	-0.0	-	119.0	128.8	-
	(-0.0)	(-0.0)	(-)	(115.2)	(180.0)	(-)
H6-C1-N2-O4	171.4	180.0	180.0	-25.9	-1.7	-0.0
	(180.0)	(179.9)	(180.0)	(-22.7)	(0.0)	(-0.0)
H7-N2-C1-X3	-148.1	-	180.0	29.7	-	0.0
	(-180.0)	(-)	(180.0)	(23.1)	(-)	(-0.0)

Table 4.5: Geometrical parameters of **2Z** and **2E** forms of FHA (TFHA) and their N-H and O-H deprotonated anions at MP2/Aug-cc-pVDZ **[L4]** theoretical level. All the bond distances are in angstrom (Å), angles and dihedrals are in degrees (°).

Parameters	2Z	2ZX3⁻	2E	2EX3⁻	2EO4⁻
C1-N2	1.294	1.335	1.287	1.338	1.301
	(1.297)	(1.317)	(1.293)	(1.311)	(1.321)
N2-O4	1.435	1.487	1.427	1.503	1.338
	(1.417)	(1.455)	(1.425)	(1.489)	(1.321)
O4-H7	0.967	0.966	0.967	0.967	-
	(0.969)	(0.968)	(0.969)	(0.967)	(-)
C1-H6	1.089	1.122	1.093	1.123	1.097
	(1.093)	(1.103)	(1.098)	(1.107)	(1.101)
C1-X3	1.352	1.271	1.359	1.273	1.415
	(1.762)	(1.725)	(1.762)	(1.728)	(1.799)
H5-X3	0.975	-	0.973	-	0.974
	(1.351)	(-)	(1.353)	(-)	(1.353)
X3-C1-N2	126.5	132.8	120.7	125.8	119.1
	(129.0)	(132.8)	(121.4)	(124.9)	(119.6)
O4-N2-C1	107.7	108.8	109.2	106.8	118.9
	(111.1)	(111.0)	(109.1)	(107.6)	(116.7)
H5-X3-C1	106.3	-	107.1	-	101.5
	(94.9)	(-)	(94.6)	(-)	(93.1)
H6-C1-N2	118.2	106.9	124.9	114.2	126.8
	(115.0)	(108.5)	(122.4)	(116.1)	(123.9)
H7-O4-N2	102.0	98.2	101.6	99.5	-
	(101.8)	(99.1)	(101.7)	(99.1)	(-)
O4-N2-C1-X3	0.0	0.0	180.0	182.3	180.0
	(-0.0)	(-0.0)	(180.0)	(180.0)	(180.0)
H7-O4-N2-C1	180.0	180.0	180.0	231.1	-
	(180.0)	(179.9)	(180.0)	(179.9)	(-)
H6-C1-N2-O4	180.0	180.0	-0.0	1.7	0.0
	(180.0)	(180.0)	(-0.0)	(0.0)	(0.0)
H5-X3-C1-N2	0.0	-	0.0	-	-0.0
	(0.0)	(-)	(-0.0)	(-)	(0.1)

The C-N bond distances in **1ZN⁻**, **1ZO⁻**, **1EN⁻** and **1EO⁻** anions are decreased significantly relative to those in their respective neutral forms while the anions **2ZX3⁻**, **2EX3⁻** and **2EO4⁻** show elongation of the distance. The most stable anionic form **1ZN⁻** of FHA has C-N bond distance of 1.332 Å which is intermediate between C-N single and double bond. The distance between carbon and X3 (X=O, S) is 1.235 Å and 1.666 Å in FHA and TFHA respectively which also indicates the presence of partial double bond character. The C=X distance elongates in **1ZN⁻**, **1ZO⁻**, **1EN⁻** and **1EO⁻** while contraction is observed in **2ZX3⁻** and **2EX3⁻**. The variations suggest stronger conjugative interactions between the lone pair of electrons present on nitrogen with the π bond of C=X in keto forms.

The deprotonation of N-H bond of **1Z** and **1E** conformations results in elongation of N-O bond while reverse is the case with deprotonation from O-H bond. All the structural isomers of FHA anions except **1EN⁻/2EX3⁻** have planar arrangement of atoms. In TFHA, all the anions are planar. The planarity facilitates the conjugative interactions between lone pairs and π bonds.

4.2.4 ATOMIC CHARGES:

The stability of any charged chemical specie is determined mainly by the ability of the species to stabilize the additional charge. In order to stabilize the charge, the anion undergoes geometrical reorientation and variation in electron delocalizations that requires change in hybridization. The changes are reflected by the variation in atomic charges, geometrical parameters and by the second order delocalization energies for orbital interactions. The atomic charges obtained using natural population analysis (NPA) incorporated within natural bond orbital (NBO) analysis are reported in Table 4.6.

Table 4.6: Atomic charges (NBO) in FHA (TFHA) at MP2/Aug-cc-pVDZ **[L4]** theoretical level.

Species	C1	N2	X3	O4	H5	H6	H7
1Z	0.696	-0.423	-0.750	-0.602	0.522	0.148	0.410
	(0.041)	(-0.338)	(-0.278)	(-0.596)	(0.524)	(0.217)	(0.430)
1ZN⁻	0.553	-0.456	-0.957	-0.733	0.500	0.093	-
	(-0.056)	(-0.296)	(-0.611)	(-0.701)	(0.492)	(0.171)	(-)
1ZO⁻	0.626	-0.341	-0.867	-0.811	-	0.066	0.328
	(-0.039)	(-0.197)	(-0.509)	(-0.753)	(-)	(0.159)	(0.339)
1E	0.719	-0.449	-0.703	-0.605	0.497	0.138	0.403
	(0.048)	(-0.401)	(-0.181)	(-0.592)	(0.499)	(0.211)	(0.415)
1EN⁻	0.622	-0.509	-0.910	-0.730	0.457	0.071	-
	(0.007)	(-0.350)	(-0.576)	(-0.712)	(0.469)	(0.162)	(-)
1EO⁻	0.653	-0.368	-0.905	-0.827	-	0.102	0.345
	(0.011)	(-0.209)	(-0.568)	(-0.781)	(-)	(0.181)	(0.365)
2Z	0.535	-0.300	-0.782	-0.683	0.534	0.187	0.509
	(-0.037)	(-0.213)	(0.004)	(-0.653)	(0.175)	(0.219)	(0.505)
2ZX3⁻	0.607	-0.516	-0.906	-0.705	-	0.069	0.451
	(-0.012)	(-0.368)	(-0.575)	(-0.667)	(-)	(0.162)	(0.460)
2E	0.541	-0.328	-0.771	-0.630	0.511	0.181	0.497

Contd.....

Species	C1	N2	X3	O4	H5	H6	H7
	(-0.026)	(-0.231)	(0.026)	(-0.632)	(0.151)	(0.215)	(0.497)
2EX3⁻	0.622	-0.509	-0.910	-0.730	-	0.071	0.457
	(0.007)	(-0.350)	(-0.576)	(-0.711)	(-)	(0.162)	(0.469)
2EO4⁻	0.292	-0.235	-0.840	-0.827	0.483	0.127	-
	(-0.283)	(-0.107)	(-0.126)	(-0.786)	(0.134)	(0.168)	(-)

The atomic charge analysis indicates that the negative charge of the anion is delocalized over all the atoms in the anions. The most stable anionic structure **1ZN⁻** in FHA possesses only -0.46 units of charge on nitrogen which is only insignificantly (0.03) different from its neutral counterpart, inspite of the fact that it is the N-H bond which is deprotonated. In **1ZN⁻** of TFHA, the charge density on nitrogen decreases slightly from -0.34 to -0.30 on N-H deprotonation. In **1ZN⁻** anion of FHA, both the carbonyl oxygen and the hydroxylic oxygen undergo variation of 0.21 and 0.13 units of charge toward higher electron density upon the anion formation. The respective variation of charges on thiocarbonyl sulfur and hydroxylic oxygen in TFHA are 0.33 and 0.11 respectively which shows comparatively large electron density shift to sulfur due to its large charge holding capacity. In the absence of significant electronegativity difference between carbon and sulfur, the thiocarbonyl carbon also holds additional electron density (atomic charge = -0.056) in **1ZN⁻** of TFHA. The enhancement in net charge density at both X3 and O4 by 0.117, 0.209 and 0.231, 0.157 respectively is observed in **1ZO⁻** anion of FHA and TFHA while charge density on nitrogen is affected only to a small extent. In all the anionic forms the charge density on X3 and O4 is higher than that of nitrogen which makes X3 and O4 relatively more nucleophilic in nature and explains the bidentate binding of TFHA to metals that occurs through utilizing S and O atoms and the observed O, O coordination mode in FHA.

4.2.5 VARIATION IN ELECTRON DELOCALIZATIONS:

The NBO analysis of all the anionic forms has been carried out to understand the role of different conjugative interactions in the stability of anions and the important orbital interactions along with second order delocalization energies $E^{(2)}$ are recorded in Table 4.7. There is appearance of $n_{X3} \rightarrow \pi^*_{C1-N2}$ orbital interaction in most stable **1ZN⁻** anion of FHA with strong second order delocalization energy $E^{(2)}$ of 119.02 kcal/mol while no such interaction was present in neutral **1Z** form. The corresponding interaction in TFHA has $E^{(2)}$ value of 54.80 kcal/mol. The interaction is expected to increase C-X bond strength with weakening of C-N bond. But the geometrical parameters suggest strengthening of C-N bond with simultaneous reduction in bond strength of C-X and N-O bond. The careful analysis of molecular orbital occupancies indicates that there is formation of C-N double bond with single bonding molecular orbital for C-X3 in all anions. The **1ZO⁻** anion has $n_{X3} \rightarrow \pi^*_{C1-N2}$ orbital interaction of 164.82 kcal/mol in case of X=O and 81.09 kcal/mol in case of X=S. The observation of all the $E^{(2)}$ values associated with **1ZO⁻** suggest relatively higher stabilization than that for **1ZN⁻** anion but the difference in C-N and C-X bond character is responsible for higher stabilization of **1ZN⁻**. The $n_{O3} \rightarrow \sigma^*_{O4-H5}$ electron delocalization in **1ZN⁻** of FHA favors the intramolecular H-bond formation with $E^{(2)}$ energy of 11.71 kcal/mol which is 5.62 kcal/mol higher than the value for **1Z**. The $n_{S3} \rightarrow \sigma^*_{O4-H5}$ electron delocalization strengthens the intramolecular H-bond O-H...S with $E^{(2)}$ value of 22.84 kcal/mol in TFHA **1ZN⁻** which is 7.04 kcal/mol higher than its neutral **1Z**. The lone pair occupancies of the anions are recorded in Table 4.8. The N-H deprotonation of **1Z** and **1E** form of both the HAs results in appearance of a pair of electrons at X3 signifying the delocalization of bonding pair of electrons available after deprotonation. The O-H deprotonation of **1Z** and **1E** form retains the additional pair of electrons at O4. In order to stabilize the additional one unit negative charge there is formation of C-N π bond which is apparent from the geometrical data and absence of lone pair of electrons at N. The deviation of the occupancies of the pair of electrons appearing at X3 or O4 in the anions is indicative of highly delocalized nature of the lone pair.

Table 4.7: Second order delocalization energies $E^{(2)}$(kcal/mol) associated with important orbital interactions of FHA (TFHA) anions obtained by NBO analysis at MP2/Aug-cc-pVDZ [L4] theoretical level.

Species	$n_{N2} \rightarrow \sigma^*_{C1-X3}$	$n_{N2} \rightarrow \pi^*_{C1-X3}$	$n_{X3} \rightarrow \sigma^*_{C1-H6}$	$n_{X3} \rightarrow \sigma^*_{O4-H5}$	$n_{X3} \rightarrow \sigma^*_{C1-N2}$	$n_{X3} \rightarrow \pi^*_{C1-N2}$	$n_{O4} \rightarrow \sigma^*_{C1-N2}$	$n_{O4} \rightarrow \pi^*_{C1-N2}$
1Z	15.37	29.99	25.94	0.87+5.22	29.97	-	-	-
	(-)	(143.97)	(13.78)	(15.80)	(15.82)	(-)	(-)	(-)
1ZN⁻	12.04	-	24.24	11.71	19.52	119.02	-	-
	(13.73)	(-)	(10.41)	(20.75+2.09)	(-)	(54.80)	(-)	(16.95)
1ZO⁻	-	-	28.16	-	28.30	164.82	10.53	-
	(-)	(-)	(10.45)	(-)	(17.27)	(81.09)	(11.26)	(53.33)
1E	-	23.50	26.64	-	32.87	-	-	-
	(1.84)	(62.73)	(13.56)	(-)	(18.99)	(-)	(-)	(-)
1EN⁻	-	-	25.42	-	22.56	125.15	-	-
	(-)	(-)	(8.72)	(-)	(13.32)	(64.32)	(-)	(7.85)
1EO-	-	-	24.68	-	24.78	149.12	-	20.44
	(-)	(-)	(10.59)	(-)	(13.34)	(65.49)	(-)	(43.52)
2Z	13.55	-	-	-	-	54.30	-	13.33
	(14.65)	(-)	(-)	(-)	(-)	(38.60)	(-)	(18.46)
2ZX3⁻	12.96	-	27.37	-	24.04	150.51	-	-
	(13.62)	(-)	(-)	(-)	(14.94)	(70.03)	(-)	(11.74)
2E	-	-	10.93	-	-	47.81	-	13.76
	(-)	(-)	(10.26)	(-)	(-)	(34.13)	(-)	(16.76)
2EX3⁻	-	-	25.42	-	22.56	125.16	-	-
	(-)	(-)	(-)	(-)	(13.32)	(64.32)	(-)	(-)
2EO4⁻	-	-	-	-	-	24.33	-	62.12
	(-)	(-)	(-)	(-)	(-)	(15.19)	(8.55)	(87.70)

Table 4.8: Lone pair occupancies of anions of FHA (TFHA) at MP2/Aug-cc-pVDZ [L4] theoretical level.

Anion	LP(1)N2	LP(1)X3	LP(2)X3	LP(3)X3	LP(1)O4	LP(2)O4	LP(3)O4
1Z	1.80	1.98	1.89	-	1.99	1.98	-
	(1.65)	(1.99)	(1.89)	(-)	(1.99)	(1.97)	(-)
1ZN⁻	1.97	1.98	1.90	1.75	1.99	1.98	-
	(1.96)	(1.99)	(1.91)	(1.83)	(1.99)	(1.96)	(-)
1E	1.86	1.99	1.89	-	2.00	1.98	-
	(1.79)	(1.99)	(1.91)	(-)	(1.99)	(1.98)	(-)
1EN⁻	1.98	1.98	1.91	1.70	1.99	1.99	-
	(1.98)	(1.99)	(1.95)	(1.80)	(2.00)	(1.98)	(-)
1ZO⁻	-	1.98	1.89	1.67	1.99	1.94	1.92
	(-)	(1.99)	(1.93)	(1.75)	(1.99)	(1.94)	(1.85)
1EO⁻	-	1.99	1.91	1.70	1.99	1.95	1.95
	(-)	(1.99)	(1.94)	(1.80)	(1.99)	(1.95)	(1.88)
2Z	1.97	1.98	1.89	-	1.99	1.97	-
	(1.96)	(1.99)	(1.89)	(-)	(1.99)	(1.95)	(-)
2ZX3⁻	1.97	1.98	1.91	1.70	2.00	1.98	-
	(1.96)	(1.99)	(1.94)	(1.78)	(2.00)	(1.97)	(-)
2E	1.97	1.98	1.90	-	2.00	1.97	-
	(1.97)	(1.99)	(1.90)	-	(2.00)	(1.96)	(-)
2EO4⁻	1.94	1.99	1.95	-	1.99	1.97	1.84
	(1.95)	(1.99)	(1.96)	(-)	(1.99)	(1.97)	(1.78)
2EX3⁻	1.98	1.98	1.91	1.70	1.99	1.99	-
	(1.98)	(1.99)	(1.95)	(1.80)	(2.00)	(1.98)	(-)

4.2.6 ISOMERISM IN FHA/TFHA ANIONS:

As suggested in several experimental studies, the HAs exist as equilibrium mixture of their several tautomeric forms. The deionization of these forms can result in anionic forms which may exist independently or together in equilibrium. There is one anticipation by Senent et al. that buffer assisted N-anion transformation into O-anion is possible and experimental evidence shows that exchange between the anions is negligible at 25°C and modest at 80°C [4,9]. For understanding whether interconversions between the several forms of anions is probable, transition states for the probable interconversions have been optimized. These probable pathways for interconversion of FHA and TFHA anions are depicted in Figure 4.2 and 4.3.

The activation barriers for all the steps involved in the pathways at B3LYP/6-31+G* [L1], B3LYP/Aug-cc-pVDZ [L2], MP2/6-31+G* [L3] and MP2/Aug-cc-pVDZ [L4] level are reported in Table 4.9 and the values at MP2/Aug-cc-pVDZ are also shown in the Figure 4.2 and 4.3 respectively for FHA and TFHA. The **1ZN⁻** and

2ZX3⁻ are generated as the result of N-H and X-H deprotonation from **1Z** and **2Z** forms respectively but can interconvert through rotation around N-O bond via rotational transition state **RTS1**. The N-O rotational transition state **RTS1** has activation barrier of 9.53 kcal/mol for **1ZN⁻** ⇌ **2ZX3⁻** in FHA while the barrier is 12.84 kcal/mol in TFHA. The **2ZX3⁻** however is 8.72 kcal/mol higher in energy than the **1ZN⁻** form in FHA while the relative energy for **2ZX3⁻** form in TFHA is 10.83 kcal/mol. This suggests that the barrier for **2ZX3⁻** ⇌ **1ZN⁻** conversion is much smaller (0.81 kcal/mol in FHA and 2.01 kcal/mol in TFHA). The activation barrier is high for reverse **1ZN⁻** ⇌ **2ZX3⁻** conversion since formation of **RTS1** not only involves N-O rotation but also involves breaking of intramolecular H-bond.

Table 4.9: Activation barriers (kcal/mol) for interconversion of anionic forms of FHA (TFHA) at B3LYP/6-31+G* [**L1**], B3LYP/Aug-cc-pVDZ [**L2**], MP2/6-31+G* [**L3**] and MP2/Aug-cc-pVDZ [**L4**] theoretical levels.

Species	Transition	L1	L2	L3	L4
RTS1	1ZN⁻ ⇌ 2ZX3⁻	8.41	8.32	9.62	9.53
		(11.38)	(11.51)	(12.13)	(12.84)
RTS2	1ZN⁻ ⇌ 1EN⁻	43.54	43.42	46.57	47.63
		(58.10)	(57.91)	(62.59)	(63.21)
RTS3	1EO⁻ ⇌ 1ZO⁻	33.93	33.09	33.46	32.04
		(48.15)	(47.63)	(49.87)	(50.04)
PTS1	2ZX3⁻ ⇌ 1ZO⁻	40.80	40.08	42.03	40.85
		(45.46)	(44.55)	(46.34)	(45.27)
PTS2	1EO⁻ ⇌ 2EO4⁻	42.80	40.64	39.13	36.36
		(36.00)	(33.91)	(33.78)	(30.50)
PTS3	1EN⁻ ⇌ 1EO⁻	38.68	36.66	36.97	36.19
		(44.60)	(42.11)	(43.12)	(42.22)

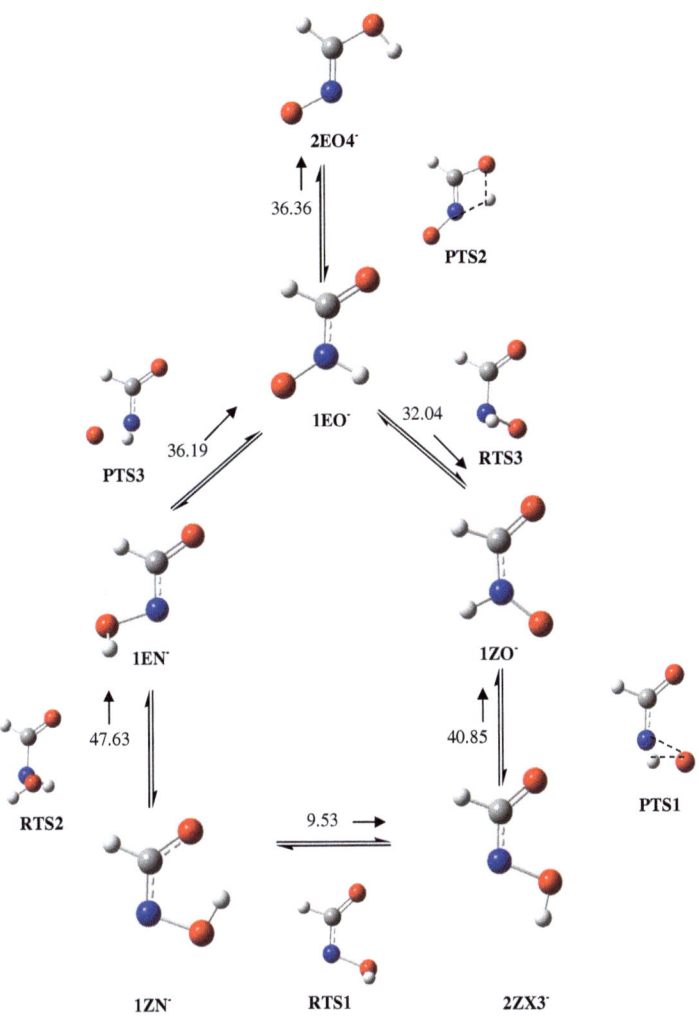

Figure 4.2: Schematic representation for intramolecular transformation amongst the tautomeric anionic forms of FHA. The values indicate the activation barriers.

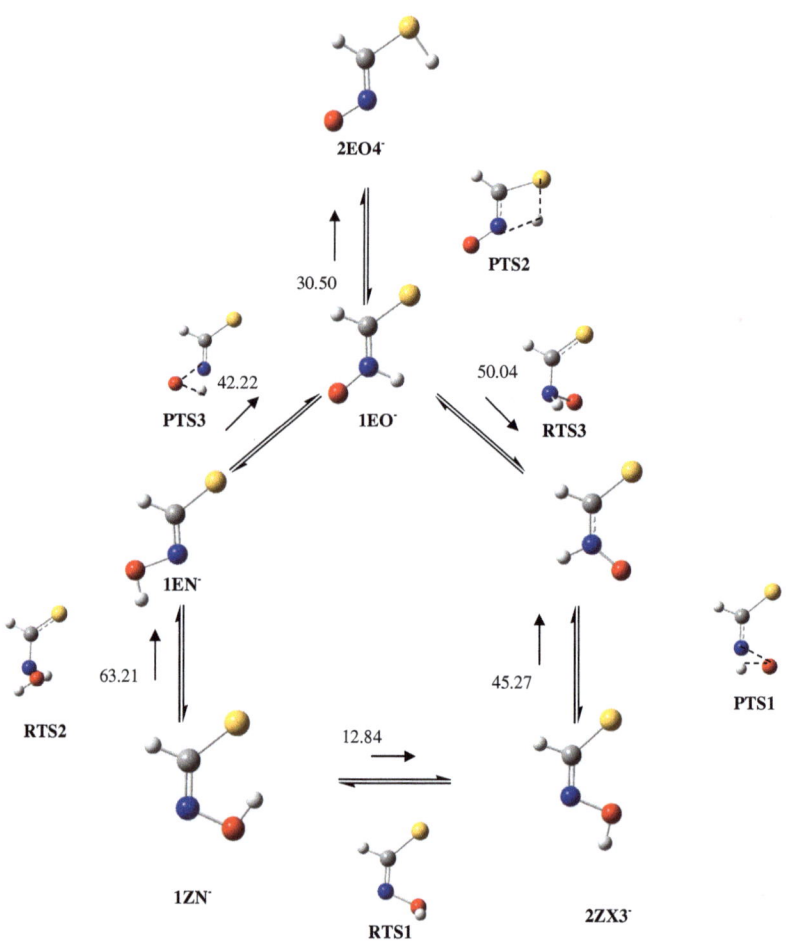

Figure 4.3: Schematic representation for intramolecular transformation amongst the tautomeric anionic forms of TFHA. The values indicate the activation barriers.

The **1EN⁻** anionic form is rotamer of **1ZN⁻** that differs from **1ZN⁻** in relative orientation of carbonyl group and NOH group and thus their interconversion involves rotation about C-N bond. The rotation around C-N bond for **1ZN⁻** ⇌ **1EN⁻** conversion has energy barrier of 47.63 kcal/mol through **RTS2** in FHA and 63.21 kcal/mol in TFHA through transition state **RTS2**. The high C-N rotational barrier for the **RTS2** also supports our earlier conclusion based on NBO analysis and geometrical parameters that there is development of C-N π bond character on N-H deprotonation. A significant difference of C-N rotation barrier (≈ 16 kcal/mol) for FHA and TFHA shows that π bond character in TFHA is relatively much stronger. The barrier calculated for the reverse process i.e. **1EN⁻** ⇌ **1ZN⁻** (calculated with respect to **1EN⁻**) is 36.44 kcal/mol in FHA and 51.90 kcal/mol in TFHA, making the interconversion probable at high temperatures only.

The **1EO⁻** anionic form, resulting from deprotonation from O-H group of **1E** is 3.07, 1.93 kcal/mol higher in energy than the **1EN⁻** conformation of FHA and in TFHA respectively. The **1EN⁻** ⇌ **1EO⁻** transformation through proton transfer transition state **PTS3** involves 1, 2 protropic shift with a barrier of 42.22 kcal/mol in TFHA and 36.19 kcal/mol in FHA respectively as it involves formation of highly strained ring.

The anion **2EO4⁻**, the deprotonated **2E** form has relative energy difference of 21.04 kcal/mol from the **1EO⁻** anionic form in TFHA and 15.79 kcal/mol in FHA. The barrier of **1EO⁻** ⇌ **2EO4⁻** (**PTS2**) is 30.50 kcal/mol for 1, 3 prototropic shift in TFHA while the similar barrier is 36.36 kcal/mol for FHA. All the rotational barriers and 1,2 proton transfer barriers are higher in TFHA but 1, 3 barrier is smaller in TFHA in comparison to FHA. Figure 4.4 is a schematic representation of the potential energy surface of interconversion in isomeric anions of FHA and TFHA anions. Tables 4.10-4.15 record the geometrical parameters of the transition states involving interconversion of different anionic species of HAs at the various theoretical levels employed.

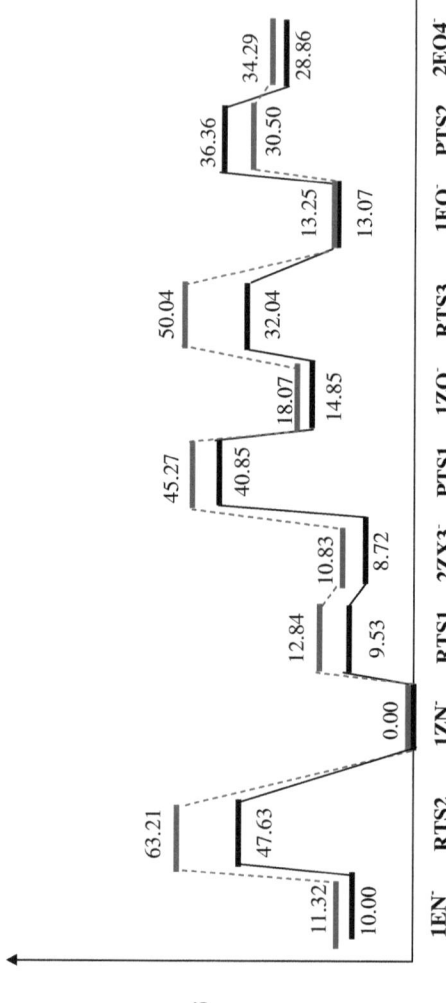

Figure 4.4: Potential energy surface indicating interconversion amongst isomeric anions of FHA (solid lines) and TFHA (Dotted lines) at MP2/Aug-cc-pVDZ.

Table 4.10: Geometrical parameters for transition state **RTS1** connecting interconversion **1ZN⁻** ⇌ **2ZX3⁻** in FHA (TFHA) at B3LYP/6-31+G* [**L1**], B3LYP/Aug-cc-pVDZ [**L2**], MP2/6-31+G* [**L3**] and MP2/Aug-cc-pVDZ [**L4**] theoretical levels. All the bond distances are in angstrom (Å), angles and dihedrals are in degrees (°).

Parameters	L1	L2	L3	L4
N2-C1	1.330	1.330	1.336	1.341
	(1.305)	(1.305)	(1.317)	(1.321)
X3-C1	1.258	1.257	1.268	1.268
	(1.726)	(1.729)	(1.708)	(1.722)
O4-N2	1.487	1.487	1.487	1.492
	(1.465)	(1.465)	(1.469)	(1.471)
H5-O4	0.970	0.965	0.974	0.968
	(0.971)	(0.966)	(0.976)	(0.969)
H6-C1	1.122	1.126	1.116	1.124
	(1.098)	(1.103)	(1.097)	(1.105)
X3-C1-N2	134.3	134.2	133.8	133.9
	(134.5)	(134.6)	(133.5)	(133.9)
O4-N2-C1	110.3	110.2	109.8	109.6
	(112.1)	(112.1)	(111.0)	(111.2)
H5-O4-N2	101.9	101.9	101.2	100.4
	(102.6)	(102.6)	(101.7)	(101.1)
H6-C1-N2	106.7	106.4	107.0	106.3
	(108.5)	(108.60)	(108.5)	(108.2)
O4-N2-C1-X3	-2.5	-2.5	-2.4	-1.9
	(-4.0)	(-3.8)	(-4.0)	(-3.4)
H5-O4-N2-C1	108.5	109.4	108.0	111.7
	(107.2)	(106.9)	(107.0)	(108.5)
H6-C1-N2-O4	177.5	177.5	177.6	177.8
	(177.1)	(177.1)	(177.0)	(177.2)

Table 4.11: Geometrical parameters for transition state **RTS2** connecting interconversion **1ZN⁻** ⇌ **1EN⁻** in FHA (TFHA) at B3LYP/6-31+G* [L1], B3LYP/Aug-cc-pVDZ [L2], MP2/6-31+G* [L3] and MP2/Aug-cc-pVDZ [L4] theoretical levels. All the bond distances are in angstrom (Å), angles and dihedrals are in degrees (°).

Parameters	L1	L2	L3	L4
N2-C1	1.378	1.380	1.385	1.392
	(1.340)	(1.342)	(1.364)	(1.368)
X3-C1	1.235	1.233	1.245	1.245
	(1.696)	(1.698)	(1.664)	(1.680)
O4-N2	1.519	1.513	1.531	1.535
	(1.482)	(1.478)	(1.525)	(1.526)
H5-O4	0.970	0.967	0.974	0.969
	(0.973)	(0.969)	(0.975)	(0.871)
H6-C1	1.142	1.147	1.132	1.141
	(1.124)	(1.128)	(0.975)	(1.128)
X3-C1-N2	129.2	129.1	128.7	128.7
	(128.9)	(128.9)	(127.8)	(128.0)
O4-N2-C1	104.2	104.6	101.1	101.1
	(114.0)	(114.1)	(105.1)	(105.5)
H5-O4-N2	101.3	101.7	100.0	99.7
	(102.8)	(103.0)	(100.3)	(100.2)
H6-C1-N2	116.7	116.6	117.0	117.0
	(119.0)	(119.3)	(119.6)	(119.9)
O4-N2-C1-X3	89.0	88.6	88.0	87.7
	(92.1)	(91.6)	(89.8)	(90.0)
H5-O4-N2-C1	108.9	105.6	111.3	103.5
	(100.2)	(96.8)	(108.3)	(97.4)
H6-C1-N2-O4	-96.4	-96.9	-95.7	-96.0
	(-95.8)	(-96.4)	(-94.8)	(-95.2)

Table 4.12: Geometrical parameters for transition state **RTS3** connecting interconversion **1ZO⁻** ⇌ **1EO⁻** in FHA (TFHA) at B3LYP/6-31+G* [**L1**], B3LYP/Aug-cc-pVDZ [**L2**], MP2/6-31+G* [**L3**] and MP2/Aug-cc-pVDZ [**L4**] theoretical levels. All the bond distances are in angstrom (Å), angles and dihedrals are in degrees (°).

Parameters	L1	L2	L3	L4
N2-C1	1.433	1.435	1.431	1.439
	(1.423)	(1.422)	(1.425)	(1.428)
X3-C1	1.223	1.220	1.229	1.228
	(1.672)	(1.674)	(1.641)	(1.655)
O4-N2	1.432	1.429	1.438	1.438
	(1.411)	(1.410)	(1.428)	(1.428)
H5-C1	1.118	1.123	1.113	1.121
	(1.098)	(1.103)	(1.098)	(1.106)
H6-N2	1.036	1.036	1.035	1.037
	(1.031)	(1.031)	(1.032)	(1.035)
X3-C1-N2	129.7	129.6	129.1	129.2
	(131.0)	(131.0)	(129.5)	(129.8)
O4-N2-C1	106.9	106.5	107.8	107.1
	(103.8)	(103.6)	(105.5)	(105.2)
H5-C1-N2	111.9	111.8	112.3	112.0
	(112.4)	(112.9)	(113.1)	(113.3)
H6-N2-C1	105.9	105.6	105.2	104.3
	(108.8)	(108.4)	(107.6)	(106.4)
O4-N2-C1-X3	-52.7	-53.2	-49.7	-50.3
	(-56.9)	(-57.0)	(-51.7)	(-51.7)
H5-C1-N2-O4	124.1	123.4	128.5	127.5
	(120.0)	(120.0)	(124.5)	(124.7)
H6-N2-C1-X3	62.1	61.4	64.0	62.4
	(59.2)	(59.0)	(62.7)	(61.8)

Table 4.13: Geometrical parameters for transition state **PTS1** connecting **1ZO⁻** ⇌ **2ZX3⁻** in FHA (TFHA) at B3LYP/6-31+G* **[L1]**, B3LYP/Aug-cc-pVDZ **[L2]**, MP2/6-31+G* **[L3]** and MP2/Aug-cc-pVDZ **[L4]** theoretical levels. All the bond distances are in angstrom (Å), angles and dihedrals are in degrees (°).

Parameters	L1	L2	L3	L4
N2-C1	1.351	1.351	1.351	1.358
	(1.318)	(1.314)	(1.325)	(1.325)
X3-C1	1.242	1.240	1.249	1.249
	(1.702)	(1.709)	(1.684)	(1.703)
O4-N2	1.535	1.529	1.526	1.525
	(1.476)	(1.467)	(1.465)	(1.455)
H5-C1	1.119	1.123	1.114	1.122
	(1.097)	(1.101)	(1.096)	(1.103)
H6-N2	1.123	1.118	1.115	1.108
	(1.129)	(1.125)	(1.124)	(1.119)
X3-C1-N2	131.6	131.3	130.7	130.6
	(130.9)	(130.5)	(129.0)	(128.7)
O4-N2-C1	115.8	116.0	116.2	116.1
	(121.1)	(122.2)	(121.0)	(122.9)
H5-C1-N2	108.6	108.4	108.9	108.5
	(110.2)	(110.6)	(110.5)	(110.4)
H6-N2-C1	126.2	127.0	128.3	127.5
	(140.7)	(146.7)	(142.1)	(151.7)
O4-N2-C1-X3	14.0	14.1	17.2	17.0
	(12.7)	(11.6)	(13.4)	(10.5
H5-C1-N2-O4	189.0	189.2	191.3	191.3
	(188.4)	(187.7)	(188.7)	(186.7)
H6-N2-C1-O4	65.6	66.3	69.7	70.0
	(76.2)	(80.1)	(80.8)	(88.7)

Table 4.14: Geometrical parameters for transition state **PTS2** connecting interconversion **1EO⁻** ⇌ **2EO4⁻** in FHA (TFHA) at B3LYP/6-31+G* **[L1]**, B3LYP/Aug-cc-pVDZ **[L2]**, MP2/6-31+G* **[L3]** and MP2/Aug-cc-pVDZ **[L4]** theoretical levels. All the bond distances are in angstrom (Å), angles and dihedrals are in degrees (°).

Parameters	L1	L2	L3	L4
N2-C1	1.300	1.302	1.317	1.325
	(1.309)	(1.310)	(1.333)	(1.341)
X3-C1	1.348	1.351	1.351	1.358
	(1.779)	(1.788)	(1.757)	(1.780)
O4-N2	1.319	1.316	1.317	1.311
	(1.293)	(1.292)	(1.300)	(1.297)
H5-C1	1.088	1.092	1.088	1.094
	(1.089)	(1.094)	(1.089)	(1.097)
H6-N2	1.405	1.387	1.401	1.383
	(1.520)	(1.511)	(1.541)	(1.552)
X3-C1-N2	108.4	107.8	108.1	107.4
	(108.1)	(107.5)	(107.8)	(107.0)
O4-N2-C1	131.6	131.6	131.4	131.7
	(126.9)	(126.3)	(125.5)	(124.7)
H5-C1-N2	128.1	128.2	127.9	128.2
	(124.9)	(125.3)	(124.6)	(125.4)
H6-N2-C1	74.1	74.0	74.4	73.8
	(81.6)	(81.1)	(80.0)	(78.9)
O4-N2-C1-X3	180.0	180.0	180.0	180.0
	(180.0)	(180.0)	(180.0)	(180.0)
H5-C1-N2-O4	0.0	-0.1	-0.0	-0.0
	(0.0)	(0.0)	(0.0)	(0.0)
H6-N2-C1-X3	-0.0	0.0	0.0	-0.0
	(0.0)	(0.0)	(-0.0)	(-0.0)

Table 4.15: Geometrical parameters of transition state **PTS3** connecting interconversion **1EN⁻** ⇌ **1EO⁻** in FHA (TFHA) at B3LYP/6-31+G* **[L1]**, B3LYP/Aug-cc-pVDZ **[L2]** MP2/6-31+G* **[L3]** and MP2/Aug-cc-pVDZ **[L4]** theoretical levels. All the bond distances are in angstrom (Å), angles and dihedrals are in degrees (°).

Parameters	L1	L2	L3	L4
N2-C1	1.347	1.348	1.349	1.359
	(1.307)	(1.305)	(1.310)	(1.311)
X3-C1	1.248	1.246	1.255	1.253
	(1.710)	(1.715)	(1.694)	(1.711)
O4-N2	1.565	1.560	1.569	1.572
	(1.504)	(1.496)	(1.498)	(1.489)
H5-O4	1.285	1.291	1.325	1.339
	(1.288)	(1.290)	(1.325)	(1.326)
H6-C1	1.114	1.119	1.110	1.117
	(1.099)	(1.104)	(1.099)	(1.107)
X3-C1-N2	126.1	126.0	126.0	125.9
	(126.9)	(127.0)	(126.7)	(126.9)
O4-N2-C1	111.2	111.5	110.4	109.9
	(117.2)	(118.1)	(118.0)	(119.7)
H5-O4-N2	45.5	45.2	44.8	44.1
	(47.2)	(47.1)	(46.7)	(46.5)
H6-C1-N2	112.9	112.7	112.3	111.9
	(113.0)	(113.0)	(112.2)	(111.6)
O4-N2-C1-X3	167.6	167.4	164.4	164.2
	(191.0)	(190.4)	(192.2)	(188.6)
H5-O4-N2-C1	239.8	240.2	237.5	240.5
	(138.0)	(142.4)	(141.9)	(152.9)
H6-C1-N2-O4	-16.4	-16.6	-20.4	-20.5
	(14.1)	(13.2)	(15.2)	(11.1)

4.2.7 INTERMOLECULAR H-BONDING WITH WATER IN ANIONS OF FHA/TFHA:

Most of the reactions of biological relevance involve water as medium. The solute-solvent interactions are important in deciding the thermodynamic properties. The highly polar water interacts with polar solutes through strong H-bond interactions. The relative energies of different isomeric forms of neutral HAs and their ionic species are expected to be altered by these interactions. In chapter 2 and 3, the analysis of the H-bond interactions between neutral FHA and TFHA with water molecule in 1:1 ratio were presented. In the present study, six anion-water aggregates have been optimized for FHA and TFHA each. The optimized aggregates of the anions of FHA and TFHA with water in 1: 1 ratio along with numbering of atoms are shown in Figure 4.5 and 4.6 respectively. The structural parameters of aggregates of anions of HAs with water are given in Table 4.16a and 4.16b.

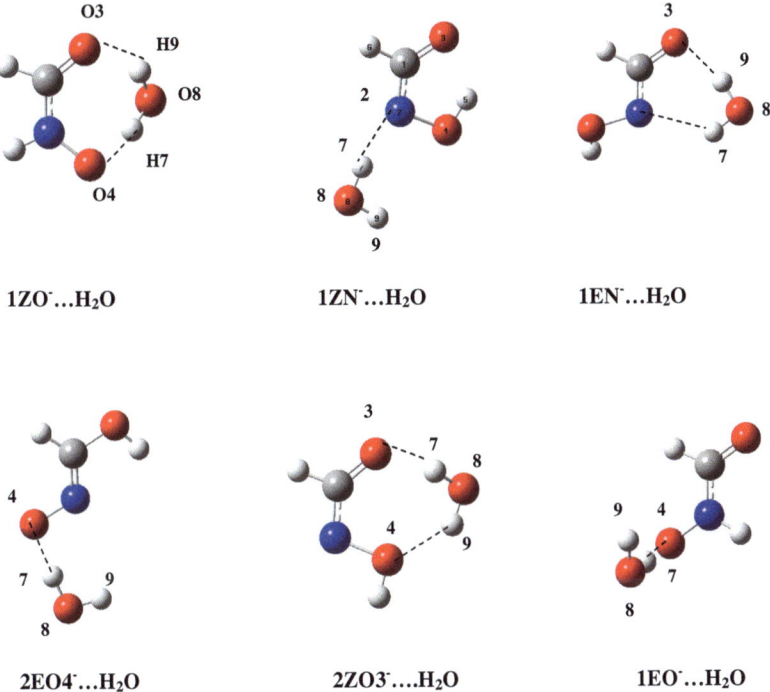

Figure 4.5: Optimized aggregates of anions of FHA with H_2O in 1: 1 ratio at MP2/Aug-cc-pVDZ level.

In all these aggregates water acts as H-bond donor and the HA anions play the role of H-bond acceptor while in the aggregates of neutral formo and thiohydroxamic acids with water, HAs acted both as H-bond donor as well as H-bond acceptor. This can be understood as the additional charge density on the anions favors their behavior as H-bond acceptor. The atomic charges and geometrical parameters important for describing the intermolecular H-bonding in aggregates of anions with water are reported in Table 4.17. The stabilization energies as the result of intermolecular H-bond are also included in the table. The analysis of these parameters suggests that in **1ZO⁻...H₂O, 1EN⁻...H₂O, 2ZX3⁻...H₂O**; water molecule bridges the two H-bond acceptor sites of the anion. In **1ZN⁻...H₂O, 2EO4⁻...H₂O** and **1EO⁻...H₂O** there is only one H-bond. The aggregation between H₂O and **1EN⁻, 2EX3⁻** resulted in same adduct as the two forms being canonical of one another result in same adduct **(Scheme 4.2)**.

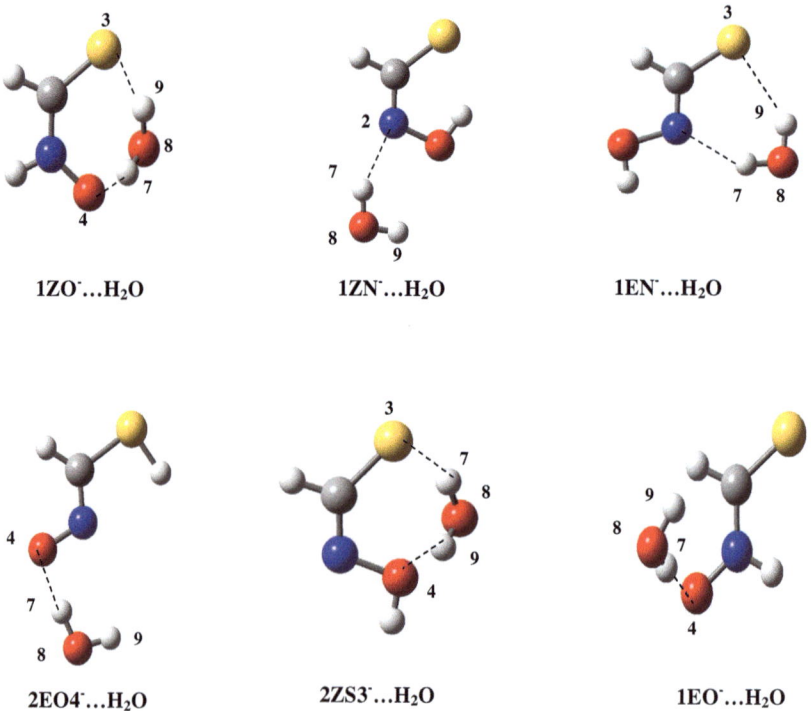

Figure 4.6: Optimized aggregates of anions of TFHA with H₂O in 1:1 ratio at MP2/Aug-cc-pVDZ

Though theoretical studies suggest N-H deprotonation being favored over O-H deprotonation, but the energy of aggregation with water is higher for **1ZO⁻...H₂O**. The lower aggregation energy in **1ZN⁻...H₂O** relative to other anionic forms is the result of presence of intramolecular H-bond present in the anion that decreases its ability to act as H-bond acceptor. The aggregation energy in the **1ZN⁻...H₂O, 2EO4⁻...H₂O, 1EO⁻...H₂O** with single H-bond interaction is 15.51, 21.95, 20.15 kcal/mol in FHA and 12.72, 19.74, 14.80 kcal/mol in TFHA while the aggregates **1ZO⁻...H₂O, 1EN⁻...H₂O, 2ZX3⁻...H₂O** with two H-bonds have stabilization energy 20.78, 18.34, 18.49 kcal/mol in FHA and 17.18, 15.07, 14.74 kcal/mol in TFHA anions. Though **2EO4⁻...H₂O** adduct has highest stabilization arising out of H-bond interactions but this anion is not detected experimentally as the stability of this anion adduct is still lower than other anion adducts. The water aggregates of FHA and TFHA anions follow the same order of relative stability as for the respective anions in gas phase i.e. the **1ZN⁻** remains the most stable anion in the aggregate form also.

Table 4.16a: Geometrical parameters for anions of FHA (TFHA) upon aggregation with water at MP2/Aug-cc-pVDZ **[L4]** theoretical level. All the bond distances are in angstrom (Å), angles and dihedrals are in degrees (°).

Parameters	1ZO⁻...H₂O	Parameters	1ZN⁻...H₂O	Parameters	1EN⁻...H₂O
N2-C1	1.347 (1.341)	N2-C1	1.330 (1.319)	N2-C1	1.333 (1.311)
X3-C1	1.259 (1.703)	X3-C1	1.286 (1.733)	X3-C1	1.280 (1.727)
O4-N2	1.339 (1.305)	O4-N2	1.451 (1.416)	O4-N2	1.489 (1.474)
H5-C1	1.118 (1.102)	H5-O4	0.991 (0.999)	H5-O4	0.967 (0.967)
H6-N2	1.023 (1.027)	H6-C1	1.109 (1.099)	H6-C1	1.117 (1.105)
H7-O4	1.746 (1.798)	H7-N2	1.806 (1.852)	H7-N2	2.288 (2.041)
O8-H7	0.996 (0.988)	O8-H7	0.995 (0.990)	O8-H7	0.972 (0.977)
H9-O8	0.973 (0.973)	H9-O8	0.966 (0.966)	H9-O8	0.990 (0.977)
X3-C1-N2	128.5 (128.3)	X3-C1-N2	126.2 (126.6)	X3-C1-N2	125.1 (125.0)
O4-N2-C1	128.4 (129.6)	O4-N2-C1	108.4 (113.1)	O4-N2-C1	107.3 (108.5)
H5-C1-N2	109.1 (109.7)	H5-O4-N2	99.9 (102.5)	H5-O4-N2	99.9 (99.9)
H6-N2-C1	114.6 (114.6)	H6-C1-N2	112.2 (112.3)	H6-C1-N2	115.2 (116.0)
H7-O4-N2	102.7 (103.4)	H7-N2-C1	150.1 (146.2)	H7-N2-C1	103.2 (114.7)
O8-H7-O4	163.4 (164.7)	O8-H7-N2	166.1 (166.5)	O8-H7-N2	128.0 (150.7)
H9-O8-H7	97.7 (98.7)	H9-O8-H7	100.7 (101.1)	H9-O8-H7	97.2 (97.5)
O4-N2-C1-X3	-0.2 (-0.2)	O4-N2-C1-X3	-0.0 (0.0)	O4-N2-C1-X3	177.5 (180.0)
H5-C1-N2-O4	179.9 (179.9)	H5-O4-N2-C1	0.0 (-0.0)	H5-O4-N2-C1	132.0 (180.0)
H6-N2-C1-X3	181.7 (180.7)	H6-C1-N2-O4	180.0 (180.0)	H6-C1-N2-O4	-1.9 (-0.0)
H7-O4-N2-C1	-47.6 (-59.6)	H7-N2-C1-X3	182.8 (179.9)	H7-N2-C1-X3	1.2 (0.2)
O8-H7-O4-N2	30.2 (26.5)	O8-H7-N2-C1	177.9 (180.0)	O8-H7-N2-C1	-3.5 (0.1)
H9-O8-H7-X3	7.5 (6.0)	H9-O8-H7-O4	0.1 (-0.0)	H9-O8-H7-X3	0.1 (-0.0)

Table 4.16b: Geometrical parameters for anions of FHA (TFHA) upon aggregation with water at MP2/Aug-cc-pVDZ **[L4]** theoretical level. All the bond distances are in angstrom (Å), angles and dihedrals are in degrees (°).

Parameters	1EO⁻...H$_2$O	Parameters	2ZX3⁻...H$_2$O	Parameters	2EO4⁻...H$_2$O
N2-C1	1.340 (1.327)	N2-C1	1.329 (1.316)	N2-C1	1.295 (1.313)
X3-C1	1.264 (1.714)	X3-C1	1.278 (1.726)	X3-C1	1.405 (1.794)
O4-N2	1.368 (1.329)	O4-N2	1.484 (1.461)	O4-N2	1.353 (1.334)
H5-C1	1.110 (1.099)	H5-O4	0.966 (0.968)	H5-X3	0.973 (1.353)
H6-N2	1.021 (1.027)	H6-C1	1.116 (1.103)	H6-C1	1.096 (1.100)
H7-O4	1.574 (1.709)	H7-X3	1.774 (2.501)	H7-O4	1.579 (1.622)
O8-H7	1.026 (1.000)	O8-H7	0.991 (0.977)	O8-H7	1.027 (1.016)
H9-O8	0.965 (0.966)	H9-O8	0.970 (0.975)	H9-O8	0.965 (0.965)
X3-C1-N2	125.8 (124.6)	X3-C1-N2	132.6 (133.2)	X3-C1-N2	119.6 (120.1)
O4-N2-C1	125.3 (127.2)	O4-N2-C1	109.2 (111.1)	O4-N2-C1	117.3 (116.0)
H5-C1-N2	111.0 (112.2)	H5-O4-N2	98.5 (99.2)	H5-X3-C1	102.5 (93.5)
H6-N2-C1	117.2 (115.6)	H6-C1-N2	107.6 (108.3)	H6-C1-N2	126.4 (123.6)
H7-O4-N2	104.4 (100.6)	H7-X3-C1	118.0 (91.9)	H7-O4-N2	100.7 (100.2)
O8-H7-O4	177.2 (171.4)	O8-H7-X3	166.1 (144.2)	O8-H7-O4	174.1 (173.3)
H9-O8-H7	102.5 (102.4)	H9-O8-H7	97.1 (97.5)	H9-O8-H7	100.6 (100.5)
O4-N2-C1-X3	180.7 (178.3)	O4-N2-C1-X3	0.6 (-0.1)	O4-N2-C1-X3	180.0 (180.0)
H5-C1-N2-O4	0.5 (-1.7)	H5-O4-N2-C1	180.0 (178.4)	H5-X3-C1-N2	-0.0 (0.0)
H6-N2-C1-O3	0.1 (-0.2)	H6-C1-N2-O4	180.9 (180.2)	H6-C1-N2-O4	-0.0 (0.0)
H7-O4-N2-C1	260.9 (293.2)	H7-X3-C1-N2	36.2 (52.2)	H7-O4-N2-C1	180.0 (180.0)
O8-H7-O4-N2	20.6 (6.1)	O8-H7-X3-C1	-16.4 (-38.0)	O8-H7-O4-N2	0.0 (0.1)
H9-O8-H7-O4	29.2 (3.4)	H9-O8-H7-O4	2.1 (-0.7)	H9-O8-H7-N2	0.0 (0.0)

Table 4.17: Important intermolecular hydrogen bonding parameters; hydrogen bond distances[a], hydrogen bond angles[b], charges on hydrogen bond acceptor and hydrogen, stabilization energies (S.E.[c]) and distortion energies (E_{Dis}[d]) of aggregates of anions of FHA (TFHA) with one water in 1:1 mole ratio at MP2/Aug-cc-pVDZ **[L4]** theoretical level.

Species	Hydrogen bond Distances[a]		Hydrogen bond Angles[b]		Atomic charges		S.E.[c]	E_{Dis}[d]
1ZO⁻…H₂O	H9…X3	2.124 (2.599)	O8-H9…X3	143.02 (142.01)	$q_H(q_X)$	0.499(-0.868) (0.491(-0.491))	20.78 (17.18)	1.07 (0.71)
	H7…O4⁻	1.740 (1.790)	O8-H7…O4⁻	163.44 (164.72)		0.535(-0.801) (0.529(-0.759))		
1ZN⁻…H₂O	H7…N2⁻	1.806 (1.850)	O8-H7…N2⁻	166.14 (166.48)	$q_H(q_N)$	0.542(-0.499) (0.536(-0.349))	15.51 (12.72)	0.70 (0.51)
1EN⁻…H₂O	H7…N2⁻	2.288 (2.041)	O8-H7…N2⁻	128.02 (150.71)	$q_H(q_N)$	0.495(-0.514) (0.510(-0.382))	18.34 (15.07)	0.85 (0.75)
	H9…O3	1.820 (2.550)	O8-H9…O3	160.22 (142.26)	$q_H(q_O)$	0.534(-0.917) (0.506(-0.560))		
2EO4⁻…H₂O	H7…O4⁻	1.579 (1.622)	O8-H7…O4⁻	174.10 (173.31)	$q_H(q_O)$	0.554(-0.814) (0.550(-0.788))	21.95 (19.74)	2.21 (1.59)
2ZX3⁻…H₂O	H7…X3⁻	1.770 (2.501)	O8-H7…X3⁻	166.13 (144.15)	$q_H(q_X)$	0.540(-0.924) (0.507(-0.569))	18.49 (14.74)	1.09 (0.70)
	H9…O4	2.180 (1.970)	O8-H9…O4	134.88 (158.86)	$q_H(q_O)$	0.492(-0.716) (0.510(-0.685))		
1EO⁻…H₂O	H7…O4⁻	1.570 (1.709)	O8-H7…O4⁻	177.23 (171.38)	$q_H(q_O)$	0.552(-0.803) (0.540(-0.780))	20.15 (14.80)	2.11 (0.90)

a- in angstrom (Å)
b- in degrees (°)
c, d -in kcal/mol

Contd……

The stabilization energies in the water aggregates of thiohydroxamate anions are smaller by approximately 2-4 kcal/mol. The low electronegativity of sulfur in comparison to oxygen is responsible for low charge density present on sulfur in these anions (Table 4.17). The charges on all atoms of anions in their aggregates with water are recorded in Table 4.18. Even the charge densities on nitrogen and oxygen are also lower in TFHA anion-H_2O aggregates. As the electrostatic interactions provide an important contribution to the H-bond stabilization energy; lower stabilization energies are the result of weaker electrostatic interactions.

Relative to the complexes of neutral forms with water, the anions on interaction with water are stabilized to larger extent as is evident from larger stabilization energies. The stabilization energies range from 3.33 to 11.30 kcal/mol and 2.60 to 10.51 kcal/mol (excluding E_{Dis}) respectively for neutral species of FHA and TFHA upon interaction with one water molecule and stabilization ranges from 15.51-21.95 kcal/mol 12.72-19.74 kcal/mol for anionic species of FHA and TFHA respectively aggregating with water in 1:1 ratio.

Relative energy difference between **1ZO⁻** and **1ZN⁻** in gas phase is 14.85 kcal/mol in FHA and 18.07 kcal/mol in TFHA. While on interaction with water relative energy difference reduces to 9.37 kcal/mol in FHA and 13.34 kcal/mol in TFHA at MP2/Aug-cc-pVDZ. **1ZN⁻** forms only one intermolecular O8-H7…N2⁻ H-bond with water in both FHA and TFHA. The intramolecular H-bond of **1ZN⁻** remains preserved upon complexation with water in **1ZN⁻…H₂O**. The strength of intramolecular H-bond in **1ZN⁻** is not affected on complexation with water as the comparison of the geometrical parameters in gas phase of **1ZN⁻** and **1ZN⁻…H₂O** shows that for both FHA and TFHA, the H-bond distance H5…X3 and angle O4-H5…X3 have the same magnitudes as for the isolated molecule.

The O-deprotonated anion **1ZO⁻** in FHA complexes with water forming two homonuclear but unsymmetrical O-H…O H-bonds. Out of these, the H-bond with negatively charged oxygen is more linear and hence expected to be stronger. **1ZO⁻ …H₂O** has more aggregation energy than **1ZN⁻…H₂O** for both FHA/TFHA. Comparison of charges shows that for **1ZN⁻…H₂O**, nitrogen bears less negative charge than oxygen or sulfur of **1ZO⁻…H₂O** in FHA and TFHA respectively. The negative charge of the **1ZO⁻** anion remains on oxygen (O4) and X mainly, with nitrogen also losing some of its negative charge density.

Table 4.18: Atomic charges (NPA) on atoms of FHA (TFHA) anions and their water aggregates at MP2/Aug-cc-pVDZ **[L4]** theoretical level.

Species	C1	N2	X3	O4	H5	H6	H7
1ZN⁻	0.553	-0.456	-0.957	-0.733	0.500	0.093	-
	(-0.056)	(-0.296)	(-0.611)	(-0.701)	(0.492)	(0.171)	(-)
1ZN⁻…H$_2$O	0.588	-0.499	-0.938	-0.723	0.507	0.103	-
	(-0.022)	(-0.349)	(-0.578)	(-0.696)	(0.498)	(0.179)	(-)
1ZO⁻	0.626	-0.341	-0.867	-0.811	-	0.066	0.328
	(-0.039)	(-0.197)	(-0.509)	(-0.753)	(-)	(0.159)	(0.339)
1ZO⁻…H$_2$O	0.641	-0.351	-0.868	-0.801	-	0.082	0.345
	(-0.010)	(-0.215)	(-0.491)	(-0.759)	(-)	(0.166)	(0.354)
1EN⁻	0.622	-0.509	-0.910	-0.730	0.457	0.071	-
	(0.007)	(0.350)	(-0.576)	(-0.712)	(0.469)	(0.162)	(-)
1EN⁻…H$_2$O	0.624	-0.514	-0.917	-0.710	0.462	0.091	-
	(0.022)	(-0.382)	(-0.560)	(-0.690)	(0.474)	(0.169)	(-)
1EO⁻	0.653	-0.368	-0.905	-0.827	-	0.102	0.345
	(0.011)	(-0.209)	(-0.568)	(-0.781)	(-)	(0.181)	(0.365)
1EO⁻…H$_2$O	0.675	-0.393	-0.877	-0.803	-	0.108	0.363
	(0.031)	(-0.247)	(-0.524)	(-0.780)	(-)	(0.189)	(0.376)
2ZX3⁻	0.607	-0.516	-0.906	-0.705	-	0.069	0.451
	(-0.012)	(-0.368)	(-0.575)	(-0.667)	(-)	(0.162)	(0.460)
2ZX3⁻…H$_2$O	0.609	-0.487	-0.924	-0.716	-	0.089	0.463
	(-0.003)	(-0.355)	(-0.569)	(-0.685)	(-)	(0.170)	(0.472)
2EX3⁻	0.622	-0.509	-0.910	-0.730	-	0.071	0.457
	(0.007)	(-0.350)	(-0.576)	(-0.711)	(-)	(0.162)	(0.469)
2EO4⁻	0.292	-0.235	-0.840	-0.827	0.483	0.127	-
	(-0.283)	(-0.107)	(-0.126)	(-0.786)	(0.134)	(0.168)	(-)
2EO4⁻…H$_2$O	0.353	-0.262	-0.829	-0.813	0.488	0.138	-
	(-0.224)	(-0.134)	(-0.104)	(-0.788)	(0.137)	(0.175)	(-)

Polar solvents like water affect the dipole moment of the solute molecules. Table 4.19 records dipole moments of isomers of FHA and TFHA, their anions and adducts with water. In gas phase and aqueous solution of both HAs, the dipole moment of ketone forms (**1Z, 1E**) is much higher than iminol forms (**2Z, 2E**). Similarly the dipole moment of **1ZW1** and **1EW1** is larger than adducts of **2ZW1** and **2EW1**.

Table 4.19: Calculated dipole moment in gas phase of the neutral and anionic forms of FHA (TFHA) and their water aggregates in the gas phase and in aqueous medium using PCM method at MP2/Aug-cc-pVDZ **[L4]** theoretical level.

Species	Gas phase	PCM (water)	Species HA...H$_2$O	Gas phase	PCM (water)
1Z	3.283	4.209	1ZW1	4.093	5.010
	(3.905)	(5.386)		(4.362)	(5.579)
2Z	0.457	0.745	2ZW1	1.805	2.158
	(0.290)	(0.486)		(1.871)	(2.291)
1E	3.478	4.428	1EW1	2.662	3.309
	(3.935)	(5.416)		(3.706)	(4.734)
2E	0.640	0.874	2EW1	2.002	2.426
	(0.342)	(0.327)		(1.731)	(2.090)
HA$^-$			HA$^-$...H$_2$O		
1ZO$^-$	5.193	6.766	1ZO$^-$...H$_2$O	5.084	7.091
	(4.962)	(7.108)		(4.994)	(6.730)
1ZN$^-$	0.950	1.717	1ZN$^-$...H$_2$O	0.569	0.403
	(0.838)	(1.509)		(1.062)	(0.586)
1EN$^-$	3.751	4.937	1EN$^-$...H$_2$O	3.788	4.758
	(4.197)	(4.693)		(3.949)	(4.368)
1EO$^-$	1.328	1.583	1EO$^-$...H$_2$O	1.310	1.808
	(0.604)	(1.300)		(0.891)	(1.778)
2EO4$^-$	4.930	6.672	2EO4$^-$...H$_2$O	5.143	6.917
	(5.724)	(8.363)		(6.131)	(8.625)
2ZO3$^-$	2.626	3.016	2ZO3$^-$...H$_2$O	3.020	3.688
	(3.501)	(3.363)		(3.350)	(3.810)

4.2.8 SOLVENT EFFECT ON DEPROTONATION PROCESSES IN FHA/TFHA:

The effect of solvent on isomeric forms of FHA and TFHA, their anions and deprotonation processes is analysed by evaluating the free energy change, considering deprotonation occurring in medium by employing PCM method and evaluation of energy changes accompanying the deprotonation of 1:1 water aggregates of FHA and TFHA in the presence of medium. The relative free energies (ΔG_{rel}) and solvation free energies (ΔG_S) of the tautomeric forms in the gas phase and aqueous medium are recorded in Table 4.20. The solvation energy is the resultant of two contributing components (i) The non electrostatic contribution involves energy change associated with dispersion, cavitation and repulsion (ii) The electrostatic component representing energy change associated with solute-solvent interactions. ΔG_S is difference between MP2 energies of isolated species and corresponding values in solution.

The ΔG_{rel} values reflect that the **1Z** tautomeric form is the most stable form in water medium for both FHA and TFHA, though the **2Z** conformer of TFHA is observed to be most stable in gas phase. The relative free energy difference $\Delta G_{rel(g)}$ of **1Z** and **1E** in FHA is 1.17 kcal/mol in gas phase and it reduces to 0.15 kcal/mol in water continuum. Similarly in case of TFHA the free energy difference $\Delta G_{rel(g)}$ between **1Z** and **1E** decreases from 4.24 in gas phase to 2.86 kcal/mol in solution. From $\Delta G_{rel(aq)}$ for both FHA and TFHA, it is concluded that ketone forms (**1Z** and **1E**) are more stabilized in aqueous phase relative to enolic forms (**2Z** and **2E**). The solvation energy for **1Z**, **2Z**, **1E**, **2E** of FHA are 10.41, 9.68, 11.50 and 11.36 kcal/mol in FHA and the respective values in case of TFHA are 9.61, 7.08, 10.88 and 7.11 kcal/mol. **1ZW1, 1EW1, 2ZW1, 2EW1** represent the most stable adducts of **1Z, 1E, 2Z** and **2E** conformers with one water molecule for FHA and TFHA are as shown in Figure 2.2 and 3.3 respectively (Chapter 2 and 3). On adduct formation with water, the solvation energy values increase for **1Z, 1E, 2Z, 2E** to 18.83, 16.33, 18.30 and 18.88 kcal/mol in FHA and 17.13, 14.05, 19.02 and 14.24 kcal/mol in TFHA.

Table 4.20: Relative free energies $\Delta G_{rel(g)}$ in gas phase, $\Delta G_{rel(aq)}$ in water, solvation free energies ΔG_S, of neutral forms and $\Delta G_S^=$ (in kcal/mol) for aggregates of molecule-water of FHA (TFHA) at MP2/Aug-cc-pVDZ **[L4]** theoretical level.

Species	Gas phase $\Delta G_{rel(g)}$	PCM (water) $\Delta G_{rel(aq)}$	$(\Delta G_S)_{MP2}$	Species	PCM water $\Delta G_{rel(aq)}$	$\Delta G_S^=$
1Z	0.00 (0.00)	0.00 (0.00)	10.41 (9.61)	1ZW1	0.00 (0.00)	18.83 (17.13)
2Z	1.12 (-1.53)	1.79 (2.85)	9.68 (7.08)	2ZW1	3.56 (3.40)	16.33 (14.05)
1E	1.17 (4.24)	0.15 (2.86)	11.50 (10.88)	1EW1	1.77 (2.24)	18.30 (19.02)
2E	5.10 (2.02)	4.57 (6.89)	11.36 (7.11)	2EW1	5.48 (7.29)	18.88 (14.24)

The solvation free energies (Table 4.21) are considerably larger for the FHA and TFHA anions in comparison to the neutral molecules as the charged species can induce higher polarization in the medium and result in stronger solute-solvent interactions. The FHA containing two electronegative oxygen atoms in comparison to only one in TFHA, has higher solvation energies for its neutral and anionic forms in comparison to the values for TFHA and its anions respectively. The solvation energies for TFHA anions are nearly 5.89-9.84 kcal/mol lower than the values for respective anions of FHA. Solvation free energies are related to dipole moment also. The **1ZO⁻** has largest solvation free energy -73.49 kcal/mol for FHA and -67.60 kcal/mol in TFHA. **1ZN⁻** has low dipole moment amongst all anions of FHA and TFHA and its solvation free energy is minimum i.e. -64.15 and -56.07 kcal/mol for both FHA and TFHA (Table 4.21). The solvation energies for the adducts of both the HAs and their anions with water are higher than solvation of respective molecules and anions.

The free energy change associated with the solvation of aggregates of HAs and their anions with water are represented as $\Delta G_S^=$. The conformer which is least stable in gas phase has maximum solvation energy when its adduct with water is present in solution e.g. for FHA, the **2E** conformer is least stable in gas phase but it's adduct **2EW1** has maximum solvation energy ($\Delta G_S^=$ =18.88 kcal/mol). In TFHA, **1E** is least stable in gas phase but it's adduct **1EW1** has maximum solvation free energy $\Delta G_S^=$ =19.02 kcal/mol. Similarly the anion **2EO4⁻** is least stable in gas phase for both FHA and TFHA but its adduct **2EO4⁻...H₂O** has maximum solvation energy. The aggregate of most stable anion of both FHA and TFHA with water i.e. **1ZN⁻...H₂O** has the lowest solvation energy amongst the aggregates of all anions with water. The water molecule present in the adduct also interacts with the medium and increases the polar character of the species, thus the observed behavior. The variations are important as they reflect the behavior of species when they bind strongly to the solvent molecules forming first coordination sphere. The Table 4.22 lists the free energy changes accompanying the deprotonation processes for the formation of respective anions in gas phase, aqueous medium and their aggregates with water. ΔG in the medium is evaluated using equation 1.5.4.3 of chapter 1 and this equation defines ΔG^{iso}. $\Delta G^{'}$ is defined by equation 1.5.4.4 of chapter 1. The free energy changes are decreased considerably in the presence of solvent, which helps in increasing the acidity. The presence of intermolecular H-bond with water in the aggregates and the solvent continuum together make deprotonation much easier as is indicated by free energy change decreasing to less than 3 kcal/mol (with the exception of **2EO4⁻...H₂O** aggregate where $\Delta G^{'}$ is 9.52 kcal/mol) in FHA. $\Delta G'$ of 0.19 kcal/mol is so small that N-H can easily ionize.

Solvation leads to significant decrease in ΔG values for the deprotonation processes in TFHA as in the case of FHA. The solution pK_a values have been calculated using equation 1.5.4.5 (chapter1) as given by Liptak et al [22]. The N-H deprotonation of **1Z** tautomer of FHA corresponds to pK_a value of 10.2 which is close to the value based on experimental (~9 for HA) [23]. For similar deprotonation process, pK_a value of TFHA is 4.2 indicating higher acidity of TFHA. The effect of the dielectric of the medium increases the acidity of both FHA and TFHA.

Table 4.21: Relative free energies $\Delta G_{rel(g)}$ in gas phase, $\Delta G_{rel(aq)}$ in water, solvation free energies ΔG_S of anionic forms and $\Delta G_S^=$ of aggregates with water of FHA (TFHA) at MP2/Aug-cc-pVDZ [L4] theoretical level. All quantities are in units of kcal/mol

Species	Gas phase anion $\Delta G_{rel(g)}$	PCM (water) $\Delta G_{rel(aq)}$	ΔG_S	Species	PCM (water) $\Delta G_{rel(aq)}$	$\Delta G_S^=$
1ZN⁻	0.00	0.00	-64.15	1ZN⁻...H₂O	0.00	-72.49
	(0.00)	(0.00)	(-56.07)		(0.00)	(-63.39)
1ZO⁻	14.75	5.48	-73.49	1ZO⁻...H₂O	2.15	-85.16
	(17.99)	(6.08)	(-67.60)		(-2.36)	(-78.64)
1EN⁻	9.59	6.44	-68.63	1EN⁻...H₂O	5.42	-77.99
	(11.04)	(7.24)	(-60.78)		(6.53)	(-68.81)
1EO⁻	13.03	7.79	-69.47	1EO⁻...H₂O	4.35	-81.25
	(13.28)	(9.32)	(-59.63)		(6.74)	(-69.53)
2EO4⁻	28.73	23.37	-70.65	2EO4⁻...H₂O	15.52	-86.84
	(33.56)	(29.15)	(-64.35)		(21.77)	(-79.04)
2ZO3⁻	8.41	2.72	-70.60	2ZO3⁻...H₂O	0.94	-80.72
	(10.61)	(3.88)	(-63.05)		(2.84)	(-71.42)

Table 4.22: The free energies of deprotonation, solution pK_a of anions and free energy changes accompanying the deprotonation in gas phase, aqueous solution and water aggregates of FHA (TFHA) at MP2/Aug-cc-pVDZ **[L4]** theoretical level.

Deprotonation Process	Free Energy Changes			
	Gas phase	PCM (Water)		
	ΔG^{iso}	ΔG	pK_a	$\Delta G'$
1Z → 1ZN⁻ + H⁺	330.37	169.13	10.2	0.19
	(314.95)	(160.99)	(4.2)	(-7.83)
1Z → 1ZO⁻ + H⁺	345.12	174.54	14.2	2.27
	(332.94)	(167.45)	(9.0)	(-5.09)
1E → 1EN⁻ + H⁺	338.80	174.17	13.9	2.59
	(321.75)	(164.35)	(6.7)	(-4.56)
1E → 1EO⁻ + H⁺	342.23	176.76	15.8	2.76
	(323.99)	(167.74)	(9.2)	(-3.04)
2Z → 2ZX⁻ + H⁺	337.66	169.24	10.3	-3.25
	(327.09)	(163.62)	(6.2)	(-6.80)
2E → 2EO4⁻ + H⁺	354.00	187.21	23.5	9.52
	(346.49)	(181.75)	(19.5)	(5.17)

4.3 CONCLUSIONS:

1. The deprotonation enthalpies of FHA and TFHA at B3LYP/6-31+G*, B3LYP/Aug-cc-pVDZ, MP2/6-31+G* and MP2/Aug-cc-pVDZ levels suggest that the formation of N-H deprotonated anion is more favored in comparison to O-H deprotonated anion.

2. The variation in geometrical parameters and NBO analysis suggests that there is C-N double bond formation along with strong intramolecular H-bond in the C(=X)N(OH) unit that is responsible for the higher stability of **1ZN⁻** anion.

3. The **1ZN⁻** anionic form and **1ZN⁻...H₂O** aggregate though have comparatively lower solvation energy but both retain their most stable anionic form status.

4. ΔG values for the deprotonation of water aggregate of HAs in aqueous medium also concludes that N-H deprotonation is the most favored process.

5. Relative acidities and pK_a values of FHA and TFHA show that TFHA is relatively more acidic than FHA.
6. The interconversion pathways among the tautomeric forms of anions of both FHA and TFHA involve high activation barriers which suggest that interconversion amongst the forms is highly improbable under normal conditions.
7. The high stabilization energy values of aggregates of HA anions with water molecule suggest that intermolecular H-bonding plays important role in the stability of anions in aqueous medium.
8. Since the free energy required for various deprotonation processes is approximately halved in aqueous phase than in the gas phase and through aggregate formation it is further reduced, thus evidently the medium facilitates the deprotonation processes in HAs both explicitly and implicitly.

4.4 REFERENCES:

1. D.A. Brown, M.V. Chidambaram, Metal Ions in Biological Systems, Marcel Dekker, New York, 1982.
2. E. Lipczynska-Kochany, H. Iwamura, J. Org. Chem. 47 (1982) 5277.
3. A. Bagno, G. Scorrano, J. Phys. Chem. 100 (1996) 1536.
4. M.L. Senent, A. Niño, A. Muñoz-Caro, S. Ibeas, B. Garcia, J.M. Leal, F. Secco, M. Venturini, J. Org. Chem. 68 (2003) 6535.
5. A. Bagno, C. Comuzzi, G. Scorrano, J. Am. Chem. Soc. 116 (1994) 916.
6. S. Böhm, O. Exner, Org. Biomol. Chem. 1 (2003) 1176.
7. M. Decouzon, O. Exner, J.-F. Gal, P.–C. Maria, J. Org. Chem. 55 (1990) 3980.
8. O. Exner, M. Hradil, J. Mollin, Collect. Czech. Chem. Commun. 58 (1993) 1109.
9. N. Mora-Diez, M.L. Senent, B. Garcia, Chem. Phys. 324 (2006) 350.
10. J.E. Yazal, Y.P. Pang, J. Phys. Chem. 104 (2000) 6499.
11. K.R. Fountain, C.J. Felkerson, J.D. Driskell, B.D. Lamp, J. Org. Chem. 68 (2003) 1810.
12. I.H. Um, H.W. Yoon, J.S. Lee, H.J. Moon, D.S. Kwon, J. Org. Chem. 62 (1997) 5939.
13. F. Filippini, R.F. Hudson, Chem. Commun. (1972) 522.
14. H. Kwart, H. Omura, J. Org. Chem. 34 (1969) 318.
15. W.B. Gruhn, M.L. Bender, J. Am. Chem. Soc. 91 (1969) 5883.
16. R. Hershfield, M.L. Bender, J. Am. Chem. Soc. 94 (1972) 1376.
17. K.K. Ghosh, Y.S. Simanenko, M.L. Satnami, S.K. Sar, Indian. J. Chem. 43B (2004) 1990.
18. K.K. Ghosh, J. Vaidya, M.L. Satnami, Int. J. Chem. Kinet. 38 (2006) 26.
19. B. García, S. Ibeas, F.J. Hoyuelos, J.M. Leal, J. Org. Chem. 66 (2001) 7986.
20. M. Remko, J. Phys. Chem. A 106 (2002) 5005.
21. R. Kakkar, A. Dua, S. Zaidi, Org. Biomol. Chem. 5 (2007) 547.
22. M.D. Liptak, K.C. Gross, P.G. Seybold, S. Feldgus, G.C. Shields, J. Am. Chem. Soc. 124 (2002) 6421.
23. A.B. Hughes, Amino Acids, Peptides and Proteins in Organic Chemistry, Vol 2, Wiley-VCH, Weinheim, 2009.

HYDROGEN BOND COOPERATIVITY IN DIMERS OF HYDROXAMIC ACIDS

5.1 INTRODUCTION:

A typical hydrogen bond (H-bond) is defined in terms of H-bond donor and H-bond acceptor. H-bonding and protonation are related phenomena that can be considered as acid base interactions, with H-bonding being relatively weak while protonation is very strong [1]. The two most crucial factors that can influence the structure and energetics of H-bonding are the acidity of the donor and the basicity of the acceptor [2]. It implies that for a given H-bond donor, the stabilization energy should be proportional to the proton affinity and for a given H-bond acceptor; stabilization energy can be anticipated to depend on acidity of H-bond donor. Several authors have correlated the bronsted basicity scales obtained in solution of closely related bases and H-bond [3].

Frank and Wen postulated an important concept of cooperativity between H-bonds in 1957 [4]. Cooperativity is a theory of H-bonding in which each of the individual H-bond making up a chain of interlinked H-bonds is more strongly bound than in the absence of others [5]. The sum of non additive many body energies has been considered as the energetic contribution of the cooperativity to the stability [6]. This non additive property arises because the ability of donor and acceptor groups is further enhanced (positive cooperativity) by an increase in polarity when H-bonds form a part of a collective ensemble [7]. Binding of oxygen to haemoglobin stands as an example of positive cooperativity. Multiple H-bonds do not always reinforce each other, this sort of weakening of an existing H-bond by adding another acceptor/donor to the interaction is referred to as negative cooperativity [8].

Different types of mechanisms for H-bond cooperativity such as conformational changes, secondary electrostatic effects and structure tightening exist for cooperative binding [9]. Classical ionic electrostatic model of H-bond include only pairwise additive interactions of localized monomers, the existence of cooperativity can be considered departure from classical ionic electrostatic model of H-bonding and thus suggest presence of significant covalent-electronic delocalization [10]. Several authors have suggested origin of cooperativity for weakly H-bonded systems to be the result of reinforcement of charge transfer occurring in H-bonds for clusters where each molecule is simultaneously donor and acceptor of H-bond.

Cooperativity effects of H-bonds are of considerable importance in understanding of structure and dynamics of liquids and solutions containing H-bonds [11,12]. Cooperativity is important in determination of stability of many systems such as DNA-duplexes, folded proteins and ligand receptor complexes. H-bond cooperativity is essential for sustaining the stable conformers of biological molecules and constructing crystal structures [13,14]. The presence of substantial cooperative effects in molecular crystals of acetic acid, urea, nitroaniline etc. is reported by Dannenberg et al. [15,16]. With the help of molecular orbital calculations, they have shown that amidic H-bonds can be extremely cooperative in some cases (α-helices) or no effect in some cases (collagen like triple helices) while in others (β sheets) cooperativity is masked by new H-bond interactions [17]. Cooperativity has also been examined in molecular systems like proton hydrates, hydrogen halides, hydrogen cyanides, hydrogen peroxide or alcohol.

Hydroxamic acids (HAs) are important class of compounds with numerous applications as being enzyme inhibitors, tumor inhibitors, chemotherapeutic agents etc [18-21]. Biological action of this class of compounds rely a part on their H-bonding ability. In order to explore the H-bond ability of thioformohydroxamic acid (TFHA) and formohydroxamic acid (FHA) and its correlation to acidity and basicity of potential sites and the rationale behind cooperativity of H-bonding in these molecules, the present study makes use of density functional and molecular orbital ab initio calculations on the protonation and deprotonation enthalpies of the tautomeric forms and their dimers.

5.2. RESULTS AND DISCUSSIONS:
5.2.1 PROTON AFFINITY:

The HAs are acid species but also behave as weak bases and are suggested to protonate in highly acidic medium [22]. The protonation of weak bases is important both as a source of information on the molecular electronic structure and as a tool to interpret the reactivity of acid catalyzed chemical reactions. The presence of O, N and X (X=O, S; FHA, TFHA) atoms possessing lone pairs of electrons in TFHA and FHA provides them the sites that can act as nucleophiles. The proton affinity (PA) is quantitative measure of the nucleophilicity of a specific site. **1Z, 1E, 2Z** and **2E** tautomeric forms of TFHA and FHA are chosen for protonation studies. **Scheme 5.1** shows diagramatic sketch of the protonated species obtained by protonation of various tautomers of FHA and TFHA at different heteroatoms. The protonation at the X3 site of **1Z** and **1E** conformation resulted in two energy minima structures i.e. **1ZXH⁺A**, **1ZXH⁺B** and **1EXH⁺A**, **1EXH⁺B** respectively as shown in **Scheme 5.1**. The two orientations **A** and **B** differ in relative position of X-H bond with respect to C-N bond. The N2 protonation of **2Z** and **2E** i.e. **2ZNH⁺** and **2ENH⁺** converge to same structure as X3 protonation of **1Z** and **1E** i.e. **1ZXH⁺A** and **1EXH⁺A**, while X3 protonation leads to **2ZXH⁺** and **2EXH⁺** respectively.

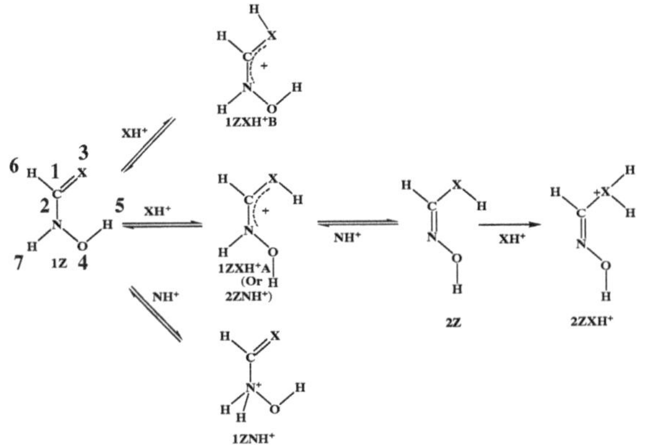

Scheme 5.1 a

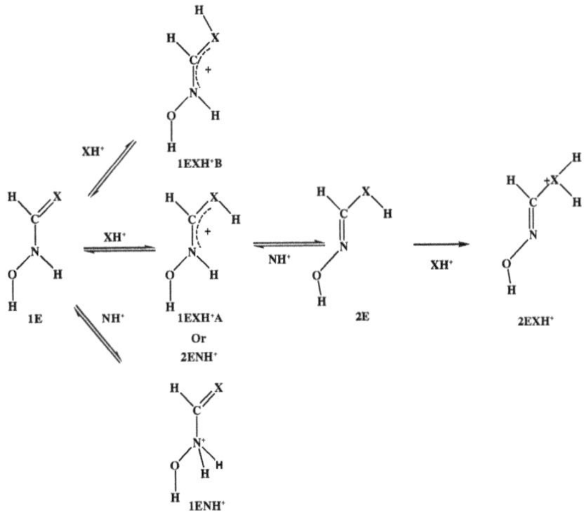

Scheme 5.1 b

Scheme 5.1: The protonation equilibria in **1Z** (scheme 5.1a) and **1E** (scheme 5.1b) of TFHA (X=S) and FHA (X=O).

The relative energy differences associated with all the protonated conformations of TFHA and FHA at B3LYP/6-31+G* [L1], MP2/6-31+G* [L3] and MP2/6-311++G**//MP2/6-31+G* [L5] theoretical levels are recorded in Table 5.1. Out of the ten protonated species optimized, the most stable protonated HA conformation is **1ZXH$^+$A** for both FHA and TFHA. It is conspicuous to note that **1ZXH$^+$A** and **1ZXH$^+$B** of TFHA differ only by 0.02 kcal/mol while the similar difference in these two corresponding protonated FHA species is 2.12 kcal/mol at [L5] theoretical level. The **1EXH$^+$A** and **1EXH$^+$B** are 4.46 and 3.47 kcal/mol higher in energy in comparison to the **1ZXH$^+$A** form in TFHA. The X3 protonation of **2E** result in least stable structure for both FHA and TFHA.

Table 5.1: Relative energy differences (kcal/mol) among protonated forms of FHA and TFHA at B3LYP/6-31+G* [L1], MP2/6-31+G* [L3] and MP2/6-311++G**//MP2/6-31+G* [L5] theoretical levels.

Species	TFHA			FHA		
	L1	L3	L5	L1	L3	L5
1ZXH$^+$A	0.0	0.00	0.00	0.00	0.00	0.00
1ZXH$^+$B	0.08	0.05	0.02	1.78	2.00	2.12
1ZNH$^+$	24.46	17.45	19.72	17.36	12.98	14.72
1EXH$^+$A	3.71	4.52	4.46	5.49	6.13	6.06
1EXH$^+$B	2.79	3.44	3.47	1.60	2.08	1.85
1ENH$^+$	26.15	18.99	21.77	18.61	14.27	16.08
2ZXH$^+$	26.04	24.48	24.03	36.32	33.25	33.45
2EXH$^+$	28.86	28.00	27.00	39.38	37.59	36.99

The geometrical parameters of protonated species are recorded in Tables 5.2-5.5 while the corresponding variations in few important bond lengths as a result of protonation are listed in Tables 5.6 and 5.7. The variation in geometrical parameters upon protonation reflects the change in strength of bonds and the hybridization of atoms present in the molecule. In order to stabilize the positive charge in protonated species, the electronic redistribution occurs such a way so as to give maximum stability. The protonation at N2 atom result in elongation of C-N bond while C-X bond is compressed at both **[L1]** and **[L3]** level (Tables 5.6 and 5.7). The elongation of C-N bond is longer in **1ZNH$^+$** of FHA in comparison to similar elongation in TFHA while the contraction of C-X bond is higher in case of TFHA. The C-N elongations on N2 protonation of **2Z** and **2E** tautomeric forms are much smaller in comparison to N2 protonation of **1Z** and **1E** form respectively in both FHA and TFHA. The compression of C-X bond distance is higher in **2ZNH$^+$** and **2ENH$^+$** relative to **1ZNH$^+$** and **1ENH$^+$** form respectively in both the acids at **[L3]** level. The X3 protonation of all the four tautomeric forms under study is accompanied by C-X bond elongations and C-N, N-O bond contractions. The C-X bond elongations on X3 protonation are much smaller in comparison to C-N bond elongation on N2 protonation of **1Z and 1E**. The magnitude of variations is higher in FHA in comparison to those in case of TFHA.

Table 5.2: Geometrical parameters of **1Z** and **1E** forms of TFHA and its X3 (X=S) and N2 protonated forms at B3LYP/6-31+G* **[L1]** and (MP2/6-31+G*) **[L3]** theoretical levels. All the bond distances are in angstrom (Å), angles and dihedrals are in degrees (°).

Parameters	1Z	1ZXH$^+$A	1ZXH$^+$B	1E	1EXH$^+$A	1EXH$^+$B	1ZNH$^+$	1ENH$^+$
C1-N2	1.329	1.306	1.304	1.352	1.303	1.304	1.533	1.525
	(1.333)	(1.308)	(1.306)	(1.366)	(1.304)	(1.303)	(1.530)	(1.516)
N2-O4	1.381	1.374	1.373	1.407	1.377	1.379	1.399	1.412
	(1.386)	(1.380)	(1.379)	(1.420)	(1.385)	(1.383)	(1.406)	(1.420)
O4-H5	0.994	0.981	0.981	0.974	0.981	0.981	0.994	0.976
	(0.997)	(0.986)	(0.986)	(0.977)	(0.985)	(0.985)	(0.995)	(0.986)
C1-H6	1.091	1.089	1.088	1.092	1.088	1.090	1.091	1.088
	(1.091)	(1.089)	(1.088)	(1.092)	(1.089)	(1.089)	(1.092)	(1.090)
C1-X3	1.666	1.697	1.704	1.647	1.712	1.704	1.588	1.583
	(1.652)	(1.683)	(1.690)	(1.631)	(1.692)	(1.698)	(1.590)	(1.591)
N2-H7	1.011	1.021	1.021	1.017	1.023	1.022	1.034	1.029
	(1.013)	(1.024)	(1.024)	(1.020)	(1.024)	(1.026)	(1.035)	(1.036)
X3-H8$^+$	-	1.354	1.351	-	1.350	1.355	-	-
	-	(1.345)	(1.343)	-	(1.346)	(1.342)	-	-
N2-H8$^+$	-	-	-	-	-	-	1.034	1.030
	-	-	-	-	-	-	(1.035)	(1.036)
X3-C1-N2	123.6	128.1	122.3	124.3	123.1	128.9	121.9	121.1
	(123.7)	(127.8)	(121.8)	(124.2)	(128.7)	(123.0)	(122.0)	(120.7)
O4-N2-C1	122.2	120.0	119.0	118.2	118.7	118.8	116.4	107.4
	(121.6)	(119.3)	(118.2)	(115.5)	(118.4)	(118.4)	(116.1)	(106.4)
H5-O4-N2	102.1	107.2	107.4	104.9	107.0	107.1	105.1	108.3
	(101.6)	(106.5)	(106.8)	(103.8)	(106.4)	(106.4)	(104.8)	(107.4)
H6-C1-N2	112.5	115.2	115.7	112.1	115.3	115.4	109.3	108.9
	(112.4)	(115.2)	(115.8)	(112.1)	(115.1)	(115.2)	(109.6)	(109.6)
H7-N2-C1	126.3	122.3	122.9	119.2	124.4	124.7	109.4	108.7

Contd......

Parameters	1Z	1ZXH+A	1ZXH+B	1E	1EXH+A	1EXH+B	1ZNH+	1ENH+
C1-X3-H8+	(126.6)	(122.7)	(123.5)	(117.0)	(125.2)	(124.8)	(109.8)	(109.1)
	-	96.5	94.1	-	94.5	98.0	-	-
H+8-N2-C1	-	(96.1)	(93.7)	-	(97.7)	(94.2)	-	-
	-	-	-	-	-	-	109.3	109.1
	-	-	-	-	-	-	(109.8)	(110.3)
O4-N2-C1-X3	0.0	-7.4	-7.6	162.9	170.0	170.7	0.1	132.6
	(0.0)	(-7.5)	(-7.7)	(158.2)	(171.4)	(170.6)	(-0.2)	(130.5)
H5-O4-N2-C1	-0.0	128.0	124.8	113.6	120.3	117.1	-0.2	172.6
	(-0.0)	(125.1)	(120.6)	(117.8)	(113.8)	(116.6)	(0.1)	(176.4)
H6-C1-N2-O4	180.0	172.6	171.9	-19.6	-9.6	-9.4	180.1	-47.6
	(180.0)	(172.8)	(172.2)	(-25.4)	(-9.1)	(-9.3)	(179.8)	(-49.1)
H7-N2-C1-X3	-180.0	189.1	188.7	20.0	5.6	6.6	121.3	11.9
	(-180.0)	(189.7)	(-170.8)	(25.3)	(7.1)	(6.2)	(120.4)	(10.7)
N2-C1-X3-H8+	-	2.3	182.3	-	-0.9	178.0	-	-
	-	(3.1)	(182.6)	-	(-1.4)	(177.7)	-	-
H+8-N2-C1-X3	-	-	-	-	-	-	238.9	252.5
	-	-	-	-	-	-	(239.2)	(249.8)

Table 5.3: Geometrical parameters of **2Z** and **2E** forms of TFHA and its X3 (X=S) and N2 protonated forms at B3LYP/6-31+G* **[L1]** and (MP2/6-31+G*) **[L3]** theoretical levels. All the bond distances are in angstrom (Å), angles and dihedrals are in degrees (°).

Parameters	2Z	2ZXH$^+$	2E	2EXH$^+$	2ZNH$^+$	2ENH$^+$
C1-N2	1.280	1.262	1.276	1.274	1.307	1.303
	(1.293)	(1.279)	(1.288)	(1.287)	(1.308)	(1.304)
N2-O4	1.410	1.365	1.414	1.351	1.374	1.379
	(1.420)	(1.384)	(1.425)	(1.364)	(1.380)	(1.385)
O4-H7	0.970	0.977	0.971	0.977	0.981	0.981
	(0.975)	(0.982)	(0.976)	(0.982)	(0.986)	(0.985)
C1-H6	1.086	1.084	1.092	1.089	1.089	1.090
	(1.085)	(1.084)	(1.091)	(1.090)	(1.089)	(1.089)
C1-X3	1.766	1.861	1.763	1.805	1.697	1.704
	(1.748)	(1.819)	(1.748)	(1.786)	(1.683)	(1.692)
X3-H5	1.348	1.358	1.351	1.358	1.354	1.355
	(1.341)	(1.351)	(1.343)	(1.351)	(1.345)	(1.346)
X3-H8$^+$	-	1.355	-	1.362	-	-
		(1.348)		(1.352)		
N2-H8$^+$	-	-	-	-	1.021	1.022
					(1.024)	(1.024)
X3-C1-N2	129.6	120.0	122.3	115.1	128.1	128.9
	(129.5)	(119.7)	(121.8)	(114.3)	(127.8)	(128.7)
O4-N2-C1	112.3	113.8	110.5	113.0	120.0	118.8
	(110.9)	(110.8)	(109.2)	(111.3)	(119.3)	(118.4)
H7-O4-N2	102.9	105.8	102.8	105.2	107.2	107.1
	(102.1)	(104.8)	(102.0)	(104.4)	(106.5)	(106.4)
H6-C1-N2	115.6	123.2	122.7	128.7	115.2	115.4
	(115.0)	(121.9)	(122.2)	(127.9)	(115.2)	(115.1)
H5-X3-C1	95.8	98.0	95.5	96.9	96.5	98.0

Contd.....

Parameters	2Z	2ZXH⁺	2E	2EXH⁺	2ZNH⁺	2ENH⁺
H8⁺-X3-C1	(95.6)	(98.3)	(94.9)	(96.3)	(96.1)	(97.6)
	-	97.0	-	101.3	-	-
H8⁺-N2-C1	-	(97.5)	-	(101.0)	-	-
	-	-	-	-	122.2	124.7
	-	-	-	-	(122.7)	(125.2)
O4-N2-C1-X3	0.0	-2.5	180.0	176.9	7.6	170.7
	(0.0)	(-2.9)	(180.0)	(176.6)	(7.5)	(171.3)
H7-O4-N2-C1	180.0	180.2	179.9	180.5	231.5	117.1
	(180.1)	(179.3)	(180.0)	(180.3)	(234.9)	(113.8)
H6-C1-N2-O4	180.0	181.6	-0.0	1.5	187.5	-9.4
	(180.0)	(181.4)	(-0.0)	(1.4)	(187.2)	(-9.0)
H5-X3-C1-N2	-0.0	49.6	0.0	-4.0	-2.3	-0.9
	(0.1)	(49.5)	(0.0)	(-2.2)	(-3.1)	(-1.4)
H8⁺-X3-C1-N2	-	146.6	-	92.6	-	-
	-	(147.9)	-	(95.7)	-	-
H8⁺-N2-C1-X3	-	-	-	-	170.6	6.6
	-	-	-	-	(170.3)	(7.1)

Table 5.4: Geometrical parameters of **1Z** and **1E** forms of FHA and its X3 (X=O) and N2 protonated forms at B3LYP/6-31+G* [**L1**] and (MP2/6-31+G*) [**L3**] theoretical levels. All the bond distances are in angstrom (Å), angles and dihedrals are in degrees (°).

Parameters	1Z	1ZXH$^+$A	1ZXH$^+$B	1E	1EXH$^+$A	1EXH$^+$B	1ZNH$^+$	1ENH$^+$
C1-N2	1.357	1.307	1.294	1.381	1.305	1.300	1.580	1.595
	(1.362)	(1.306)	(1.294)	(1.388)	(1.304)	(1.299)	(1.570)	(1.580)
N2-O4	1.401	1.379	1.362	1.412	1.378	1.374	1.398	1.406
	(1.410)	(1.384)	(1.364)	(1.423)	(1.384)	(1.381)	(1.404)	(1.412)
O4-H5	0.985	0.981	0.982	0.973	0.981	0.981	0.989	0.981
	(0.988)	(0.986)	(0.987)	(0.976)	(0.985)	(0.985)	(0.992)	(0.985)
C1-H6	1.101	1.088	1.088	1.104	1.088	1.090	1.097	1.099
	(1.099)	(1.087)	(1.087)	(1.101)	(1.087)	(1.089)	(1.096)	(1.098)
C1-X3	1.227	1.286	1.307	1.216	1.293	1.297	1.177	1.173
	(1.236)	(1.290)	(1.313)	(1.225)	(1.298)	(1.302)	(1.191)	(1.187)
N2-H7	1.013	1.021	1.019	1.018	1.023	1.024	1.034	1.035
	(1.016)	(1.023)	(1.022)	(1.020)	(1.025)	(1.026)	(1.036)	(1.037)
X3-H8$^+$	-	0.985	0.977	-	0.980	0.977	-	-
	-	(0.990)	(0.981)	-	(0.984)	(0.982)	-	-
N2-H8$^+$	-	-	-	-	-	-	1.034	1.033
	-	-	-	-	-	-	(1.036)	(1.035)
X3-C1-N2	122.0	124.0	117.5	123.0	126.1	118.4	116.2	117.7
	(122.0)	(123.8)	(116.6)	(122.7)	(126.1)	(117.9)	(116.5)	(117.8)
O4-N2-C1	117.0	117.6	125.7	115.6	118.0	118.4	112.6	113.6
	(115.9)	(116.9)	(125.0)	(113.5)	(117.5)	(118.0)	(112.2)	(112.9)
H5-O4-N2	102.1	107.3	108.1	104.5	106.9	106.9	105.7	107.7
	(101.8)	(106.6)	(107.5)	(103.6)	(106.3)	(106.2)	(105.4)	(107.4)
H6-C1-N2	113.3	119.0	118.9	112.3	118.1	118.3	112.6	109.5
	(113.5)	(119.2)	(119.3)	(112.6)	(118.1)	(118.4)	(112.6)	(109.5)
H7-N2-C1	122.4	123.0	123.3	116.2	125.0	122.7	110.4	109.8

Contd.....

Parameters	1Z	1ZXH⁺A	1ZXH⁺B	1E	1EXH⁺A	1EXH⁺B	1ZNH⁺	1ENH⁺
C1-X3-H8⁺	(120.6)	(123.5)	(123.7)	(114.7)	(125.7)	(123.1)	(110.9)	(110.4)
	-	113.3	114.4	-	117.3	113.9	-	-
	-	(112.3)	(113.7)	-	(116.6)	(113.0)	-	-
H⁺8-N2-C1	-	-	-	-	-	-	110.4	109.2
	-	-	-	-	-	-	(110.9)	(109.7)
O4-N2-C1-X3	-11.0	7.9	-0.0	159.6	171.0	170.2	0.0	206.7
	(-13.2)	(7.8)	(-0.0)	(157.7)	(172.0)	(171.0)	(-0.2)	(202.3)
H5-O4-N2-C1	6.04	-124.0	0.0	117.5	114.4	119.6	-0.04	77.9
	(7.7)	(-121.9)	(0.0)	(119.9)	(111.2)	(116.1)	(0.5)	(80.1)
H6-C1-N2-O4	172.1	187.2	180.0	-23.4	-9.2	-9.5	180.0	26.4
	(170.9)	(186.7)	(180.0)	(-25.9)	(-8.4)	(-8.9)	(179.8)	(22.3)
H7-N2-C1-X3	-154.1	-190.0	-180.0	26.5	6.8	5.8	120.3	-26.0
	(-150.0)	(-190.2)	(-180.0)	(30.1)	(7.2)	(6.2)	(119.6)	(-30.6)
N2-C1-X3-H8⁺	-	-3.6	180.0	-	-1.1	178.6	-	-
	-	(-4.3)	(180.0)	-	(-1.2)	(178.4)	-	-
H⁺8-N2-C1-X3	-	-	-	-	-	-	239.7	91.3
	-	-	-	-	-	-	(240.0)	(87.3)

199

Table 5.5: Geometrical parameters of **2Z** and **2E** forms of FHA and its X3 (X=O) and N2 protonated forms at B3LYP/6-31+G* **[L1]** and (MP2/6-31+G* **[L3]** theoretical levels. All the bond distances are in angstrom (Å), angles and dihedrals are in degrees (°).

Parameters	2Z	2ZXH⁺	2E	2EXH⁺	2ZNH⁺	2ENH⁺
C1-N2	1.280 (1.290)	1.253 (1.268)	1.274 (1.283)	1.253 (1.265)	1.307 (1.306)	1.305 (1.304)
N2-O4	1.429 (1.436)	1.373 (1.390)	1.418 (1.427)	1.352 (1.366)	1.379 (1.384)	1.378 (1.384)
O4-H7	0.969 (0.974)	0.977 (0.981)	0.970 (0.974)	0.981 (0.981)	0.981 (0.986)	0.981 (0.985)
C1-H6	1.084 (1.082)	1.081 (1.082)	1.088 (1.087)	1.085 (1.085)	1.088 (1.086)	1.088 (1.087)
C1-X3	1.344 (1.348)	1.493 (1.478)	1.350 (1.355)	1.481 (1.472)	1.286 (1.290)	1.293 (1.298)
X3-H5	0.977 (0.982)	0.996 (1.004)	0.975 (0.980)	0.992 (0.998)	0.985 (0.990)	0.980 (0.984)
X3-H8⁺	-	0.986 (0.991)	-	0.987 (0.993)	-	-
N2-H8⁺	-	-	-	-	1.021 (1.023)	1.023 (1.025)
X3-C1-N2	126.8 (126.7)	117.9 (118.0)	121.1 (120.8)	111.5 (111.1)	124.0 (123.8)	126.1 (126.1)
O4-N2-C1	108.6 (107.7)	112.7 (110.6)	110.6 (109.3)	115.3 (113.1)	117.6 (116.9)	118.0 (117.5)
H7-O4-N2	103.0 (102.3)	106.0 (105.1)	102.5 (101.8)	105.1 (104.2)	107.3 (106.6)	106.9 (106.3)
H6-C1-N2	118.6 (118.4)	128.8 (127.8)	125.2 (125.2)	136.4 (135.9)	119.0 (119.2)	118.1 (118.1)
H5-X3-C1	107.8	108.8	108.3	110.0	113.3	117.3

Contd......

Parameters	2Z	2ZXH⁺	2E	2EXH⁺	2ZNH⁺	2ENH⁺
H8⁺-X3-C1	(107.2)	(107.6)	(107.7)	(109.1)	(112.3)	(116.6)
	-	115.8	-	114.7	-	-
H8⁺-N2-C1	-	(114.8)	-	(113.9)	-	-
	-	-	-	-	123.0	125.0
	-	-	-	-	(123.5)	(125.7)
O4-N2-C1-X3	0.0	-2.9	180.0	176.1	7.9	189.0
	(0.0)	(-3.1)	(180.0)	(176.0)	(7.8)	(188.0)
H7-O4-N2-C1	180.1	179.9	180.0	180.1	236.0	245.6
	(179.9)	(179.6)	(180.0)	(179.8)	(238.1)	(248.8)
H6-C1-N2-O4	180.0	181.9	-0.0	1.5	187.2	9.2
	(180.0)	(181.8)	(0.0)	(1.5)	(186.7)	(8.4)
H5-X3-C1-N2	0.0	5.4	0.0	-4.4	-3.6	1.1
	(-0.0)	(5.5)	(-0.0)	(-4.1)	(-4.3)	(1.2)
H8⁺-X3-C1-N2	-	129.5	-	120.4	-	-
	-	(131.9)	-	(118.5)	-	-
H8⁺-N2-C1-X3	-	-	-	-	170.0	-6.8
	-	-	-	-	(169.8)	(-7.3)

Table 5.6: Variations in geometrical parameters of protonated TFHA at B3LYP/6-31+G* [L1] and (MP2/6-31+G*) [L3] theoretical levels.

Species	ΔC-N	ΔN-O	ΔC-X
1ZXH⁺A	-0.023	-0.007	0.031
	(-0.025)	(-0.006)	(0.031)
1ZXH⁺B	-0.025	-0.008	0.038
	(-0.023)	(-0.002)	(0.024)
1EXH⁺A	-0.049	-0.030	0.065
	(-0.062)	(-0.035)	(0.061)
1EXH⁺B	-0.048	-0.028	0.057
	(-0.063)	(-0.037)	(0.067)
2ZXH⁺	-0.018	-0.045	0.095
	(-0.014)	(-0.036)	(0.071)
2EXH⁺	-0.002	-0.063	0.042
	(-0.001)	(-0.061)	(0.038)
1ZNH⁺	0.204	0.018	-0.078
	(0.197)	(0.020)	(-0.062)
1ENH⁺	0.173	0.005	-0.064
	(0.150)	(0.000)	(-0.040)
2ZNH⁺	0.027	-0.036	-0.069
	(0.015)	(-0.040)	(-0.065)
2ENH⁺	0.027	-0.035	-0.059
	(0.016)	(-0.040)	(-0.056)

Table 5.7: Variations in geometrical parameters of protonated FHA at B3LYP/6-31+G* [L1] and (MP2/6-31+G*) [L3] theoretical levels.

Species	ΔC-N	ΔN-O	ΔC-X
1ZXH⁺A	-0.050	-0.022	0.059
	(-0.056)	(-0.026)	(0.054)
1ZXH⁺B	-0.063	-0.039	0.080
	(-0.068)	(-0.046)	(0.077)
1EXH⁺A	-0.076	-0.034	0.077
	(-0.084)	(-0.039)	(0.073)
1EXH⁺B	-0.081	-0.038	0.081
	(-0.089)	(-0.042)	(0.077)
2ZXH⁺	-0.027	-0.056	0.149
	(-0.022)	(-0.046)	(0.130)
2EXH⁺	-0.021	-0.066	0.131
	(-0.018)	(-0.061)	(0.117)
1ZNH⁺	0.223	-0.003	-0.050
	(0.208)	(-0.006)	(-0.045)
1ENH⁺	0.214	-0.006	-0.043
	(0.192)	(-0.011)	(-0.038)
2ZNH⁺	0.027	-0.050	-0.057
	(0.016)	(-0.052)	(-0.058)
2ENH⁺	0.031	-0.040	-0.058
	(0.021)	(-0.043)	(-0.057)

It is discussed in previous chapters that **1Z** and **2Z** of FHA and TFHA both are stabilized by intramolecular H-bonding. To look for the effect of protonation on intramolecular H-bonding, the H-bond distances and angles important for H-bonding along with the sum of angles around N2 at [L3] theoretical level are listed in Table 5.8. The analysis of geometrical parameters of X-protonated **1Z** species suggests the presence of intramolecular H-bonding in **1ZXH$^+$A** of both FHA and TFHA. Intramolecular H-bond although preserved in **1ZXH$^+$A** and **1ZXH$^+$B** of FHA and **1ZXH$^+$A** of TFHA (absent in **1ZXH$^+$B** of TFHA) but gets slightly weakened relative to the intramoleclar H-bond of **1Z**, but intramolecular H-bond on N2 protonation is approximately of the same strength as that of **1Z** as evident from negligible change in H-bonding parameters. In **1ZXH$^+$A**, the X-H bond acts as H-bond donor while O4 of hydroxyl group acts as H-bond acceptor while in **1ZXH$^+$B** orientation of FHA only, it is O-H bond that acts as H-bond donor while X3 of X-H acts as acceptor.

The **2ZXH$^+$** protonated species of the two acids is also stabilized by intramolecular H-bonding. There are no intramolecular H-bonding interactions in **1EXH$^+$A** and **1EXH$^+$B** species as suggested by their geometrical structures. The relative energy difference between these two protonated species is 0.99 kcal/mol and 4.21 kcal/mol in TFHA and FHA respectively at [L5] theoretical level. The sum of angles around N shows the **1ZXH$^+$B** protonated species of FHA is planar around N that favors delocalization of lone pair of electrons at N2. The **1ZXH$^+$A** species shows deviation of 2.4° and 2.1° from 360° for the sum of angles around nitrogen in FHA and TFHA respectively. The protonation at N of **1Z** leads to pyramidalization around N2.

Table 5.8: Hydrogen bond distances (Å) and angles (°) important for intramolecular hydrogen bonding in the protonated species of TFHA (FHA) at MP2/6-31+G* [L3] theoretical level.

Species	H-bond distances		H-bond angles		Sum of angles around N
		FHA		FHA	FHA
1Z	H5...X3	2.307	O4-H5...X3	124.4	360.0
		(2.046)		(117.8)	(346.5)
1ZXH$^+$A	H8...O4	2.310	O4-H8...X3	108.6	357.9
		(2.206)		(106.3)	357.6
1ZXH$^+$B	H5...X3	-	O4-H5...X3	-	357.9
		(2.189)		(108.1)	(360.0)

Contd.....

Species	H-bond distances		H-bond angles		Sum of angles around N
		FHA		FHA	FHA
1ZNH$^+$	H5...X3	2.309 (2.027)	O4-H5...X3	124.9 (117.9)	320.9 (322.6)
2Z	H5...O4	2.399 (2.020)	X3-H5...O4	108.6 (111.5)	-
2ZXH$^+$	H5...O4	2.399 (1.896)	X3-H5...O4	86.9 (111.8)	-

Sum of van der Waal radii $r_H+r_S=3.05$Å, $r_H+r_O=2.6$Å

The proton affinities associated with the formation of various protonated species at **[L1]**, **[L3]** and **[L5]** theoretical levels are listed in Table 5.9. The proton affinity (PA) of X3 site of the **1Z** and **1E** tautomeric forms of TFHA at **[L5]** theoretical level differ by 1.31 kcal/mol while PA of the N2 site is much lower than X3 site and differ by 1.82 kcal/mol for **1Z** and **1E** conformers. The O3 site in **1Z** and **1E** forms of FHA has PA values that differ by 6.10 kcal/mol. The proton affinities of all the sites of FHA are lower than the respective values for TFHA sites. The chalcogen site in **1Z** and **1E** conformation of both FHA and TFHA is more basic than the N2 site as seen from Table 5.9 and besides it the basicity of S3 site in TFHA is higher in comparison to O3 site of FHA. In **2Z** and **2E** conformers however it is the N2 site which is more basic than the X3 site.

Table 5.9: Proton affinities (kcal/mol) of X3 and N2 sites of different tautomeric forms of FHA and TFHA at B3LYP/6-31+G* **[L1]**, MP2/6-31+G* **[L3]** and MP2/6-311++G**// MP2/6-31+G* **[L5]** theoretical levels.

Species	TFHA			FHA		
	L1	L3	L5	L1	L3	L5
1ZXH$^+$A	199.08	195.81	202.67	192.88	189.66	196.11
1ZXH$^+$B	199.0	196.10	202.95	190.88	187.31	193.65
1ZNH$^+$	174.56	178.68	183.28	175.25	176.17	181.28
1EXH$^+$A	198.11	193.73	201.36	187.79	183.37	190.01
1EXH$^+$B	198.99	194.75	202.29	191.68	187.40	194.20
1ENH$^+$	175.59	180.31	185.10	174.36	174.88	179.67
2ZNH$^+$	198.42	193.30	199.30	196.34	191.05	196.81
2ZXH$^+$	172.12	168.62	175.07	159.77	155.10	163.10
2ENH$^+$	197.65	192.13	198.06	194.25	189.07	194.42
2EXH$^+$	172.42	168.53	175.41	160.18	157.38	163.20

The NBO analysis provides the quantitative picture of electron delocalizations responsible for the various conjugative interactions present in the protonated species.

The NBO analysis has been carried out and the second order delocalization energies $E^{(2)}$, which are representative of stabilization energies associated with various electronic delocalizations obtained at [L3] theoretical level for protonated species are given in Table 5.10. NBO indicates that protonation of the S3 site both in case of **1Z** and **1E** conformers of TFHA results in replacement of C-X π bond by C-N π bond. The $n_{N2} \rightarrow \sigma^*_{C1-X3}/\pi^*_{C1-X3}$ orbital interactions that are the most important conjugative interactions for the neutral **1Z** and **1E** are absent in N2as well as X3 protonated **1Z** and **1E** species for both HAs. In the absence of these interactions, the $n_{X3} \rightarrow \sigma^*_{C1-N2}$ and $n_{X3} \rightarrow \pi^*_{C1-N2}$ orbital interactions appear in **1ZXH⁺A** and **1ZXH⁺B** species, while $n_{X3} \rightarrow \sigma^*_{C1-N2}$ are significant in **1ZNH⁺**. The variation in $E^{(2)}$ values for all orbital interactions indicates that protonation of the tautomeric forms is not accompanied by enhancement of conjugative interactions, but it is the formation of C=N π bond which is important for the stability of **1ZXH⁺A, 1ZXH⁺B, 1EXH⁺A** and **1EXH⁺B**. Inspite of the stability difference (4.46 kcal/mol) between **1ZXH⁺A** and **1EXH⁺A** in TFHA, the PA values of these species of TFHA differ only by 1.31 kcal/mol at [L5] theoretical level. The **1EXH⁺A** of FHA is 6.06 kcal/mol less stable in comparison to **1ZXH⁺A** form of FHA and the presence of intramolecular H-bond in **1ZXH⁺A** species is responsible for higher stability of this species. Though **1EXH⁺B** of TFHA is not stabilized by H-bonding interactions but has PA comparable to X-protonation of **1Z** resulting in **1ZXH⁺B** at [L5] theoretical level as protonation in the latter involves rupture of intramolecular interactions.

The PA at N2 of **1Z** conformer of TFHA is 183.28 kcal/mol while that of **1E** is 185.10 kcal/mol. The **1ZNH⁺** and **1ENH⁺** species are 19.72 and 21.77 kcal/mol less stable than the most stable **1ZXH⁺A** species in case of TFHA. The large difference in the stability of N-protonated species relative to X3 protonated species is the result of loss of conjugative interactions in **1ZNH⁺** due to non availability of lone pair of electrons. The proton affinities of O3 and N2 sites in four conformers of FHA under study are 2.20-12.21 kcal/mol lower than the corresponding values in TFHA.

The high PA values for N-protonation of **2Z** and **2E** forms and indistinguishability of **2ZNH⁺** and **2ENH⁺** from **1ZXH⁺A** and **1EXH⁺A** respectively suggest that **2Z** and **2E** tautomerize to **1Z** and **1E** forms in the presence of H⁺. Lone pair occupancies also reflect the extent of electron delocalization involving the lone pair. The occupancies are listed in Table 5.11 and are supportive of conclusions drawn from the NBO analysis.

Table 5.10: The important second order $E^{(2)}$ delocalization energies (kcal/mol) for the various orbital interactions present in protonated TFHA (FHA) obtained by NBO analysis at MP2/6-31+G* **[L3]** theoretical level.

Species	$n_{N2}\rightarrow\sigma^*_{C1-X3}$	$n_{N2}\rightarrow\pi^*_{C1-X3}$	$n_{X3}\rightarrow\sigma^*_{C1-N2}$	$n_{X3}\rightarrow\pi^*_{C1-N2}$	$n_{X3}\rightarrow\sigma^*_{C1-H6}$	$n_{O4}\rightarrow\sigma^*_{C1-N2}$	$n_{O4}\rightarrow\pi^*_{C1-N2}$	$n_{N2}\rightarrow\sigma^*_{C1-H6}$
1Z	- (14.98)	145.34 (32.68)	15.92 (29.25)	- (-)	13.09 (23.91)	6.56 (-)	- (-)	- (-)
1ZXH+A	- (-)	- (-)	6.38+7.31 (6.47)	56.87 (3.17)	- (-)	6.85 (4.16)	- (-)	- (-)
1ZXH+B	- (-)	- (-)	- (-)	52.69 (76.36)	5.71 (8.72)	7.35 (7.30)	- (18.82)	- (-)
1ZNH+	- (-)	- (-)	34.35 (59.61)	- (-)	13.59 (20.58)	5.60 (-)	- (-)	- (-)
1E	- (12.11)	49.47 (20.90)	18.93 (32.25)	- (-)	12.75 (24.38)	- (-)	- (-)	- (-)
1EXH+A	- (-)	- (-)	6.22 (8.87)	56.61 (3.62)	- (-)	7.18 (6.41)	- (-)	- (-)
1EXH+B	- (-)	- (-)	- (-)	47.36 (35.07)	5.51 (8.77)	7.43 (-)	- (7.32)	- (-)
1ENH+	- (-)	- (-)	19.17 (64.76)	- (-)	5.46+13.15 (20.72)	- (5.94)	- (-)	- (-)
2Z	14.57 (13.50)	- (-)	34.64 (8.63)	- (54.42)	- (-)	18.29 (-)	- (13.22)	- (-)
2ZXH+	17.91 (-)	- (-)	- (6.47)	4.48 (-)	- (-)	- (4.16)	29.70 (-)	- (-)
2E	- (3.82)	- (-)	- (6.12)	34.78 (48.37)	- (-)	- (-)	- (13.80)	9.69 (10.60)
2EXH+	- (11.87)	- (-)	- (5.80)	- (-)	- (-)	- (9.82)	37.11 (10.82)	10.36 (12.07)

Table 5.11: Lone pair occupancies of neutral and protonated TFHA (FHA) at MP2/6-31+G* **[L3]** theoretical level.

Species	LP(1)N2	LP(1)X3	LP(2)X3	LP(1)O4	LP(2)O4
1Z	1.66	1.99	1.89	1.99	1.97
	(1.80)	(1.98)	(1.90)	(1.99)	(1.98)
1ZXH$^+$A	-	1.98	1.72	1.99	1.97
	(-)	(1.98)	(-)	(1.99)	(1.97)
1ZXH$^+$B	-	1.99	1.73	1.98	1.97
	(-)	(1.98)	(1.84)	(1.99)	(1.95)
1ZNH$^+$	-	1.99	1.84	1.99	1.98
	(-)	(1.98)	(1.85)	(1.99)	(1.98)
1E	1.81	1.99	1.91	1.99	1.98
	(1.87)	(1.98)	(1.90)	(1.98)	(1.90)
1EXH$^+$A	-	1.98	1.75	1.99	1.98
	(-)	(1.97)	(1.81)	(1.99)	(1.98)
1EXH$^+$B	-	1.99	1.76	1.99	1.98
	(-)	(1.98)	(1.82)	(1.99)	(1.98)
1ENH$^+$	-	1.99	1.87	1.99	1.99
	(-)	(1.98)	(1.85)	(1.99)	(1.98)
2Z	1.96	1.99	1.89	1.99	1.96
	(1.97)	(1.98)	(1.89)	(1.99)	(1.97)
2ZXH$^+$	1.95	1.98	-	1.99	1.93
	(1.94)	(1.98)	(-)	(1.98)	(1.94)
2E	1.97	1.99	1.90	2.00	1.96
	(1.96)	(1.98)	(1.90)	(2.00)	(1.97)
2EXH$^+$	1.97	1.99	-	2.00	1.91
	(1.95)	(1.98)	(-)	(2.00)	(1.92)

The protonated species are stabilized by the spread of the charge to different sites through electron delocalizations. The atomic charges recorded in Table 5.12 also reflect that the positive charge is not localized at the site of protonation. Larger spread of positive charge is reflected in protonated TFHA where the sulfur due to high polarizability and low electronegativity holds the positive charge whether protonation occurs at X3 or N2, while in case of protonation of FHA, large fraction of positive charge remains at incoming proton.

Table 5.12: Atomic charges (NPA) of the neutral and protonated species of TFHA (FHA) at MP2/6-31+G* **[L3]** theoretical level.

Species	C1	N2	X3	O4	H5	H6	H7	H+8
1Z	-0.019	-0.347	-0.251	-0.603	0.531	0.242	0.447	-
	(0.636)	(-0.426)	(-0.732)	(-0.616)	(0.535)	(0.173)	(0.430)	(-)
1ZXH+A	0.067	-0.302	0.275	-0.568	0.552	0.300	0.482	0.194
	(0.706)	(-0.348)	(-0.685)	(-0.572)	(0.557)	(0.273)	(0.490)	(0.580)
1ZXH+B	0.071	-0.298	0.280	-0.557	0.550	0.295	0.482	0.177
	(0.632)	(-0.272)	(-0.738)	(-0.527)	(0.553)	(0.267)	(0.499)	(0.586)
1ZNH+	-0.227	-0.391	0.333	-0.543	0.539	0.288	0.501	0.501
	(0.657)	(-0.437)	(-0.494)	(-0.538)	(0.562)	(0.248)	(0.501)	(0.501)
1E	-0.016	-0.412	-0.145	-0.616	0.518	0.240	0.431	-
	(0.661)	(-0.454)	(-0.685)	(-0.626)	(0.517)	(0.166)	(0.422)	(-)
1EXH+A	0.083	-0.305	0.279	-0.545	0.546	0.308	0.472	0.161
	(0.711)	(-0.349)	(-0.688)	(-0.539)	(0.548)	(0.278)	(0.473)	(0.566)
1EXH+B	0.089	-0.300	0.240	-0.546	0.547	0.302	0.476	0.193
	(0.696)	(-0.325)	(-0.705)	(-0.543)	(0.548)	(0.264)	(0.485)	(0.581)
1ENH+	-0.227	-0.388	0.337	-0.541	0.549	0.290	0.490	0.489
	(0.652)	(-0.440)	(-0.468)	(-0.540)	(0.549)	(0.244)	(0.494)	(0.510)
2Z	-0.099	-0.197	0.025	-0.677	0.176	0.248	0.524	-
	(0.478)	(-0.282)	(-0.783)	(-0.703)	(0.547)	(0.214)	(0.528)	(-)
2ZXH+	-0.151	-0.009	0.477	-0.650	0.246	0.300	0.568	0.219
	(0.370)	(0.066)	(-0.766)	(-0.658)	(0.636)	(0.286)	(0.579)	(0.619)
2E	-0.082	-0.217	0.046	-0.651	0.150	0.238	0.515	-
	(0.488)	(-0.312)	(-0.772)	(-0.650)	(0.524)	(0.206)	(0.516)	(-)
2EXH+	-0.185	-0.044	0.506	-0.575	0.235	0.293	0.550	0.221
	(0.362)	(-0.120)	(-0.744)	(-0.566)	(0.620)	(0.278)	(0.553)	(0.615)

5.2.2 GAS PHASE ACIDITIES:

Eight anionic forms each of FHA and TFHA that result from N-H, O-H and X-H deprotonation have been optimized. **2EX3⁻** and **1EN⁻** converge to same structure and **2ZO4⁻** does not converge at [L1] and [L3] levels for TFHA. For FHA, it converged only at [L1]. The anionic forms are shown in **Scheme 5.2**.

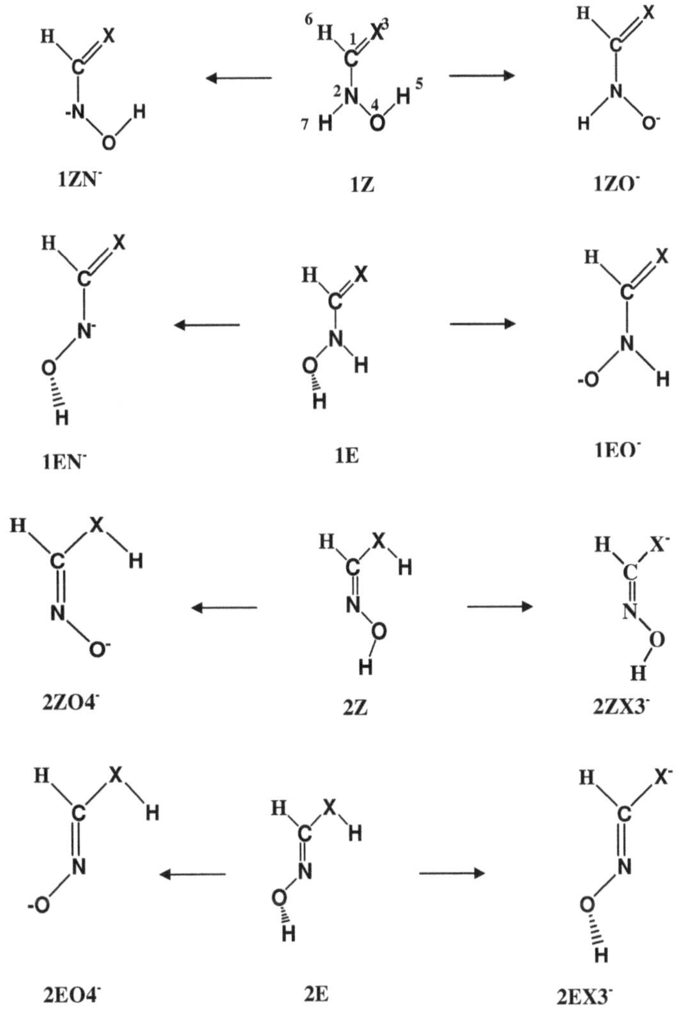

Scheme 5.2: The deprotonation of different isomers of hydroxamic acids

The geometrical parameters of the anions at B3LYP/6-31+G* **[L1]** and (MP2/6-31+G*) **[L3]** theoretical levels are listed in the Table 5.13-5.16 while the variation in important bond distances on deprotonation are recorded in Table 5.17-5.18.

Table 5.13: Geometrical parameters of the anionic and neutral forms of TFHA at B3LYP/6-31+G* **[L1]** and (MP2/6-31+G*) **[L3]** theoretical levels. All the bond distances are in angstrom (Å), angles and dihedrals are in degrees (°).

Parameters	1Z	1ZN⁻	1ZO⁻	1E	1EN⁻	1EO⁻
C1-N2	1.329	1.300	1.328	1.349	1.297	1.315
	(1.333)	(1.315)	(1.338)	(1.366)	(1.306)	(1.322)
N2-O4	1.381	1.416	1.309	1.407	1.486	1.326
	(1.386)	(1.422)	(1.302)	(1.420)	(1.488)	(1.320)
O4-H5	0.994	0.994	-	0.974	0.968	-
	(0.997)	(1.001)	(-)	(0.977)	(0.973)	(-)
C1-H6	1.091	1.093	1.097	1.092	1.100	1.091
	(1.091)	(1.091)	(1.096)	(1.092)	(1.099)	(1.091)
C1-X3	1.666	1.748	1.705	1.647	1.731	1.725
	(1.652)	(1.730)	(1.691)	(1.631)	(1.716)	(1.713)
N2-H7	1.011	-	1.028	1.017	-	1.025
	(1.013)	(-)	(1.029)	(1.020)	(-)	(1.028)
C1-X3...H5	2.325	2.243				
	(2.307)	(2.189)				
X3-C1-N2	123.6	128.5	129.6	124.3	125.4	125.3
	(123.7)	(128.2)	(128.5)	(124.2)	(125.0)	(124.6)
O4-N2-C1	122.2	113.7	130.6	118.2	108.4	128.3
	(121.6)	(112.7)	(130.3)	(115.5)	(107.7)	(128.6)
H5-O4-N2	102.1	104.5	-	104.9	100.0	-
	(101.6)	(103.3)	(-)	(103.8)	(99.3)	(-)
H6-C1-N2	112.5	112.4	110.0	112.1	116.4	112.8
	(112.4)	(112.1)	(110.0)	(112.1)	(116.2)	(112.7)
H7-N2-C1	126.3	-	114.5	119.2	-	115.4
	(126.6)	(-)	(114.6)	(117.0)	(-)	(115.2)
C1-X3...H5-O4	123.8	127.8				
	(124.4)	(129.8)				
O4-N2-C1-X3	0.0	0.0	-0.0	162.9	180.0	180.0
	(0.0)	(0.0)	(0.0)	(158.2)	(180.0)	(180.0)
H5-O4-N2-C1	-0.0	0.1	-	113.6	180.0	-
	(-0.0)	(0.0)	(-)	(117.8)	(180.0)	(-)
H6-C1-N2-O4	180.0	180.0	180.0	-19.6	-0.0	-0.0
	(180.0)	(180.0)	(180.0)	(-25.4)	(0.0)	(0.0)
H7-N2-C1-X3	-180.0	-	180.0	20.0	-	-0.0
	(-180.0)	(-)	(180.0)	(25.3)	(-)	(0.0)

Table 5.14: Geometrical parameters of the anionic and neutral forms of TFHA at B3LYP/6-31+G* **[L1]** and (MP2/6-31+G*) **[L3]** theoretical levels. All the bond distances are in angstrom (Å), angles and dihedrals are in degrees (°).

Parameters	2Z	2ZX3⁻	2E	2EX3⁻	2EO4⁻
C1-N2	1.280	1.302	1.276	1.296	1.299
	(1.288)	(1.313)	(1.288)	(1.306)	(1.312)
N2-O4	1.410	1.450	1.414	1.486	1.316
	(1.425)	(1.454)	(1.425)	(1.488)	(1.327)
O4-H7	0.970	0.969	0.971	0.968	-
	(0.976)	(0.974)	(0.976)	(0.973)	(-)
C1-H6	1.086	1.097	1.092	1.099	1.095
	(1.091)	(1.096)	(1.091)	(1.098)	(1.094)
C1-X3	1.766	1.728	1.763	1.731	1.805
	(1.748)	(1.711)	(1.748)	(1.716)	(1.783)
H5-X3	1.348	-	1.351	-	1.354
	(1.343)	(-)	(1.343)	(-)	(1.344)
X3-C1-N2	129.6	133.6	122.3	125.4	120.6
	(121.8)	(132.7)	(121.8)	(124.9)	(120.2)
O4-N2-C1	112.3	112.2	110.5	108.4	118.3
	(109.2)	(111.0)	(109.2)	(107.7)	(116.9)
H5-X3-C1	95.8	-	95.5	-	93.7
	(94.9)	(-)	(94.9)	(-)	(93.5)
H6-C1-N2	115.6	108.8	122.7	116.4	124.0
	(122.2)	(108.8)	(122.2)	(116.2)	(123.3)
H7-O4-N2	102.9	99.9	102.8	100.0	-
	(102.0)	(99.3)	(102.0)	(99.3)	(-)
O4-N2-C1-X3	0.0	-0.0	180.0	180.0	180.0
	(-0.0)	(-0.0)	(180.0)	(180.0)	(180.0)
H7-O4-N2-C1	180.0	180.0	179.9	180.0	-
	(180.0)	(180.0)	(180.0)	(180.0)	(-)
H6-C1-N2-O4	180.0	180.0	-0.0	0.0	-0.0
	(180.0)	(180.0)	(-0.0)	(-0.0)	(-0.0)
H5-X3-C1-N2	-0.0	-	0.0	-	0.0
	(0.0)	(-)	(0.0)	(-)	(-0.3)

Table 5.15: Geometrical parameters of the anionic and neutral forms of FHA at B3LYP/6-31+G* **[L1]** and (MP2/6-31+G*) **[L3]** theoretical levels. All the bond distances are in angstrom (Å), angles and dihedrals are in degrees (°).

Parameters	1Z	1ZN⁻	1ZO⁻	1E	1EN⁻	1EO⁻
C1-N2	1.357	1.318	1.342	1.381	1.324	1.330
	(1.362)	(1.326)	(1.343)	(1.388)	(1.330)	(1.330)
N2-O4	1.401	1.455	1.337	1.412	1.504	1.358
	(1.410)	(1.457)	(1.338)	(1.423)	(1.502)	(1.361)
O4-H5	0.985	0.987	-	0.973	0.967	-
	(0.988)	(0.995)	(-)	(0.976)	(0.972)	(-)
C1-H6	1.101	1.108	1.119	1.104	1.119	1.106
	(1.099)	(1.104)	(1.116)	(1.101)	(1.115)	(1.103)
C1-X3	1.227	1.280	1.250	1.216	1.264	1.266
	(1.236)	(1.291)	(1.259)	(1.225)	(1.273)	(1.274)
N2-H7	1.013	-	1.026	1.018	-	1.023
	(1.016)	(-)	(1.025)	(1.020)	(-)	(1.023)
C1-X3...H5	2.069	1.937				
	(2.046)	(1.893)				
X3-C1-N2	122.0	128.1	129.7	123.0	126.1	126.0
	(122.0)	(127.5)	(129.4)	(122.7)	(125.9)	(125.7)
O4-N2-C1	117.0	108.8	129.8	115.6	107.1	127.6
	(115.9)	(108.1)	(129.6)	(113.5)	(106.7)	(127.1)
H5-O4-N2	102.1	102.0	-	104.5	99.5	-
	(101.8)	(100.9)	(-)	(103.6)	(99.3)	(-)
H6-C1-N2	113.3	111.8	109.0	112.3	114.6	111.8
	(113.5)	(112.1)	(109.1)	(112.6)	(114.6)	(111.9)
H7-N2-C1	122.4	-	113.9	116.2	-	115.6
	(120.6)	(-)	(114.3)	(114.7)	(-)	(116.1)
C1-X3...H5-O4	116.9	122.4				
	(117.8)	(124.5)				
O4-N2-C1-X3	-11.0	0.06	-0.0	159.6	181.9	180.0
	(-13.2)	(0.0)	(-0.0)	(157.7)	(182.2)	(180.0)
H5-O4-N2-C1	6.0	-0.0	-	117.5	208.3	-
	(7.7)	(-0.0)	(-)	(119.9)	(219.8)	(-)
H6-C1-N2-O4	172.1	180.0	180.0	-23.4	1.6	-0.0
	(170.9)	(180.0)	(180.0)	(-25.9)	(1.7)	(-0.0)
H7-N2-C1-X3	-154.1	-	180.0	26.5	-	0.0
	(-150.0)	(-)	(180.0)	(30.1)	(-)	(0.0)

Table 5.16: Geometrical parameters of the anionic and neutral forms of FHA at B3LYP/6-31+G* **[L1]** and (MP2/6-31+G*) **[L3]** theoretical levels. All the bond distances are in angstrom (Å), angles and dihedrals are in degrees (°).

Parameters	2Z	2ZX3⁻	2E	2EX3⁻	2EO4⁻
C1-N2	1.280	1.324	1.274	1.324	1.281
	(1.290)	(1.329)	(1.283)	(1.330)	(1.294)
N2-O4	1.429	1.484	1.418	1.504	1.333
	(1.436)	(1.483)	(1.427)	(1.502)	(1.344)
O4-H7	0.969	0.967	0.970	0.967	-
	(0.974)	(0.972)	(0.974)	(0.972)	(-)
C1-H6	1.084	1.120	1.088	1.119	1.091
	(1.082)	(1.115)	(1.087)	(1.115)	(1.090)
C1-X3	1.344	1.260	1.350	1.264	1.410
	(1.348)	(1.270)	(1.355)	(1.273)	(1.409)
H5-X3	0.977	-	0.975	-	0.977
	(0.982)	(-)	(0.980)	(-)	(0.981)
X3-C1-N2	126.8	134.5	121.1	126.1	119.7
	(126.7)	(132.8)	(120.8)	(125.9)	(119.2)
O4-N2-C1	108.6	109.0	110.6	107.1	120.6
	(107.7)	(109.1)	(109.3)	(106.7)	(118.7)
H5-X3-C1	107.8	-	108.3	-	102.5
	(107.2)	(-)	(107.7)	(-)	(101.9)
H6-C1-N2	118.6	105.9	125.2	114.6	127.0
	(118.4)	(107.5)	(125.2)	(114.6)	(126.9)
H7-O4-N2	103.0	97.6	102.5	99.5	-
	(102.3)	(98.2)	(101.8)	(99.3)	(-)
O4-N2-C1-X3	0.0	-0.0	180.0	178.1	179.9
	(0.0)	(-0.0)	(180.0)	(177.8)	(180.0)
H7-O4-N2-C1	180.1	180.0	180.0	151.7	-
	(179.9)	(180.0)	(180.0)	(140.2)	(-)
H6-C1-N2-O4	180.0	180.0	-0.0	-1.6	0.0
	(180.0)	(180.0)	(0.0)	(-1.7)	(-0.0)
H5-X3-C1-N2	0.0	-	0.0	-	0.0
	(-0.0)	(-)	(-0.0)	(-)	(-0.0)

As can be seen from Tables 5.17 and 5.18, the C-N, N-O and C-X bond distances show significant variation with the anion formation. In both FHA and TFHA, the C-N bond contracts while the C-X bond elongates in 1ZN⁻, 1ZO⁻, 1EO⁻ and 1EN⁻. C-N bond contraction is more is 1ZN⁻ and 1EN⁻ of FHA than TFHA. C-X elongation occurs on N2-H7 and O4-H5 deprotonation of TFHA and FHA and C-X elongation is more in N and O deprotonated 1Z and 1E of TFHA than FHA.

Table 5.17: Variation in geometrical parameters of FHA at B3LYP/6-31+G* [L1] and (MP2/6-31+G*) [L3] theoretical levels as a result of deprotonation.

Deprotonated Species	Δ C-N	Δ N-O	Δ C-X
1ZN⁻	-0.039 (-0.036)	0.054 (0.047)	0.053 (0.055)
1ZO⁻	-0.015 (-0.019)	-0.064 (-0.072)	0.023 (0.023)
1EN⁻	-0.057 (-0.058)	0.092 (0.079]	0.048 (0.048)
1EO⁻	-0.051 (-0.058)	-0.054 (-0.062)	0.050 (0.049)
2ZX3⁻	0.044 (0.039)	0.055 (0.047)	-0.084 (-0.078)
2EO4⁻	0.007 (0.011)	-0.085 (-0.083)	0.060 (0.054)
2EX3⁻	0.050 (0.047)	0.086 (0.075)	-0.086 (-0.082)

Table 5.18: Variation in geometrical parameters of TFHA at B3LYP/6-31+G* [L1] and (MP2/6-31+G*) [L3] theoretical levels as a result of deprotonation.

Deprotonated Species	Δ C-N	Δ N-O	Δ C-X
1ZN⁻	-0.029 (-0.018)	0.035 (0.036)	0.082 (0.078)
1ZO⁻	-0.001 (0.005)	-0.072 (-0.084)	0.039 (0.039)
1EN⁻	-0.052 (-0.060)	0.079 (0.060)	0.084 (0.085)
1EO⁻	-0.034 (-0.044)	-0.081 (-0.100)	0.078 (0.082)
2ZX3⁻	0.022 (0.025)	0.040 (0.029)	-0.038 (-0.037)
2EO4⁻	0.023 (0.024)	-0.098 (-0.098)	0.042 (0.035)
2EX3⁻	0.020 (0.018)	0.072 (0.063)	-0.032 (-0.032)

The relative energies of the deprotonated species are evaluated at [L5] theoretical level with respect to the most stable 1ZN⁻ form, are reported in Table 5.19. The energy difference in the anionic forms ranges from 8.17 to 28.10 kcal/mol in FHA and 10.48 to 32.18 kcal/mol in TFHA. The relative energy differences are larger in TFHA over FHA. The enthalpy change associated with the deprotonation processes are listed in Table 5.19.

Table 5.19: Deprotonation enthalpies (kcal/mol) and relative energies (kcal/mol) associated with the formation of various anionic species of TFHA (FHA) at MP2/6-311++G**// MP2/6-31+G* [L5] theoretical level.

Deprotonated Species	Deprotonation enthalpies	Relative energies
1ZN⁻	327.68	0.00
	(342.73)	(0.00)
1ZO⁻	346.94	19.48
	(358.83)	(16.25)
1EN⁻	335.54	11.07
	(351.56)	(9.29)
1EO⁻	338.98	14.13
	(355.94)	(13.44)
2EO4⁻	360.57	32.18
	(366.36)	(28.10)
2EX3⁻	339.31	11.06
	(347.43)	(9.29)
2ZX3⁻	342.20	10.48
	(349.88)	(8.17)

Deprotonation enthalpies range from 327.68 to 360.57 kcal/mol in TFHA and for FHA the range is from 342.73 to 366.36 kcal/mol. The range is broader in TFHA. TFHA is more acidic over FHA as relatively lower deprotonation enthalpies are required for deprotonation. N2 deprotonation is favored for both TFHA and FHA over O4 deprotonation by an amount of 19.26 and 16.10 kcal/mol respectively. The N-H and O-H deprotonation affinities for **1Z** of FHA by Bagno et al. are 350.34 and 362.95 kcal/mol at MP2/6-311++G** level [23]. The values are 7.61 and 4.12 kcal/mol higher than our values evaluated at [L5] theoretical level.

NBO analysis (Table 5.20) of **1ZN⁻** indicates the presence of C-N double bond and C-X single bond both in case of TFHA and FHA. In addition the $n_{X3} \rightarrow \pi^*_{C1-N2}$ orbital interaction with $E^{(2)}$ value of 55.92 kcal/mol and 119.51 kcal/mol for TFHA and FHA respectively stabilize the anion. The molecular orbitals of **1ZO**⁻ anion of TFHA and FHA indicate the presence of C-N pi bond and absence of C-X pi bond. The geometrical parameters suggest C-N and C-X bond to be intermediate in strength of the bonds present in **1Z** and **1ZN**⁻. The $n_{X3} \rightarrow \pi^*_{C1-N2}$ orbital interaction tend to stabilize the **1ZO**⁻ of TFHA with $E^{(2)}$ value 85.29 kcal/mol and of FHA with $E^{(2)}$ of 195.71 kcal/mol. Therefore the difference in stabilities of **1ZN**⁻ and **1ZO**⁻ of TFHA and FHA results from the difference in strengths of C-N, C-X and N-O bonds. Atomic charge analysis (Table 5.21) of the anions indicates the additional one unit negative charge of the anion is more delocalized in **1ZN**⁻ rather than **1ZO**⁻ anion. The higher acidity associated with N-H deprotonation of **1Z** of TFHA in comparison to that in case of FHA can be rationalized as the higher stability of **1ZN**⁻ anion of TFHA is due to larger charge holding ability of sulfur and smaller relative stability of **1ZN**⁻ with respect to **1Z** with C-S π bond being replaced by C-N π bond.

Acidic as well as the basic characteristics of TFHA are stronger than FHA as evident from the smaller deprotonation enthalpies and higher proton affinities of all sites of TFHA over FHA. The H-bonding interactions depend not only on the basicity of H-bond acceptor but also on H-bond donor ability. In order to understand the correlation between H-bond energies with the acidity and basicity of the sites, intermolecular H-bonding between various tautomeric forms is studied.

Table 5.20: The second order delocalization energies $E^{(2)}$ in kcal/mol for the various orbital interactions present in neutral and deprotonated species of TFHA (FHA) obtained by NBO analysis at MP2/6-31+G* [L3] theoretical level. (The interactions of relevant neutral molecules are also included for comparison.

Species	$n_{N2} \to \sigma^*_{C1-X3}$	$n_{N2} \to \pi^*_{C1-X3}$	$n_{X3} \to \sigma^*_{O4-H5}$	$n_{X3} \to \sigma^*_{C1-N2}$	$n_{X3} \to \pi^*_{C1-N2}$	$n_{O4} \to \sigma^*_{C1-N2}$	$n_{O4} \to \pi^*_{C1-N2}$
1Z	-	145.34	13.71	15.92	-	6.56	-
	(14.98)	(32.68)	(5.36)	(29.25)	(-)	(-)	(-)
1ZN⁻	14.11	-	20.34	-	55.92	-	16.72
	(12.55)	(-)	(11.21)	(18.75)	(119.51)	(-)	(9.04)
1ZO⁻	-	-	-	16.71	85.29	11.04	51.77
	(-)	(-)	(-)	(-)	(27.57+168.14)	(10.48)	(26.34)
1E	-	49.47	-	18.93	-	-	-
	(12.11)	(20.90)	(-)	(32.25)	(-)	(-)	(-)
1EN⁻	-	-	-	12.89	65.50	-	7.94
	(-)	(-)	(-)	(20.91+11.19)	(8.39+91.27)	(-)	(-)
1EO⁻	-	-	-	13.01	67.16	-	-
	(-)	(-)	(-)	(24.29)	(147.78)	(-)	(20.21)
2Z	14.57	-	-	39.47	-	18.29	-
	(13.50)	(-)	(-)	(8.63)	(54.42)	(-)	(13.22)
2ZX3⁻	13.70	-	-	14.29	72.12	-	11.86
	(13.13)	(-)	(-)	(23.40)	(151.70)	(-)	(6.74)
2E	-	-	-	-	34.78	-	16.87
	(-)	(-)	(-)	(6.12)	(48.37)	(-)	(13.80)
2EX3⁻	-	-	-	12.86	65.34	-	7.94
	(-)	(-)	(-)	(20.91+11.19)	(8.39+91.27)	(-)	(-)
2EO4⁻	-	-	-	-	15.85	8.59	83.92
	(7.33)	(-)	(-)	(-)	(24.92)	(7.92)	(59.42)

Table 5.21: Atomic charges (NPA) in TFHA (FHA) at MP2/6-31+G* **[L3]** theoretical level.

Species	C1	N2	X3	O4	H5	H6	H7
1Z	-0.019	-0.347	-0.251	-0.603	0.531	0.242	0.447
	(0.636)	(-0.426)	(-0.732)	(-0.616)	(0.535)	(0.173)	(0.430)
1ZN⁻	-0.113	-0.277	-0.594	-0.709	0.496	0.197	-
	(0.493)	(-0.433)	(-0.941)	(-0.744)	(0.507)	(0.118)	(-)
1ZO⁻	-0.086	-0.213	-0.487	-0.755	-	0.179	0.362
	(0.577)	(-0.353)	(-0.849)	(-0.812)	(-)	(0.086)	(0.351)
1E	-0.016	-0.412	-0.145	-0.616	0.518	0.240	0.431
	(0.661)	(-0.454)	(-0.685)	(-0.626)	(0.517)	(0.166)	(0.422)
1EN⁻	-0.045	-0.332	-0.561	-0.732	-0.486	0.184	-
	(0.564)	(-0.486)	(-0.896)	(-0.753)	(0.476)	(0.094)	(-)
1EO⁻	-0.039	-0.221	-0.551	-0.782	-	0.207	0.386
	(0.594)	(-0.370)	(-0.891)	(-0.827)	(-)	(0.128)	(0.366)
2Z	-0.099	-0.197	0.025	-0.677	0.176	0.248	0.524
	(0.478)	(-0.282)	(-0.783)	(-0.703)	(0.547)	(0.214)	(0.528)
2ZX3⁻	-0.068	-0.357	-0.557	-0.684	-	0.188	0.477
	(0.551)	(-0.501)	(-0.890)	(-0.721)	(-)	(0.093)	(0.468)
2E	-0.082	-0.217	0.046	-0.651	0.150	0.238	0.516
	(0.488)	(-0.312)	(-0.772)	(-0.650)	(0.524)	(0.206)	(0.516)
2EX3⁻	-0.045	-0.332	-0.562	-0.732	-	0.184	0.486
	(0.564)	(-0.486)	(-0.896)	(-0.752)	(-)	(0.094)	(0.476)
2EO4⁻	-0.320	-0.102	-0.106	-0.792	0.133	0.188	-
	(0.253)	(-0.224)	(-0.842)	(-0.831)	(0.495)	(0.150)	(-)

5.2.3 DIMERIZATION AND THE STABILIZATION ENERGIES:

An extensive conformational search on the potential energy surface (PES) of dimeric thioformohydroxamic acid $(H(C=S)NHOH)_2$ in the gas phase has been performed at **[L1]**, **[L3]** and **[L5]** theoretical levels and a rich variety of H-bonded stationary points are located. Twenty one different dimeric structures of TFHA are optimized out of which twelve are thione-thione, four are thiol-thiol and five are thione-thiol dimers. Of the nine thione homodimers, four comprise of association of **1Z** conformer with **1Z** and five are homodimers of **1E**. There are two homodimers each of **2Z** and **2E** and the rest eight are mixed dimers. The optimized structures of all the dimers are shown in Figure 5.1 and the corresponding aggregation energies (S.E.) at **[L5]** theoretical level are reported in Table 5.22. The geometrical data analysis of the dimers suggests that all the dimers except one (**1E-1E (5)**) are bonded through two H-bonds.

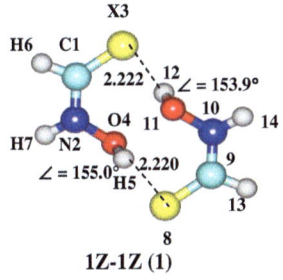

1Z-1Z (1)

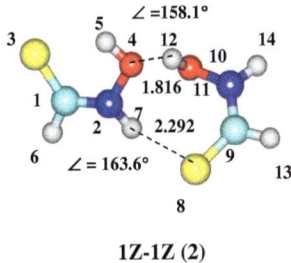

1Z-1Z (2)

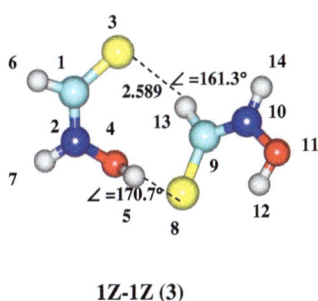

1Z-1Z (3)

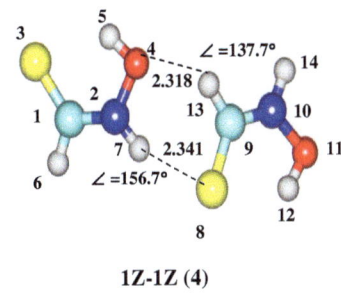

1Z-1Z (4)

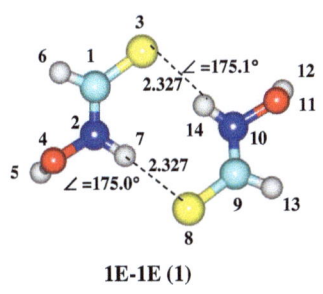

1E-1E (1)

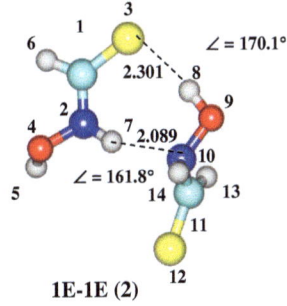

1E-1E (2)

Contd…..

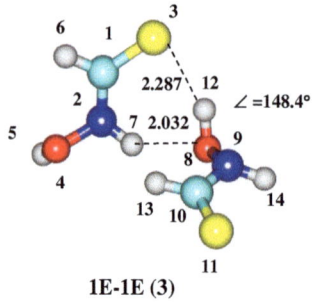

1E-1E (3)

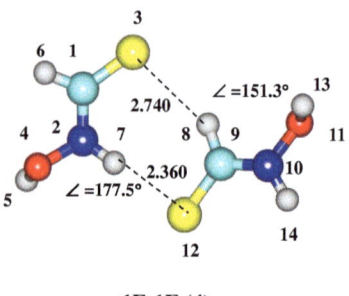

1E-1E (4)

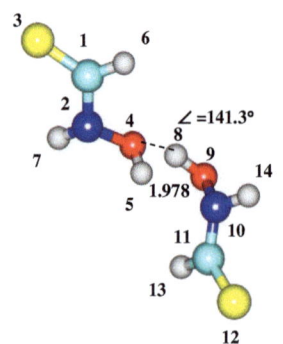

1E-1E (5)

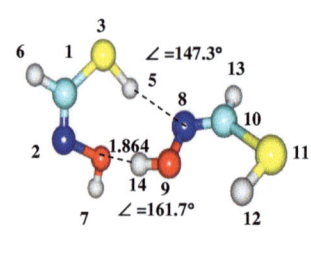

2Z-2Z (1)

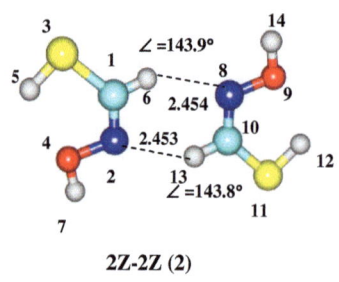

2Z-2Z (2)

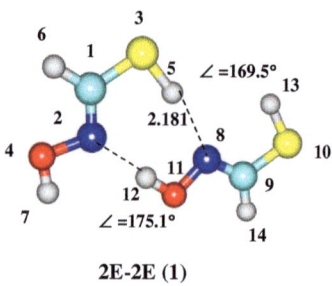

2E-2E (1)

Contd…..

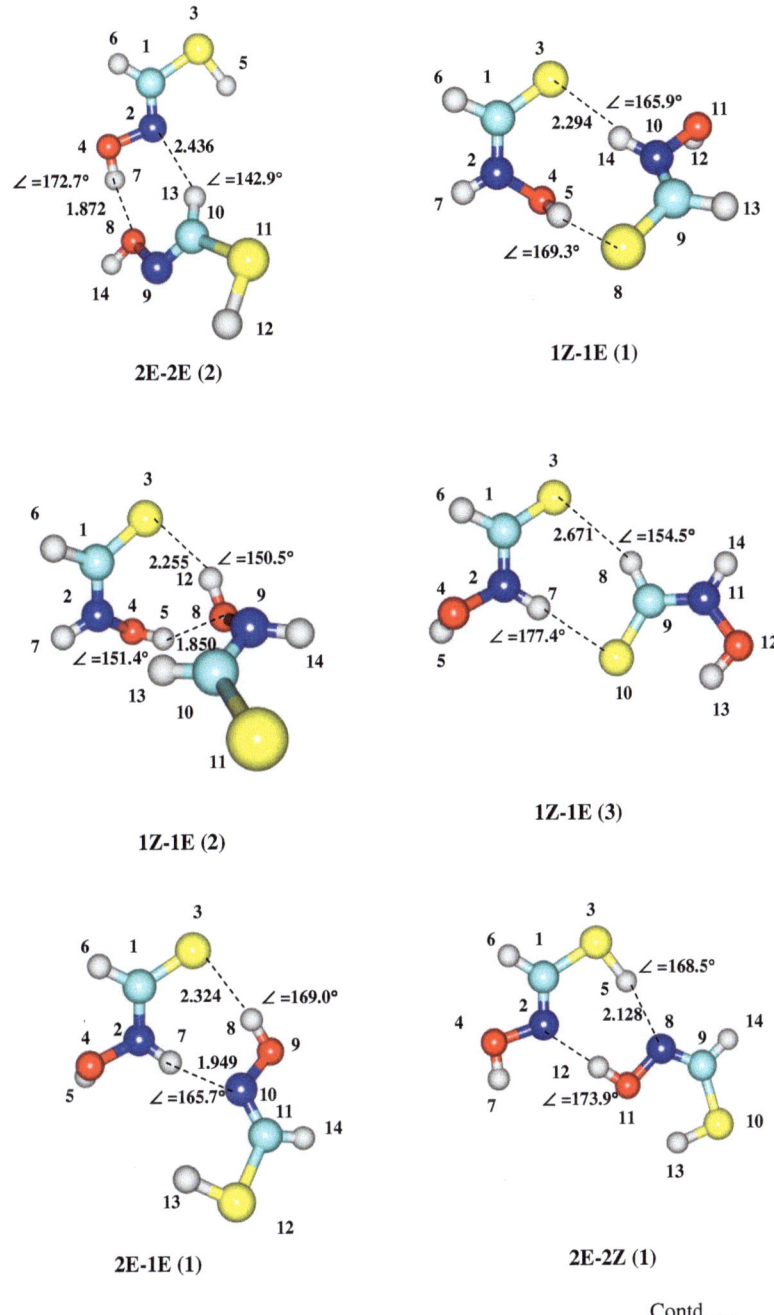

2E-2E (2)　　1Z-1E (1)

1Z-1E (2)　　1Z-1E (3)

2E-1E (1)　　2E-2Z (1)

Contd.....

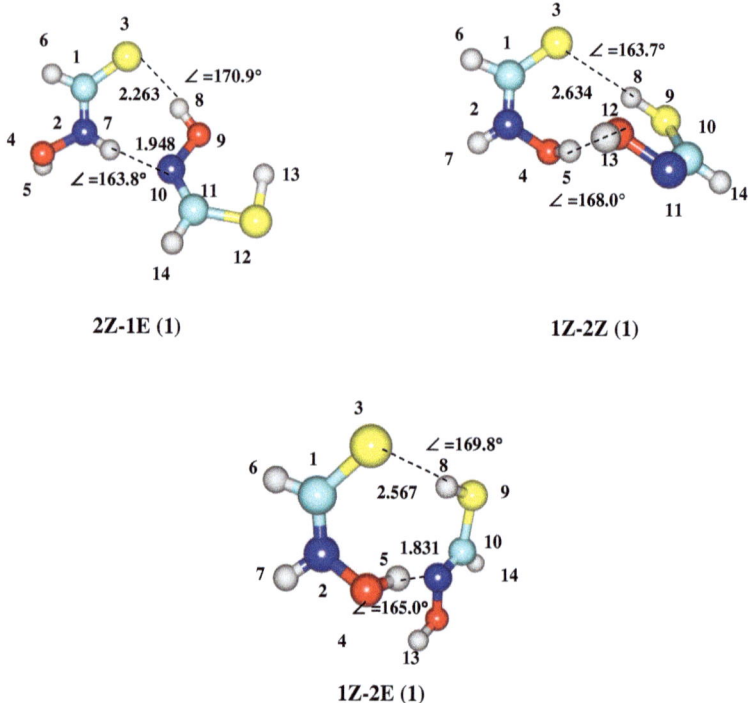

Figure 5.1: The optimized dimers of TFHA. H-bond distances and A-H...B angle are also included. A-H is H-bond donor and B is H-bond acceptor.

Table 5.22: Stabilization energy (S.E.), distortion energy (E_{Dis}), gas phase stabilization energy $(S.E.)_g$ that includes distortion energy, free energy of solvation (ΔG_{sol}), free energy of association in medium ($\Delta\Delta G_{Assoc}$) and stabilization energy in aqueous phase $(S.E.)_{aq}$ of TFHA (FHA) dimers in aqueous solution at MP2/6-311++G**//MP2/6-31+G* [L5] theoretical level. All energies are in kcal/mol.

Sr. No.	Species	S.E. (-ive)	E_{Dis} (-ive)	$(S.E.)_g$ (-ive)	ΔG_{sol} (-ive)	$\Delta\Delta G_{Assoc}$[a]	$(S.E.)_{aq}$[b]
1	1Z	-	-	-	9.51 (11.13)	-	-
2	1E	-	-	-	10.24 (11.21)	-	-
3	2Z	-	-	-	6.52 (9.80)	-	-
4	2E	-	-	-	7.26 (11.77)	-	-
5	1Z-1Z (1)	13.36 (15.52)	8.89 (3.35)	22.25 (18.87)	10.67 (12.91)	8.35 (9.35)	-13.90 (-9.52)
6	1Z-1Z (2)	11.24 (12.51)	4.00 (1.74)	15.24 (14.25)	9.76 (11.95)	9.26 (10.31)	-5.98 (-3.94)
7	1Z-1Z (3)	8.99 (11.02)	3.86 (2.01)	12.85 (13.03)	9.13 (16.27)	9.89 (5.99)	-2.96 (-7.04)
8	1Z-1Z (4)	7.37 (8.18)	0.25 (0.74)	7.62 (8.92)	10.60 (13.67)	8.42 (8.59)	0.80 (-0.33)
9	1E-1E (1)	10.34 (12.22)	1.90 (1.53)	12.24 (13.75)	10.39 (11.86)	10.09 (10.56)	-2.15 (-3.19)
10	1E-1E (2)	9.05	0.85	9.90	9.14	11.34	1.44
11	1E-1E (3)	7.15 (7.69)	0.79 (0.50)	7.94 (8.19)	10.54 (13.67)	10.09 (8.75)	2.54 (0.56)
12	1E-1E (4)	6.42 (7.38)	0.75 (0.61)	7.17 (7.99)	14.98 (14.85)	5.50 (7.57)	-1.67 (-0.42)
13	1E-1E (5)	4.73	0.24	4.97	14.01	6.47	1.74
14	2Z-2Z (1)	5.92	0.69	6.61	5.00	8.04	1.43
15	2Z-2Z (2)	3.21 (3.12)	0.06 (0.10)	3.27 (3.22)	7.57 (13.23)	5.47 (6.37)	2.20 (3.15)
16	2E-2E (1)	8.09 (12.20)	0.25 (1.02)	8.34 (13.22)	4.62 (11.60)	9.90 (11.94)	1.56 (-1.28)
17	2E-2E (2)	5.63 (5.46)	0.16 (0.19)	5.79 (5.65)	7.19 (16.45)	7.33 (7.09)	1.70 (1.44)
18	1Z-1E (1)	13.45 (15.29)	5.34 (2.91)	18.79 (18.20)	7.66 (12.20)	12.09 (10.14)	-6.70 (-8.06)
19	1Z-1E (2)	8.56 (9.19)	3.81 (1.58)	12.37 (10.77)	10.02 (11.58)	9.73 (10.76)	-2.64 (-0.01)
20	1Z-1E (3)	7.11	0.79	7.90	12.58	7.17	-0.73

Contd…..

Sr. No.	Species	S.E. (-ive)	E_{Dis} (-ive)	$(S.E.)_g$ (-ive)	ΔG_{sol} (-ive)	$\Delta\Delta G_{Assoc}{}^a$	$(S.E.)_{aq}{}^b$
		(8.37)	(0.70)	(9.07)	(14.36)	(7.98)	(-1.09)
21	2E-1E (1)	9.92	0.93	10.85	6.69	10.81	-0.04
		(10.00)	(0.64)	(10.64)	(13.16)	(9.82)	(-0.82)
22	2E-2Z (1)	9.14	0.60	9.74	4.68	9.10	-0.64
		(14.36)	(1.38)	(15.74)	(8.43)	(13.14)	(-2.60)
23	2Z-1E (1)	11.32	1.03	12.35	6.74	10.02	-2.33
		(12.35)	(1.11)	(13.46)	(10.20)	(10.81)	(-2.65)
24	1Z-2Z (1)	6.18	3.35	9.53	9.91	6.12	-3.41
		(11.60)	(3.24)	(14.84)	(13.48)	(7.45)	(-7.39)
25	1Z-2E (1)	8.88	3.42	12.30	10.64	6.13	-6.17
		(15.77)	(2.99)	(18.76)	(13.21)	(9.69)	(-9.07)

$\Delta\Delta G_{Assoc}{}^a = (\Delta G_{sol})_{dimer} - [(\Delta G_{sol})_{monomer1} + (\Delta G_{sol})_{monomer2}]$
$(S.E.)_{aq}{}^b = (S.E.)_g + \Delta\Delta G_{Assoc}$

The geometrical parameters of the optimized dimers of TFHA at B3LYP/6-31+G* [L1] and MP2/6-31+G* [L3] theoretical levels are listed in Table 5.23-5.29.

Table 5.23: Geometrical parameters of dimers of TFHA at B3LYP/6-31+G* [L1] and (MP2/6-31+G*) [L3] theoretical levels. All the bond distances are in angstrom (Å), angles and dihedrals are in degrees (°).

Parameters	1Z-1Z (1)	Parameters	1Z-1Z (2)	Parameters	1Z-1Z (3)
N2-C1	1.336 (1.345)	N2-C1	1.325 (1.331)	N2-C1	1.337 (1.346)
S3-C1	1.659 (1.643)	S3-C1	1.670 (1.655)	S3-C1	1.650 (1.640)
O4-N2	1.379 (1.389)	O4-N2	1.390 (1.398)	O4-N2	1.370 (1.380)
H5-O4	0.997 (0.996)	H5-O4	1.002 (1.005)	H5-O4	0.990 (0.990)
H6-C1	1.093 (1.093)	H6-C1	1.091 (1.091)	H6-C1	1.090 (1.090)
H7-N2	1.015 (1.018)	H7-N2	1.030 (1.031)	H7-N2	1.010 (1.010)
S8-H5	2.244 (2.220)	S8-H7	2.336 (2.292)	S8-H5	2.220 (2.200)
C9-S8	1.660 (1.643)	C9-S8	1.660 (1.644)	C9-S8	1.670 (1.660)
N10-C9	1.336 (1.345)	N10-C9	1.335 (1.342)	N10-C9	1.310 (1.320)
O11-N10	1.378 (1.389)	O11-N10	1.380 (1.389)	O11-N10	1.380 (1.380)
H12-O11	0.998 (0.996)	H12-O11	0.989 (0.990)	H12-O11	0.980 (0.990)
H13-C9	1.093 (1.093)	H13-C9	1.094 (1.093)	H13-C9	1.090 (1.090)
H14-N10	1.015 (1.018)	H14-N10	1.014 (1.016)	H14-N10	1.010 (1.010)
S3-C1-N2	128.9 (128.0)	S3-C1-N2	123.4 (123.5)	S3-C1-N2	129.1 (128.3)
O4-N2-C1	121.9 (119.8)	O4-N2-C1	119.8 (119.0)	O4-N2-C1	123.7 (121.4)
H5-O4-N2	108.4 (106.7)	H5-O4-N2	102.0 (101.1)	H5-O4-N2	106.7 (105.0)
H6-C1-N2	110.5 (110.7)	H6-C1-N2	112.6 (112.5)	H6-C1-N2	109.8 (110.1)
H7-N2-C1	120.6 (119.7)	H7-N2-C1	128.8 (128.3)	H7-N2-C1	121.4 (120.6)
S8-H5-O4	151.4 (154.5)	S8-H7-N2	162.7 (163.6)	S8-H5-O4	171.4 (170.7)
C9-S8-H5	96.7 (93.2)	C9-S8-H7	106.7 (104.1)	C9-S8-H5	91.3 (86.7)
N10-C9-S8	128.8 (128.0)	N10-C9-S8	129.3 (128.6)	N10-C9-S8	123.3 (123.5)
O11-N10-C9	122.0 (119.7)	O11-N10-C9	124.0 (121.7)	O11-N10-C9	123.4 (123.0)
H12-O11-N10	108.4 (106.7)	H12-O11-N10	107.2 (105.6)	H12-O11-N10	103.1 (102.5)
H13-C9-N10	110.5 (110.7)	H13-C9-N10	110.2 (110.5)	H13-C9-S8	122.9 (123.1)
H14-N10-C9	120.5 (119.7)	H14-N10-C9	121.5 (121.0)	H14-N10-C9	125.4 (126.0)
O4-N2-C1-S3	-14.2 (-16.5)	O4-N2-C1-S3	-2.2 (-5.2)	O4-N2-C1-S3	-9.8 (-13.8)

Contd.....

Parameters	1Z-1Z (1)	Parameters	1Z-1Z (2)	Parameters	1Z-1Z (3)
H5-O4-N2-C1	107.9 (114.6)	H5-O4-N2-C1	-0.4 (4.4)	H5-O4-N2-C1	88.4 (97.2)
H6-C1-N2-O4	166.9 (165.2)	H6-C1-N2-O4	178.1 (176.1)	H6-C1-N2-O4	171.4 (168.3)
H7-N2-C1-S3	-169.6 (-163.4)	H7-N2-C1-S3	-171.8 (-166.7)	H7-N2-C1-S3	-172.6 (-165.4)
S8-H5-O4-N2	178.3 (172.3)	S8-H7-N2-O4	5.2 (15.7)	S8-H5-O4-N2	-115.9 (-91.7)
C9-S8-H5-O4	72.2 (75.7)	C9-S8-H7-N2	-19.7 (-38.2)	C9-S8-H5-O4	32.7 (1.3)
N10-C9-S8-H5	32.4 (30.1)	N10-C9-S8-H7	-23.1 (-19.3)	N10-C9-S8-H5	-171.0 (-162.8)
O11-N10-C9-S8	14.4 (16.5)	O11-N10-C9-S8	-11.9 (-14.7)	O11-N10-C9-S8	0.3 (0.5)
H12-O11-N10-S3	1.6 (2.8)	H12-O11-N10-O4	10.3 (14.8)	H12-O11-N10-C9	0.7 (2.8)
H13-C9-N10-O11	193.3 (194.8)	H13-C9-N10-O11	169.9 (167.8)	H13-C9-S8-S3	-11.4 (-8.8)
H14-N10-C9-S8	169.8 (163.2)	H14-N10-C9-S8	186.9 (192.9)	H14-N10-C9-H13	-1.2 (-0.8)

Table 5.24: Geometrical parameters of dimers of TFHA at B3LYP/6-31+G* [L1] and (MP2/6-31+G*) [L3] theoretical levels. All the bond distances are in angstrom (Å), angles and dihedrals are in degrees (°).

Parameters	1Z-1Z (4)	Parameters	1E-1E (1)	Parameters	1E-1E (2)
N2-C1	1.300 (1.331)	N2-C1	1.328 (1.338)	N2-C1	1.334 (1.343)
S3-C1	1.663 (1.656)	S3-C1	1.660 (1.650)	S3-C1	1.663 (1.646)
O4-N2	1.359 (1.393)	O4-N2	1.400 (1.410)	O4-N2	1.399 (1.410)
H5-O4	0.959 (1.000)	H5-O4	0.970 (0.978)	H5-O4	0.975 (0.979)
H6-C1	1.078 (1.091)	H6-C1	1.090 (1.092)	H6-C1	1.091 (1.091)
H7-N2	1.001 (1.026)	H7-N2	1.030 (1.033)	H7-N2	1.022 (1.026)
S8-H7	2.590 (2.341)	S8-H7	2.350 (2.310)	H8-S3	2.281 (2.301)
C9-S8	1.668 (1.662)	C9-S8	1.660 (1.640)	O9-H8	0.995 (0.993)
N10-C9	1.298 (1.325)	C10-C9	1.320 (1.330)	N10-O9	1.400 (1.414)
O11-N10	1.355 (1.384)	O11-N10	1.400 (1.410)	C11-N10	1.355 (1.372)
H12-O11	0.958 (0.994)	H12-O11	0.970 (0.970)	S12-C11	1.650 (1.631)
H13-C9	1.077 (1.091)	H13-C9	1.090 (1.090)	H13-C11	1.092 (1.091)
H14-N10	0.995 (1.014)	H14-N10	1.030 (1.030)	H14-N10	1.019 (1.022)
S3-C1-N2	127.5 (123.7)	S3-C1-N2	125.7 (125.5)	S3-C1-N2	125.5 (125.3)

Contd.....

Parameters	1Z-1Z (4)	Parameters	1E-1E (1)	Parameters	1E-1E (2)
O4-N2-C1	122.7 (120.0)	O4-N2-C1	118.4 (116.9)	O4-N2-C1	118.9 (117.1)
H5-O4-N2	105.5 (101.3)	H5-O4-N2	105.2 (104.3)	H5-O4-N2	105.3 (104.3)
H6-C1-N2	112.2 (112.4)	H6-C1-N2	112.8 (112.8)	H6-C1-N2	112.7 (112.7)
H7-N2-C1	125.9 (126.9)	H7-N2-C1	123.9 (122.8)	H7-N2-C1	121.8 (120.3)
S8-H7-N2	151.6 (156.7)	S8-H7-N2	175.1 (175.1)	H8-S3-C1	105.3 (104.0)
C9-S8-H7	91.9 (86.7)	C9-S8-H7	104.5 (103.5)	O9-H8-S3	171.7 (170.2)
N10-C9-S8	125.1 (123.4)	N10-C9-S8	125.7 (125.6)	N10-O9-H8	104.3 (103.1)
O11-N10-C9	124.2 (122.8)	O11-N10-C9	118.5 (116.8)	C11-N10-O9	118.3 (115.6)
H12-O11-N10	106.1 (102.6)	H12-O11-N10	105.2 (104.3)	S12-C11-N10	124.4 (124.0)
H13-C9-S8	122.0 (123.3)	H13-C9-N10	122.8 (112.7)	H13-C11-N10	111.9 (112.2)
H14-N10-C9	124.4 (126.0)	H14-N10-C9	123.9 (122.6)	H14-N10-C11	118.2 (116.4)
O4-N2-C1-S3	0.0 (-5.3)	O4-N2-C1-S3	170.0 (167.1)	O4-N2-C1-S3	167.4 (163.8)
H5-O4-N2-C1	0.1 (3.6)	H5-O4-N2-C1	115.2 (116.5)	H5-O4-N2-C1	117.2 (118.9)
H6-C1-N2-O4	180.0 (170.6)	H6-C1-N2-O4	-11.1 (-15.1)	H6-C1-N2-O4	-14.1 (-18.7)
H7-N2-C1-S3	-181.1 (-166.9)	H7-N2-C1-S3	12.9 (20.1)	H7-N2-C1-S3	13.3 (18.8)
S8-H7-N2-O4	-0.0 (-26.0)	O8-H7-N2-O4	28.3 (-0.8)	H8-S3-C1-N2	-2.1 (-7.4)
C9-S8-H7-N2	-0.0 (4.5)	C9-S8-H7-N2	160.2 (-173.2)	O9-H8-S3-C1	3.5 (2.1)
N10-C9-S8-H7	179.8 (185.2)	N10-C9-S8-H7	-3.9 (-4.5)	N10-O9-H8-H7	-5.6 (-3.7)
O11-N10-C9-S8	-0.0 (-0.1)	O11-N10-C9-S8	170.1 (167.0)	C11-N10-O9-H8	113.6 (114.4)
H12-O11-N10-C9	0.2 (0.6)	O12-O11-N10-C9	115.0 (116.6)	S12-C11-N10-O9	165.7 (162.4)
H13-C9-S8-O4	-0.2 (-2.5)	H13-C9-N10-O11	-11.0 (-15.1)	H13-C11-N10-O9	-17.5 (-22.1)
H14-N10-C9-S8	180.0 (180.8)	H14-N10-C9-S3	0.4 (-2.5)	H14-N10-C11-H13	-159.6 (-154.4)

Table 5.25: Geometrical parameters of dimers of TFHA at B3LYP/6-31+G* **[L1]** and (MP2/6-31+G*) **[L3]** theoretical levels. All the bond distances are in angstrom (Å), angles and dihedrals are in degrees (°).

Parameters	1E-1E (3)	Parameters	1E-1E (4)	Parameters	1E-1E (5)
N2-C1	1.336 (1.347)	N2-C1	1.336 (1.346)	N2-C1	1.359 (1.370)
S3-C1	1.663 (1.646)	S3-C1	1.661 (1.643)	S3-C1	1.641 (1.628)
O4-N2	1.400 (1.412)	O4-N2	1.404 (1.414)	O4-N2	1.413 (1.421)
H5-O4	0.975 (0.979)	H5-O4	0.975 (0.978)	H5-O4	0.974 (0.979)
H6-C1	1.091 (1.091)	H6-C1	1.092 (1.092)	H6-C1	1.092 (1.092)
H7-N2	1.022 (1.025)	H7-N2	1.030 (1.030)	H7-N2	1.018 (1.020)
O8-H7	2.056 (2.032)	S8-H7	2.801 (2.740)	H8-O4	1.911 (1.978)
N9-O8	1.400 (1.414)	C9-S8	1.094 (1.093)	O9-H8	0.980 (0.982)
C10-N9	1.346 (1.361)	N10-C9	1.338 (1.349)	N10-O9	1.402 (1.418)
S11-C10	1.650 (1.633)	O11-N10	1.400 (1.410)	C11-N10	1.348 (1.368)
H12-O8	0.996 (0.993)	H12-O11	1.660 (1.642)	S12-C11	1.650 (1.631)
H13-C10	1.093 (1.093)	H13-C9	0.975 (0.978)	H13-C11	1.092 (1.092)
H14-N9	1.016 (1.020)	H14-N10	1.016 (1.019)	H14-N10	1.017 (1.021)
S3-C1-N2	125.7 (125.6)	S3-C1-N2	125.4 (125.2)	S3-C1-N2	123.9 (124.0)
O4-N2-C1	118.8 (116.7)	O4-N2-C1	118.4 (116.5)	O4-N2-C1	117.3 (115.0)
H5-O4-N2	105.3 (104.3)	H5-O4-N2	105.1 (104.1)	H5-O4-N2	105.7 (105.3)
H6-C1-N2	112.5 (112.5)	H6-C1-N2	112.3 (112.3)	H6-C1-N2	112.1 (112.2)
H7-N2-C1	122.2 (120.4)	H7-N2-C1	123.1 (121.5)	H7-N2-C1	118.9 (117.0)
O8-H7-N2	143.3 (145.4)	S8-H7-N2	102.7 (101.9)	H8-O4-N2	127.2 (146.4)
N9-O8-H7	144.6 (125.7)	C9-S8-H7	150.6 (151.3)	O9-H8-O4	164.2 (141.3)
C10-N9-O8	120.0 (117.1)	N10-C9-S8	113.7 (113.4)	N10-O9-H8	104.1 (102.7)
S11-C10-N9	124.5 (124.4)	O11-N10-C9	119.7 (117.5)	C11-N10-O9	119.0 (115.8)
H12-O8-H7	85.5 (84.6)	H12-O11-N10	122.6 (122.7)	S12-C11-N10	124.6 (124.2)
H13-C10-N9	112.2 (112.4)	H13-C9-N10	105.4 (104.3)	H13-C11-N10	112.0 (112.0)
H14-N9-C10	120.3 (118.2)	H14-N10-C9	121.3 (119.7)	H14-N10-C11	119.4 (116.7)
O4-N2-C1-S3	167.9 (164.0)	O4-N2-C1-S3	168.9 (165.0)	O4-N2-C1-S3	161.5 (157.4)

Contd.....

Parameters	1E-1E (3)	Parameters	1E-1E (4)	Parameters	1E-1E (5)
H5-O4-N2-C1	111.5 (115.3)	H5-O4-N2-C1	113.6 (116.3)	H5-O4-N2-C1	116.6 (119.1)
H6-C1-N2-O4	-13.6 (-18.8)	H6-C1-N2-O4	-12.8 (-17.9)	H6-C1-N2-O4	-21.2 (-26.4)
H7-N2-C1-S3	14.1 (21.2)	H7-N2-C1-S3	15.0 (21.7)	H7-N2-C1-S3	21.2 (25.5)
O8-H7-N2-O4	-173.1 (-151.1)	O8-H7-N2-O4	-5.3 (-16.2)	O8-O4-N2-C1	-96.4 (-89.3)
N9-O8-H7-N2	125.7 (95.9)	C9-S8-H7-N2	-21.6 (-15.7)	O9-H8-O4-N2	-147.7 (-158.0)
C10-N9-O8-H7	0.1 (15.5)	N10-C9-S8-H7	205.0 (209.4)	N10-O9-H8-H5	-16.4 (-6.4)
S11-C10-N9-O8	165.2 (159.3)	O11-N10-C9-S8	-15.8 (-21.2)	C11-N10-O9-H8	111.6 (115.8)
H12-O8-H7-S3	1.5 (5.1)	H12-O11-N10-C9	3.8 (6.4)	S12-C11-N10-C9	165.3 (160.3)
H13-C10-N9-O8	-17.0 (-24.4)	H13-C9-N10-O11	104.1 (109.6)	H13-C11-N10-O9	-17.0 (-23.5)
H14-N9-C10-S11	15.4 (22.1)	H14-N10-C9-S3	192.3 (197.2)	H14-N10-C11-H13	-163.1 (-156.7)

Table 5.26: Geometrical parameters of dimers of TFHA at B3LYP/6-31+G* [**L1**] and (MP2/6-31+G*) [**L3**] theoretical levels. All the bond distances are in angstrom (Å), angles and dihedrals are in degrees (°).

Parameters	2Z-2Z (1)	Parameters	2Z-2Z (2)	Parameters	2E-2E (1)
N2-C1	1.281 (1.293)	N2-C1	1.281 (1.294)	N2-C1	1.279 (1.290)
S3-C1	1.758 (1.742)	S3-C1	1.764 (1.746)	S3-C1	1.753 (1.741)
O4-N2	1.425 (1.435)	O4-N2	1.412 (1.422)	O4-N2	1.414 (1.424)
H5-S3	1.355 (1.346)	H5-S3	1.348 (1.341)	H5-S3	1.364 (1.353)
H6-C1	1.087 (1.086)	H6-C1	1.088 (1.087)	H6-C1	1.091 (1.091)
H7-O4	0.971 (0.977)	H7-O4	0.970 (0.975)	H7-O4	0.971 (0.976)
N8-H5	2.253 (2.290)	N8-H6	2.503 (2.454)	N8-H5	2.164 (2.181)
O9-N8	1.399 (1.411)	O9-N8	1.412 (1.422)	C9-N8	1.277 (1.289)
C10-N8	1.281 (1.294)	C10-N8	1.281 (1.294)	S10-C9	1.763 (1.749)
S11-C10	1.765 (1.747)	S11-C10	1.764 (1.746)	O11-N8	1.399 (1.412)
H12-S11	1.349 (1.341)	H12-S11	1.348 (1.341)	H12-O11	0.989 (0.990)
H13-C10	1.086 (1.085)	H13-C10	1.088 (1.087)	H13-S10	1.351 (1.343)
H14-O9	0.981 (0.985)	H14-O9	0.970 (0.975)	H14-C9	1.091 (1.091)
S3-C1-N2	132.3 (131.4)	S3-C1-N2	129.3 (129.1)	S3-C1-N2	123.7 (123.2)

Contd.....

Parameters	2Z-2Z (1)	Parameters	2Z-2Z (2)	Parameters	2E-2E (1)
O4-N2-C1	112.5 (110.7)	O4-N2-C1	112.4 (111.2)	O4-N2-C1	111.0 (109.6)
H5-S3-C1	99.1 (98.2)	H5-S3-C1	96.0 (95.8)	H5-S3-C1	96.4 (95.8)
H6-C1-N2	114.1 (113.8)	H6-C1-N2	115.5 (114.9)	H6-C1-N2	121.1 (120.7)
H7-O4-N2	102.3 (101.7)	H7-O4-N2	103.0 (102.2)	H7-O4-N2	103.5 (102.8)
N8-H5-S3	156.2 (147.3)	N8-H6-C1	144.7 (143.9)	N8-H5-S3	167.9 (169.5)
O9-N8-H5	100.9 (99.2)	O9-N8-H6	146.2 (147.0)	C9-N8-H5	142.6 (143.4)
C10-N8-H5	143.4 (143.8)	C10-N8-H6	101.3 (99.9)	C10-C9-O8	123.3 (122.5)
S11-C10-N8	129.0 (128.8)	S11-C10-N8	129.3 (129.1)	O11-N8-C9	111.8 (110.3)
H12-S11-C10	95.5 (95.1)	H12-S11-C10	96.0 (95.8)	H12-O11-N8	103.3 (102.2)
H13-C10-N8	116.0 (115.4)	H13-C10-N8	115.5 (114.9)	H13-S10-C9	96.0 (95.3)
H14-O9-N8	102.7 (101.6)	H14-O9-N8	103.1 (102.2)	H14-C9-N8	122.0 (121.7)
O4-N2-C1-S3	0.1 (0.7)	O4-N2-C1-S3	-0.0 (-0.2)	O4-N2-C1-S3	180.0 (180.0)
H5-S3-C1-N2	-3.6 (-1.3)	H5-S3-C1-N2	0.1 (-2.8)	H5-S3-C1-N2	0.0 (0.1)
H6-C1-N2-O4	180.0 (180.7)	H6-C1-N2-O4	180.0 (179.0)	H6-C1-N2-O4	0.0 (0.0)
H7-O4-N2-C1	190.3 (192.8)	H7-O4-N2-C1	180.0 (177.7)	H7-O4-N2-C1	180.0 (180.0)
N8-H5-S3-C1	246.0 (252.3)	N8-H6-C1-N2	-4.3 (-16.4)	N8-H5-S3-C1	-0.9 (0.2)
O9-H8-H5-S3	90.5 (76.3)	O9-H8-H6-C1	183.7 (176.2)	C9-N8-H5-S3	181.0 (179.7)
C10-N8-H5-S3	292.0 (287.7)	C10-N8-H6-C1	5.7 (15.4)	S10-C9-N8-H5	-0.3 (-0.2)
S11-C10-N8-H5	157.3 (146.7)	S11-C10-N8-H6	178.8 (169.1)	S11-N8-C9-S10	180.0 (180.0)
H12-S11-C10-N8	-0.6 (1.9)	H12-S11-C10-N8	-0.1 (2.8)	H12-O11-N8-N2	0.0 (-0.0)
H13-C10-N8-O9	180.3 (180.7)	H13-C10-N8-N2	0.4 (-7.0)	H13-S10-C9-N8	-0.0 (-0.0)
H14-O9-N8-S3	25.4 (36.2)	H14-O9-N8-C10	180.1 (182.3)	H14-C9-N8-O11	0.0 (0.0)

Table 5.27: Geometrical parameters of dimers of TFHA at B3LYP/6-31+G* [L1] and (MP2/6-31+G*) [L3] theoretical levels. All the bond distances are in angstrom (Å), angles and dihedrals are in degrees (°).

Parameters	2E-2E (2)	Parameters	1Z-1E (1)	Parameters	1Z-1E (2)
N2-C1	1.276 (1.289)	N2-C1	1.333 (1.342)	N2-C1	1.334 (1.346)
S3-C1	1.764 (1.749)	S3-C1	1.660 (1.643)	S3-C1	1.664 (1.644)
O4-N2	1.404 (1.415)	O4-N2	1.377 (1.387)	O4-N2	1.381 (1.392)
H5-S3	1.351 (1.343)	H5-O4	1.005 (1.002)	H5-O4	0.988 (0.990)
H6-C1	1.092 (1.091)	H6-C1	1.093 (1.093)	H6-C1	1.093 (1.094)
H7-O4	0.980 (0.984)	H7-N2	1.014 (1.017)	H7-N2	1.013 (1.016)
O8-H7	1.894 (1.872)	S8-H5	2.156 (2.145)	O8-H5	1.821 (1.850)
N9-O8	1.428 (1.439)	C9-S8	1.670 (1.655)	N9-O8	1.400 (1.414)
C10-N9	1.277 (1.289)	N10-C9	1.328 (1.333)	C10-N9	1.346 (1.360)
S11-C10	1.760 (1.746)	O11-N10	1.401 (1.408)	S11-C10	1.649 (1.633)
H12-O11	1.351 (1.343)	H12-O11	0.975 (0.979)	H12-O8	0.999 (0.994)
H13-C10	1.093 (1.092)	H13-C9	1.091 (1.091)	H13-C10	1.093 (1.092)
H14-O8	0.970 (0.977)	H14-N10	1.036 (1.035)	H14-N9	1.016 (1.019)
S3-C1-N2	122.8 (122.0)	S3-C1-N2	128.9 (128.1)	S3-C1-N2	129.3 (128.5)
O4-N2-C1	111.2 (109.9)	O4-N2-C1	123.4 (121.1)	O4-N2-C1	125.2 (122.6)
H5-S3-C1	95.8 (95.1)	H5-O4-N2	107.2 (105.1)	H5-O4-N2	107.3 (105.1)
H6-C1-N2	122.4 (122.0)	H6-C1-N2	110.5 (110.7)	H6-C1-N2	110.2 (110.3)
H7-O4-N2	103.2 (102.4)	H7-N2-C1	121.5 (121.0)	H7-N2-C1	121.6 (120.7)
O8-H7-O4	175.1 (172.7)	S8-H5-O4	168.6 (169.3)	O8-H5-O4	155.8 (151.4)
N9-O8-H7	127.7 (115.7)	C9-S8-H5	97.9 (92.5)	N9-S8-H5	136.0 (125.9)
C10-N9-O8	110.3 (109.0)	N10-C9-S8	125.9 (125.7)	C10-N9-O8	120.3 (117.3)
O11-C10-N9	121.7 (121.1)	O11-N10-C9	117.9 (117.1)	S11-C10-N9	124.6 (124.5)
H12-O11-C10	95.8 (95.1)	H12-O11-N10	105.1 (104.3)	H12-O8-N9	107.5 (104.8)
H13-O10-N9	123.2 (122.7)	H13-C9-S8	121.2 (121.5)	H13-C10-N9	112.2 (112.2)
H14-O8-N9	103.0 (101.9)	H14-N10-C9	123.8 (124.0)	H14-C9-C10	120.6 (118.8)
O4-N2-C1-S3	180.0 (180.1)	O4-N2-C1-S3	-10.8 (-13.9)	O4-N2-C1-S3	-12.2 (-17.1)

Contd.....

Parameters	2E-2E (2)	Parameters	1Z-1E (1)	Parameters	1Z-1E (2)
H5-S3-C1-N2	0.7 (2.2)	H5-O4-N2-C1	88.7 (96.3)	H5-O4-N2-C1	65.6 (79.6)
H6-C1-N2-O4	-0.1 (0.0)	H6-C1-N2-O4	170.7 (168.5)	H6-C1-N2-O4	169.8 (165.9)
H7-O4-N2-C1	178.4 (178.2)	H7-N2-C1-S3	173.3 (-166.8)	H7-N2-C1-S3	-174.9 (-168.1)
O8-H7-O4-N2	-9.1 (23.1)	S8-H5-O4-N2	206.9 (234.6)	O8-H5-O4-N2	-82.7 (-64.1)
N9-O8-H7-O4	40.1 (43.4)	C9-S8-H5-O4	44.9 (6.0)	N9-O8-H5-O4	147.1 (106.4)
C10-N9-O8-H7	-21.8 (-46.8)	N10-C9-S8-H5	13.1 (18.0)	C10-N9-O8-H5	-5.1 (5.3)
S11-C10-N9-O8	180.4 (180.3)	O11-N10-C9-S8	171.1 (171.2)	S11-C10-N9-O8	165.8 (160.5)
H12-S11-C10-N9	0.5 (0.3)	H12-O11-N10-C9	117.7 (116.4)	H12-O8-N9-C10	17.0 (21.3)
H13-O10-N9-O4	-11.6 (-29.9)	H13-C9-S8-H5	194.4 (199.7)	H13-C10-N9-O8	-16.4 (-23.3)
H14-O8-N9-C10	181.6 (182.0)	H14-N10-C9-S3	-12.1 (-11.7)	H14-N9-C10-S11	15.4 (22.3)

Table 5.28: Geometrical parameters of dimers of TFHA at B3LYP/6-31+G* [L1] and (MP2/6-31+G*) [L3] theoretical levels. All the bond distances are in angstrom (Å), angles and dihedrals are in degrees (°).

Parameters	1Z-1E (3)	Parameters	2E-1E (1)	Parameters	2E-2Z (1)
N2-C1	1.334 (1.343)	N2-C1	1.333 (1.345)	N2-C1	1.280 (1.291)
S3-C1	1.662 (1.645)	S3-C1	1.664 (1.646)	S3-C1	1.752 (1.740)
O4-N2	1.402 (1.411)	O4-N2	1.404 (1.415)	O4-N2	1.416 (1.425)
H5-O4	0.975 (0.978)	H5-O4	0.975 (0.978)	H5-S3	1.366 (1.355)
H6-C1	1.092 (1.092)	H6-C1	1.092 (1.092)	H6-C1	1.091 (1.091)
H7-N2	1.029 (1.029)	H7-N2	1.033 (1.032)	H7-O4	0.971 (0.976)
H8-S3	2.731 (2.671)	H8-S3	2.323 (2.324)	N8-H5	2.102 (2.128)
C9-H8	1.094 (1.092)	O9-H8	0.991 (0.991)	C9-N8	1.281 (1.294)
S10-C9	1.676 (1.661)	N10-O9	1.396 (1.409)	S10-C9	1.764 (1.746)
N11-C9	1.321 (1.324)	C11-N10	1.277 (1.289)	O11-N8	1.392 (1.405)
O12-N11	1.381 (1.385)	S12-C11	1.762 (1.748)	H12-O11	0.991 (0.992)
H13-O12	0.990 (0.994)	H13-S12	1.351 (1.344)	H13-S10	1.349 (1.342)
H14-N11	1.011 (1.014)	H14-C11	1.091 (1.091)	H14-C9	1.086 (1.085)
S3-C1-N2	125.6 (125.5)	S3-C1-N2	124.8 (124.8)	S3-C1-N2	124.1 (123.6)

Contd.....

Parameters	1Z-1E (3)	Parameters	2E-1E (1)	Parameters	2E-2Z (1)
O4-N2-C1	118.6 (116.9)	O4-N2-C1	118.9 (116.7)	O4-N2-C1	110.8 (109.5)
H5-O4-N2	105.2 (104.2)	H5-O4-N2	105.1 (104.0)	H5-S3-C1	96.5 (96.0)
H6-C1-N2	112.3 (112.2)	H6-C1-N2	113.2 (113.1)	H6-C1-N2	120.8 (120.4)
H7-N2-C1	123.4 (122.4)	H7-N2-C1	122.2 (119.9)	H7-O4-N2	103.5 (102.9)
H8-S3-C1	102.6 (100.1)	N8-S3-C1	99.1 (99.7)	N8-H5-S3	167.1 (168.5)
C9-H8-S3	154.1 (154.5)	C9-H8-S3	171.1 (169.0)	C9-H8-H5	138.5 (139.8)
S10-C9-H8	122.9 (122.9)	N10-O9-H8	103.8 (102.4)	S10-C9-N8	128.7 (128.6)
N11-C9-H8	113.8 (113.6)	C11-N10-O9	112.1 (110.9)	O11-N8-C9	113.8 (112.1)
O12-N11-C9	123.2 (122.8)	S12-C11-N10	123.4 (122.4)	H12-O11-N8	103.3 (102.2)
H13-O12-N11	102.9 (102.3)	H13-S12-C11	96.2 (95.4)	H13-S10-C9	95.3 (95.1)
H14-N11-C9	125.6 (125.9)	H14-C11-N10	121.8 (121.5)	H14-C9-N8	116.2 (115.6)
O4-N2-C1-S3	169.3 (166.1)	O4-N2-C1-S3	169.5 (164.9)	O4-N2-C1-S3	180.0 (180.0)
H5-O4-N2-C1	113.0 (115.6)	H5-O4-N2-C1	108.0 (113.7)	H5-S3-C1-N2	0.1 (-0.0)
H6-C1-N2-O4	-12.2 (-16.8)	H6-C1-N2-O4	-12.1 (-17.8)	H6-C1-N2-O4	-0.0 (-0.0)
H7-N2-C1-S3	13.6 (19.7)	H7-N2-C1-S3	15.4 (23.4)	H7-O4-N2-C1	179.8 (180.1)
H8-S3-C1-N2	-5.7 (-22.3)	N8-S3-C1-N2	-3.5 (-5.2)	N8-H5-S3-C1	-5.3 (0.3)
C9-H8-S3-C1	-6.7 (2.8)	O9-H8-S3-C1	-11.8 (-20.7)	C9-N8-H5-S3	185.6 (179.5)
S10-C9-H8-H7	0.4 (11.5)	N10-O9-H8-H7	-10.6 (-18.5)	S10-C9-N8-H5	179.9 (180.5)
N11-C9-H8-S3	186.7 (200.4)	C11-N10-O9-H8	185.0 (188.0)	O11-N8-C9-S10	-0.0 (0.0)
O12-N11-C9-H8	179.6 (179.7)	S12-C11-N10-O9	179.8 (180.5)	H12-O11-N8-N2	0.2 (-0.0)
H13-O12-N11-C9	-0.4 (-2.8)	H13-O12-C11-N10	-4.1 (4.4)	H13-S10-C9-N8	-0.0 (0.2)
H14-N11-C9-H8	0.6 (-1.1)	H14-C11-N10-O9	0.0 (0.6)	H14-C9-N8-O11	180.0 (180.0)

Table 5.29: Geometrical parameters of dimers of TFHA at B3LYP/6-31+G* **[L1]** and (MP2/6-31+G*) **[L3]** theoretical levels. All the bond distances are in angstrom (Å), angles and dihedrals are in degrees (°).

Parameters	2Z-1E (1)	Parameters	1Z-2Z (1)	Parameters	1Z-2E (1)
N2-C1	1.331 (1.341)	N2-C1	1.338 (1.350)	N2-C1	1.336 (1.346)
S3-C1	1.666 (1.649)	S3-C1	1.655 (1.638)	S3-C1	1.657 (1.640)
O4-N2	1.403 (1.413)	O4-N2	1.382 (1.394)	O4-N2	1.380 (1.391)
H5-O4	0.975 (0.978)	H5-O4	0.990 (0.989)	H5-O4	1.000 (0.997)
H6-C1	1.092 (1.092)	H6-C1	1.094 (1.094)	H6-C1	1.094 (1.094)
H7-N2	1.035 (1.034)	H7-N2	1.013 (1.016)	H7-N2	1.013 (1.016)
H8-S3	2.278 (2.263)	H8-S3	2.619 (2.634)	H8-S3	2.509 (2.567)
O9-H8	0.993 (0.994)	S9-H8	1.360 (1.349)	S9-H8	1.365 (1.350)
N10-O9	1.390 (1.403)	C10-S9	1.755 (1.741)	C10-S9	1.748 (1.738)
C11-N10	1.282 (1.295)	N11-C10	1.282 (1.294)	N11-C10	1.280 (1.289)
S12-C11	1.761 (1.743)	O12-N11	1.431 (1.438)	O12-N11	1.416 (1.425)
H13-S12	1.349 (1.342)	H13-O12	0.971 (0.977)	H13-O12	0.972 (0.977)
H14-C11	1.086 (1.086)	H14-C10	1.088 (1.087)	H14-C10	1.091 (1.091)
S3-C1-N2	125.1 (125.2)	S3-C1-N2	128.3 (127.5)	S3-C1-N2	128.7 (127.4)
O4-N2-C1	116.7 (116.8)	O4-N2-C1	123.9 (121.0)	O4-N2-C1	124.2 (121.5)
H5-O4-N2	105.2 (104.1)	H5-O4-N2	106.5 (104.1)	H5-O4-N2	107.7 (105.0)
H6-C1-N2	113.1 (113.0)	H6-C1-N2	110.1 (110.4)	H6-C1-N2	110.1 (110.5)
H7-N2-C1	122.5 (120.9)	H7-N2-C1	122.0 (120.9)	H7-N2-C1	121.4 (121.0)
H8-S3-C1	99.8 (100.1)	H8-S3-C1	110.5 (108.3)	H8-S3-C1	112.4 (106.4)
O9-H8-S3	171.3 (170.9)	S9-H8-S3	160.3 (163.7)	S9-H8-S3	166.6 (169.8)
N10-O9-H8	104.1 (102.7)	C10-S9-H8	99.2 (98.4)	C10-S9-H8	96.8 (96.3)
C11-N10-O9	113.7 (112.3)	N11-C10-S9	132.3 (131.3)	N11-C10-S9	124.2 (123.5)
S12-C11-N10	128.6 (128.3)	O12-N11-C10	111.9 (110.2)	O12-N11-C10	111.0 (109.8)
H13-S12-C11	95.4 (94.9)	H13-O12-N11	102.3 (101.6)	H13-O12-N11	103.5 (102.8)
H14-C11-N10	116.3 (115.6)	H14-C10-N11	114.0 (113.8)	H14-C10-N11	120.7 (120.5)
O4-N2-C1-S3	169.3 (166.2)	O4-N2-C1-S3	-11.3 (-15.4)	O4-N2-C1-S3	11.5 (15.1)

Contd......

Parameters	2Z-1E (1)	Parameters	1Z-2Z (1)	Parameters	1Z-2E (1)
H5-O4-N2-C1	110.1 (113.9)	H5-O4-N2-C1	67.9 (78.6)	H5-O4-N2-C1	-71.0 (-78.5)
H6-C1-N2-O4	-11.7 (-16.3)	H6-C1-N2-O4	171.5 (168.4)	H6-C1-N2-O4	189.6 (-192.3)
H7-N2-C1-S3	13.5 (21.2)	H7-N2-C1-S3	-171.7 (-163.6)	H7-N2-C1-S3	-187.7 (-194.6)
H8-S3-C1-N2	-5.7 (-5.6)	H8-S3-C1-N2	-3.7 (-2.2)	H8-S3-C1-N2	34.0 (48.1)
C9-N8-S3-C1	-14.8 (-23.8)	S9-H8-S3-C1	124.7 (115.0)	S9-H8-S3-C1	110.4 (103.1)
N10-O9-H8-H7	-3.6 (-13.0)	C10-S9-H8-S3	187.9 (187.0)	C10-S9-H8-S3	236.9 (224.3)
C11-N10-O9-H8	182.8 (187.6)	N11-C10-S9-H8	0.0 (4.7)	N11-C10-S9-H5	2.3 (1.4)
S12-C11-N10-O9	0.2 (0.9)	O12-N11-C10-H5	-32.7 (-33.6)	O12-N11-C10-S9	179.4 (179.5)
H13-S12-C11-N10	0.4 (4.6)	H13-O12-N11-C10	191.6 (191.7)	H13-O12-N11-C10	173.9 (175.9)
H14-C11-N10-O9	180.1 (181.2)	H14-C10-N11-O12	180.5 (181.3)	H14-C10-N11-O12	-0.8 (-0.5)

As a result of the intermolecular interactions of two isomeric units in dimer, structural changes (distortions) in the monomeric units in comparison to the isolated monomers can be anticipated. The distortion energy is calculated as the difference between the energy of relaxed monomers and the distorted monomers as present in the dimeric units. The distortion energies (E_{Dis}) associated with the dimer formation are also listed in the Table 5.22. Since the dimers are formed at the expense of this energy, after incorporating the distortion energy in the stabilization energy, the total stabilization energies (S.E.)$_g$ are also included in the Table 5.22. **1Z-1Z (1)** is the most stabilized dimer having two non covalent C=S...HON bonds of almost equal length of 2.22 Å and bridging hydrogen having angle of 153.9°. **1Z-1Z (1)** dimer of TFHA has stabilization of 13.36 kcal/mol at **[L5]** theoretical level as the result of aggregation and distortion energy of 8.89 kcal/mol. Addition of distortion energy to stabilization energy gives the total stabilization (S.E.)$_g$ of 22.25 kcal/mol. The similar **1Z-1Z (1)** dimer of FHA has been reported in earlier studies with two C=O...H-O H-bonds with aggregation energy 15.52 kcal/mol and distortion energy 3.35 kcal/mol at **[L5]** theoretical level [24]. Hence total stabilization after incorporation of the distortion energy is 18.87 kcal/mol which is 3.38 kcal/mol lower than that in TFHA. It is important to note that the energy change (E_{Dis}) accompanying the distortion of monomeric unit to dimeric unit in **1Z-1Z (1)** is 5.54 kcal/mol higher in case of TFHA relative to **1Z-1Z (1)** of FHA.

The presence of intramolecular H-bond in **1Z** conformer of TFHA and FHA has been concluded in the earlier studies [25]. The dimerization of **1Z** conformer of FHA and TFHA involves the breakage of the intramolecular H-bonding. Is the higher distortion energy in TFHA resulting from stronger intramolecular H-bond in TFHA or are there some other factors that contribute to higher distortion energy? In order to explore the reasons, the structural parameters, conjugative interactions, lone pair occupancies of the monomers and dimers have been analysed. The structural parameters of **1Z** conformers of TFHA and FHA respectively indicate total planarity of the conformation in case of TFHA while deviation from planarity is observed in case of **1Z** of FHA (dihedral O-N-C-O = -13.2°). The planarity of **1Z** of TFHA favors strong conjugative interactions involving lone pairs of electrons present on N, S, O and π bond of the thiocarbonyl group. The dimerization of **1Z** conformer of TFHA is accompanied by variation in C-N, C-S and N-O bond distances. The similar variation in **1Z** of FHA is comparatively smaller in magnitude. Thus the structural changes including the loss of planarity of O-N-C-S diminish the conjugative interactions between lone pair of N and C=X group and hence result in larger distortion energy. The results are also supported by the NBO analysis.

Homodimers **1Z-1Z (1)**, **1E-1E (1)**, **2Z-2Z (2)** are centrosymmetric dimers having a pair of geometrically equivalent monomers with equivalent H-bond donors and acceptors possessing equal H-bond lengths, bridging H-bond angles and charges on the interacting atoms. From the symmetric nature of the dimers, it can be suggested that the average single H-bond energy involving C=S...H-O is 11.12 kcal/mol in **1Z-1Z (1)** while C=S...N-H hydrogen bond energy is 6.12 kcal/mol in **1E-1E (1)** of TFHA at **[L5]** theoretical level. The similar H-bond energies in case of centrosymmetric homodimers of FHA for C=O...H-O and the C=O...H-N have been evaluated as 9.43 and 6.87 kcal/mol in **1Z-1Z (1)** and **1E-1E (1)** respectively at the same theoretical level. The H-O...S H-bond strength in case of H_2C (=S)...H_2O with the similar H-bond distance and O-H...S angle as in **1Z-1Z (1)** of TFHA has been evaluated at **[L5]** theoretical level to be 2.83 kcal/mol. The O-H...O H-bond strength in $H_2C(=O)...H_2O$ with the same H-bond distance and angle O-H...O corresponding to **1Z-1Z (1)** of FHA is 3.32 kcal/mol. The stabilization energies in **1Z-1Z (1)** dimers of FHA and TFHA are approximately three to four times higher than the respective H-bond strengths of thioformaldehyde and formaldehyde with water. The H-bond interactions between nitrogen of **2Z** conformer with C-H as H-bond donor from CH_4 results in stabilization of 0.78 kcal/mol. The **2Z-2Z (2)** dimer that involves interactions between similar functionalities results in stabilization energy of 3.27 kcal/mol. The comparison shows that the average H-bond strengths suggested by the centrosymmetric dimers are much higher. The reasons for the difference explored in the section 5.2.4.

The variation in stabilization energy of the centrosymmetric dimers **1Z-1Z (1)**, **1E-1E (1)** and **2Z-2Z (2)** as a result of change in O-H...S, N-H...S and C-H...N distances respectively in the three dimers keeping the configuration of the monomeric units unchanged as in the dimeric unit is depicted in Figure 5.2. The similar plot for the analogous dimers of FHA are also included in the Figure. The decrease in distance between the two units from equilibrium distance decreases the stabilization. The relative energy difference between energies of dimer and equilibrium monomer at the larger distances (>6Å) indicates the distortion energy of the two monomeric units from the equilibrium geometry to the geometry of the dimeric form. As can be seen there is sharp decrease in potential energy with the decrease in intermolecular distance to less than sum of van der Waals radii. The interactions between monomeric units start at comparatively longer distance in TFHA relative to FHA.

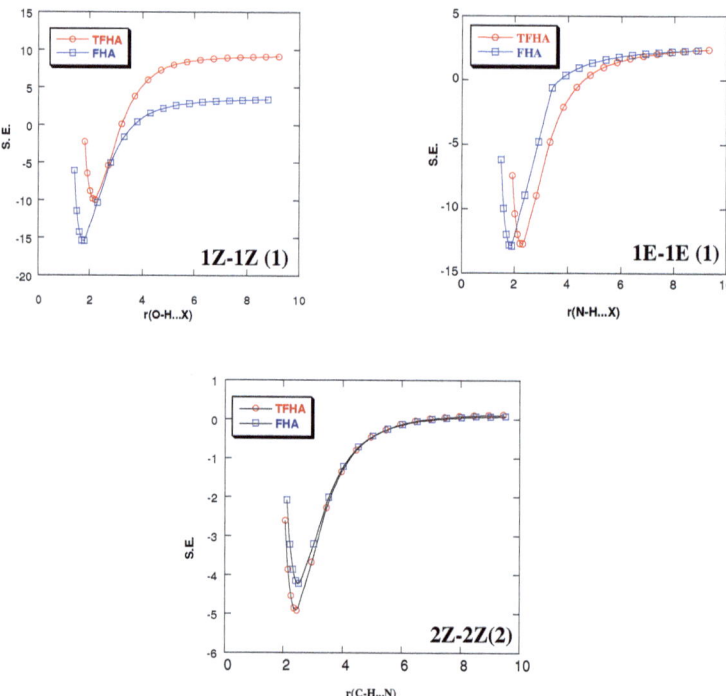

Figure 5.2: Variation of stabilization energies (kcal/mol) of centrosymmetric dimers of FHA and TFHA at [**L5**] theoretical level as a function of varying distance (Å) from equilibrium geometry of dimer. Counterpoise correction not applied.

As can be seen from the figure, stabilizing interactions in **1Z-1Z (1)** dimer become operative at a distance longer than the sum of van der Waals radii in both TFHA and FHA. The variation in stabilization energy becomes pronounced in case of FHA as the distance between monomeric units approaches toward equilibrium distance. The plot of stabilization energy vs. N-H...X distance in **1E-1E (1)** dimer of TFHA and FHA indicates that interactions between the two units start at a larger distance in case of TFHA relative to FHA. The stabilization is stronger in TFHA in comparison to FHA when the separation between the monomeric units is longer than the equilibrium distance but at a distance close to the equilibrium, the **1E-1E (1)** dimer of FHA gains nearly the same stabilization as in case of TFHA with oxygen being highly electronegative and small in size allow further proximity and higher electrostatic interactions. The stabilization energy in case of **2Z-2Z (2)** dimer of TFHA is nearly the same as that in case of FHA at larger C-H...N distances but the stabilization energy is stronger for TFHA in comparison to that of FHA near to the energy minima.

The **1Z-1Z (2)** dimer has stabilization energy of 15.24 kcal/mol with C=S...H-N and O-H...O hydrogen bonding interactions and the stabilization energy is 7.01 kcal/mol lower than that in case of **1Z-1Z (1)** dimer. The difference in stabilization energies for the **1Z-1Z (1)**, **1Z-1Z (2)**, **1Z-1Z (3)** and **1Z-1Z (4)** highlights the importance of nature of H-bond donor and H-bond acceptor affecting the overall stability of the dimer formed.

The stabilization energies of dimers involving **1E** conformers are comparatively lower. The centrosymmetric **1E-1E (1)** dimer differs from **1Z-1Z (1)** with respect to H-bond donor, the O-H bonds act as H-bond donors in **1Z-1Z (1)** while the hydrogen bond donors in case of **1E-1E (1)** are N-H bonds. The stabilization energies associated with **1E-1E (1)** is 12.24 kcal/mol and 13.75 kcal/mol for TFHA and FHA respectively at [L5] theoretical level. The N-H...S H-bond strength evaluated from the interaction of thioformaldehyde and ammonia is 1.54 kcal/mol and in case of formaldehyde with NH_3, the N-H...O strength is 1.74 kcal/mol at [L5] theoretical level. Clearly the stabilization energies associated with dimerization are much higher in comparison to the anticipated values from the model molecules. The stabilization energies associated with dimers of **2Z** and **2E** conformers are lower than those of the dimers of **1Z** and **1E**. In the dimers of **2Z** and **2E**, one of the H-bond interaction results from the weak H-bond donor (S-H or C-H). Interestingly, the dimer **2Z-2Z (2)** of TFHA and similar dimer of FHA show approximately same stabilization of 3.2 kcal/mol with H-bond donor and acceptor involved in the dimerization being the same. Though **2Z** conformer of TFHA is predicted to be the most stable conformer in the isolated state, but the dimerization energies suggest that **1Z** conformer with C=S as H-bond acceptor and highly polar O-H as H-bond donor results in strong H-bonds and thereby stabilizing the **1Z** conformer more than the other tautomeric forms. The stabilization energies associated with eleven dimers of TFHA are lower than the corresponding values for the dimers of FHA except for the **1Z-1Z (1), 1Z-1Z (2), 1Z-1E (1), 1Z-1E (2), 2E-1E (1), 2E-2E (2)** and **2Z-2Z (2)**. The difference in total stabilization energy of the dimers of TFHA and FHA is less than ±2 kcal/mol in thirteen dimers.

When results of protonation-deprotonation phenomena are linked to the dimerization studies, it is observed that the most stable dimer involves X3 as H-bond acceptor site which is also most basic site of the molecule but the H-bond donor in most stable dimer is O-H which is not the most acidic site. The **1E-1E (1)** dimer that involves C=X as H-bond acceptors and N-H as H-bond donors for the two H-bonds has 10.01 and 5.12 kcal/mol lower stabilization energy than the most stable **1Z-1Z (1)** dimer for TFHA and FHA respectively. Though the stabilization energies associated with the dimerization process show dependence on the nature of H-bond donor and H-bond acceptor but the acidity and basicity of these solely do not control the interaction energy.

5.2.4 HYDROGEN BOND COOPERATIVITY:

As discussed in previous section, the stabilization energies associated with the dimerization are much higher than the H-bond strengths evaluated for similar functionalities in model molecules placed at the similar distance and angle. This section explores the reasons for larger stabilization observed in the dimers.

There are a number of reports in literature where the stabilization resulting from two or more than two H-bonded units in a molecular cluster is higher than the additive sum of individual H-bond energies, the additional stabilization is named as cooperativity [26]. The cooperativity has been assigned in clusters bound by H-bonds to the polarization of bonds arising out of the H-bond formation making the next H-bond comparatively stronger. Tsuzuki et al. in their study on intermolecular H-bonding of water-crownophane complex with single water having four H-bonds with the two hydrogen donating phenolic hydroxy groups and the two hydrogen accepting oxygen atoms of the polyoxyethylene chain assigned the cooperative enhancement of the stabilization energy to induced polarization [27]. The additional stabilization in the H-bonded dimers of FHA and TFHA indicates cooperativity. Is this cooperativity arising due to induced polarization of the bonds or is the result of some other factors? This section analyses the factors responsible for the observed cooperativity.

The atomic charge analysis (Table 5.30) of the dimeric units in comparison to that of the monomer units reflects the variation in polarization of the bonds. The increase in polarization of C-N bond in dimers of TFHA and FHA suggests that the strength of bond should increase and hence decrease in bond distance is expected. The N2 and X3 both being electronegative carry negative charge density in monomers of both TFHA and FHA. This negative charge density increases upon dimer formation thereby elongation of the bond should occur. But elongation of N-O is reflected only in **1Z-1Z (1)** of TFHA, the bond is shortened in the similar dimer of FHA. The enhanced polarization of C-X bond in case of **1Z-1Z (1)** also anticipates the shortening of the bond, however the elongation occurs. Thus the polarizations of the bonds cannot explain the geometrical variation and hence it is not the sole criterion for the cooperativity.

The variation in conjugative interactions involving lone pair of electrons present on N, O and the thiocarbonyl functionality can also contribute to the strengthening of the dimeric unit. The second order delocalization energies $E^{(2)}$ associated with various electron delocalizations are recorded in the Table 5.31. Prior to dimerization, there are strong conjugative interactions between the lone pair of electrons present on N2 and the C=S π bond as suggested by $E^{(2)}$ value of 145.35 kcal/mol in **1Z** conformer of TFHA at **[L3]** theoretical level. The similar interactions in **1Z** of FHA (Table 5.31) are comparatively weak ($E^{(2)}$ 14.98 and 32.68 for $n_{N2} \rightarrow \sigma^*_{C1-O3}$ and $n_{N2} \rightarrow \pi^*_{C1-O3}$ respectively). There is loss of $n_{N2} \rightarrow \pi^*_{C1-S3}$ conjugative interactions on dimerization of **1Z** to **1Z-1Z (1)** as reflected by $E^{(2)}$ value of 45.36 kcal/mol in TFHA while **1Z-1Z (1)** of FHA shows enhancement in the second order stabilization energy ($E^{(2)}$ for $n_{N2} \rightarrow \pi^*_{C1-O3}$ 83.9 kcal/mol for each unit). Though the $E^{(2)}$ values for the $n_{N2} \rightarrow \pi^*_{C1-O3}$ orbital interaction are larger for **1Z-1Z (1)** dimer of FHA in comparison to that in similar dimer in TFHA but the lone pair occupancies of electrons present on N2 are comparatively lower in the latter suggesting higher delocalization of N2 lone pair in the dimer **1Z-1Z (1)** of TFHA. Inspite of the lower $E^{(2)}$ values for the orbital interactions $n_{N2} \rightarrow \sigma^*_{C1-S3}$ in TFHA, the C-N bond distance in **1Z-1Z (1)** dimer of TFHA is shorter in comparison to that in **1Z-1Z (1)** of FHA. The C-N bond distance in **1Z-1Z (1)** of TFHA is 1.345 Å while in similar dimer of FHA is 1.359Å at **[L3]** theoretical level.

Table 5.30: Atomic Charges (NPA) in dimers of TFHA (FHA) at MP2/6-31+G* [L3] theoretical level.

Species	U*	C1	N2	X3	O4	H5	H6	H7
1Z		-0.019	-0.348	-0.251	-0.603	0.531	0.242	0.447
		(0.636)	(-0.426)	(-0.732)	(-0.616)	(0.535)	(0.173)	(0.430)
1Z-1Z (1)	1	0.011	-0.374	-0.208	-0.626	0.531	0.235	0.430
		(0.671)	(-0.430)	(-0.752)	(-0.635)	(0.561)	(0.162)	(0.422)
	2	0.011	-0.374	-0.206	-0.626	0.531	0.235	0.430
		(0.671)	(-0.430)	(-0.752)	(-0.635)	(0.561)	(0.162)	(0.423)
1Z-1Z (2)	1	-0.004	-0.369	-0.255	-0.649	0.537	0.248	0.462
		(0.640)	(-0.438)	(-0.750)	(-0.659)	(0.550)	(0.180)	(0.477)
	2	0.012	-0.372	-0.215	-0.617	0.551	0.238	0.436
		(0.670)	(-0.431)	(-0.757)	(-0.627)	(0.551)	(0.166)	(0.430)
1Z-1Z (3)	1	0.004	-0.375	-0.231	-0.620	0.526	0.232	0.431
		(0.659)	(-0.435)	(-0.749)	(-0.637)	(0.559)	(0.155)	(0.423)
	2	0.008	-0.329	-0.304	-0.595	0.533	0.268	0.453
		(0.635)	(-0.404)	(-0.807)	(-0.608)	(0.539)	(0.228)	(0.442)
1Z-1Z (4)	1	-0.009	-0.363	-0.266	-0.635	0.532	0.245	0.459
		(0.636)	(-0.440)	(-0.754)	(-0.648)	(0.540)	(0.174)	(0.471)
	2	0.002	-0.330	-0.295	-0.594	0.534	0.267	0.453
		(0.639)	(-0.407)	(-0.792)	(-0.605)	(0.538)	(0.206)	(0.442)
1E		-0.016	-0.412	-0.145	-0.616	0.518	0.240	0.431
		(0.661)	(-0.454)	(-0.685)	(-0.626)	(0.517)	(0.166)	(0.422)
1E-1E (1)	1	0.037	-0.406	-0.238	-0.603	0.519	0.244	0.448
		(0.680)	(-0.462)	(-0.761)	(-0.615)	(0.516)	(0.171)	(0.471)
	2	0.037	-0.406	-0.238	-0.603	0.519	0.244	0.448
		(0.680)	(-0.462)	(-0.761)	(-0.615)	(0.516)	(0.171)	(0.471)
1E-1E (2)	1	0.030	-0.401	-0.216	-0.600	0.523	0.247	0.452
	2	-0.017	-0.450	-0.144	-0.633	0.526	0.246	0.437
1E-1E (3)	1	0.021	-0.407	-0.199	-0.605	0.522	0.248	0.457
		(0.676)	(-0.453)	(-0.737)	(-0.614)	(0.519)	(0.178)	(0.447)
	2	-0.014	-0.398	-0.154	-0.666	0.532	0.230	0.434
		(0.660)	(-0.443)	(-0.693)	(-0.678)	(0.558)	(0.158)	(0.422)
1E-1E (4)	1	0.021	-0.412	-0.234	-0.609	0.517	0.240	0.447
		(0.670)	(-0.469)	(-0.736)	(-0.624)	(0.515)	(0.164)	(0.463)
	2	0.023	-0.395	-0.213	-0.602	0.520	0.261	0.436
		(0.669)	(-0.444)	(-0.741)	(-0.616)	(0.518)	(0.207)	(0.424)
1E-1E (5)	1	-0.028	-0.403	-0.117	-0.655	0.539	0.241	0.432
	2	-0.018	-0.431	-0.146	-0.626	0.539	0.240	0.433
2Z		-0.099	-0.197	0.025	-0.677	0.176	0.248	0.524
		(0.478)	(-0.282)	(-0.783)	(-0.703)	(0.547)	(0.214)	(0.528)
2Z-2Z (1)	1	-0.083	-0.209	0.033	-0.700	0.192	0.251	0.531
	2	-0.092	-0.221	0.021	-0.693	0.180	0.244	0.547
2Z-2Z (2)	1	-0.088	-0.228	0.028	-0.674	0.175	0.265	0.522
		(0.485)	(-0.311)	(-0.780)	(-0.700)	(0.546)	(0.233)	(0.527)
	2	-0.088	-0.228	0.028	-0.674	0.175	0.265	0.522
		(0.485)	(-0.311)	(-0.780)	(-0.700)	(0.546)	(0.233)	(0.526)
2E		-0.082	-0.217	0.046	-0.651	0.150	0.238	0.516
		(0.488)	(-0.312)	(-0.772)	(-0.650)	(0.524)	(0.206)	(0.516)

Contd…..

Species	U*	C1	N2	X3	O4	H5	H6	H7
2E-2E (1)	1	-0.043	-0.274	0.037	-0.639	0.177	0.242	0.516
		(0.536)	(-0.369)	(-0.800)	(-0.641)	(0.548)	(0.205)	(0.514)
	2	-0.071	-0.245	0.050	-0.671	0.142	0.239	0.541
		(0.520)	(-0.361)	(-0.767)	(-0.661)	(0.524)	(0.210)	(0.543)
2E-2E (2)	1	-0.080	-0.233	0.045	-0.670	0.144	0.238	0.542
		(0.490)	(-0.328)	(-0.773)	(-0.671)	(0.521)	(0.205)	(0.541)
	2	-0.071	-0.223	0.056	-0.681	0.150	0.258	0.527
		(0.495)	(-0.315)	(-0.767)	(-0.677)	(0.525)	(0.226)	(0.527)
1Z-1E (1)	1	0.011	-0.371	-0.213	-0.635	0.517	0.235	0.433
		(0.669)	(-0.429)	(-0.758)	(-0.638)	(0.560)	(0.162)	(0.426)
	2	0.043	-0.397	-0.243	-0.601	0.519	0.245	0.457
		(0.682)	(-0.457)	(-0.771)	(-0.613)	(0.516)	(0.175)	(0.477)
1Z-1E (2)	1	0.003	-0.387	-0.203	-0.607	0.545	0.240	0.439
		(0.664)	(-0.440)	(-0.747)	(-0.615)	(0.543)	(0.170)	(0.432)
	2	-0.017	-0.395	-0.152	-0.668	0.538	0.227	0.438
		(0.658)	(-0.440)	(-0.690)	(-0.680)	(0.565)	(0.154)	(0.426)
1Z-1E (3)	1	0.025	-0.406	-0.245	-0.605	0.518	0.241	0.450
		(0.672)	(-0.468)	(-0.744)	(-0.622)	(0.515)	(0.166)	(0.465)
	2	0.003	-0.333	-0.301	-0.597	0.533	0.264	0.452
		(0.634)	(-0.410)	(-0.793)	(-0.610)	(0.538)	(0.219)	(0.439)
2E-1E (1)	1	0.026	-0.415	-0.210	-0.614	0.519	0.242	0.466
		(0.680)	(-0.463)	(-0.741)	(-0.621)	(0.517)	(0.173)	(0.454)
	2	-0.059	-0.262	0.053	-0.658	0.145	0.242	0.525
		(0.501)	(-0.352)	(-0.771)	(-0.664)	(0.520)	(0.209)	(0.551)
2E-2Z (1)	1	-0.040	-0.281	0.038	-0.639	0.183	0.241	0.515
		(0.546)	(-0.380)	(-0.795)	(-0.641)	(0.555)	(0.205)	(0.512)
	2	-0.085	-0.231	0.023	-0.696	0.180	0.244	0.549
		(0.511)	(-0.343)	(-0.779)	(-0.712)	(0.551)	(0.215)	(0.556)
2Z-1E (1)	1	0.031	-0.410	-0.220	-0.609	0.518	0.243	0.471
		(0.682)	(-0.460)	(-0.752)	(-0.617)	(0.517)	(0.174)	(0.461)
	2	-0.076	-0.248	0.029	-0.686	0.184	0.244	0.529
		(0.497)	(-0.332)	(-0.783)	(-0.715)	(0.552)	(0.210)	(0.565)
1Z-2Z (1)	1	-0.014	-0.388	-0.194	-0.621	0.547	0.231	0.431
		(0.671)	(-0.436)	(-0.746)	(-0.628)	(0.550)	(0.163)	(0.428)
	2	-0.083	-0.215	0.037	-0.701	0.186	0.251	0.532
		(0.511)	(-0.311)	(-0.795)	(-0.720)	(0.575)	(0.210)	(0.529)
1Z-2E (1)	1	-0.002	-0.378	-0.218	-0.632	0.549	0.232	0.431
		(0.673)	(-0.429)	(-0.767)	(-0.636)	(0.549)	(0.165)	(0.429)
	2	-0.034	-0.288	0.057	-0.639	0.165	0.243	0.516
		(0.560)	(-0.395)	(-0.789)	(-0.644)	(0.569)	(0.204)	(0.511)

U* - Monomer unit

Table 5.31: The second order delocalization energies $E^{(2)}$ in kcal/mol for the important orbital interactions in dimers of TFHA (FHA) from NBO analysis at MP2/6-31+G* [L3] theoretical level.

Species	$n_{N2} \to \sigma^*_{C1-X3}$	$n_{N2} \to \pi^*_{C1-X3}$	$n_{X3} \to \sigma^*_{C1-H6}$	$n_{X3} \to \sigma^*_{O4-H5}$	$n_{X3} \to \sigma^*_{C1-N2}$	$n_{X3} \to \pi^*_{C1-N2}$	$n_{O4} \to \pi^*_{C1-N2}$
1Z	-	145.34	13.09	13.71	15.92	-	-
	(14.98)	(32.68)	(23.91)	(5.36)	(29.25)	(-)	(-)
1Z-1Z (1)	17.79	45.36	13.93	-	14.33	-	-
	(-)	(83.99)	(24.83)	(-)	(26.48)	(-)	(-)
	21.88	39.27	13.94	-	14.27	-	-
	(-)	(83.90)	(24.83)	(-)	(26.48)	(-)	(-)
1Z-1Z (2)	-	140.49	13.00	20.86	14.57	-	-
	(-)	(44.53)	(23.22)	(-)	(27.87)	(-)	(-)
	108.92	-	14.21	-	13.94	-	-
	(-)	(-)	(24.78)	(-)	(26.68)	(-)	(-)
1Z-1Z (3)	-	110.65	12.88	-	18.03	-	-
	(-)	(62.05)	(25.08)	(-)	(30.02)	(-)	(-)
	-	154.78	2.06+5.53	-	16.23	-	-
	(-)	(66.39)	(15.59)	(-)	(26.33)	(-)	(-)
1Z-1Z (4)	-	150.03	12.92	17.21	15.01	-	-
	(-)	(48.13)	(23.49)	(-)	(28.00)	(-)	(-)
	-	159.11	9.29	9.75	16.19	-	-
	(-)	(67.82)	(18.56)	(-)	(27.30)	(-)	(-)
1E	-	49.47	12.75	-	18.93	-	-
	(12.11)	(20.90)	(24.38)	(-)	(32.25)	(-)	(-)
1E-1E (1)	-	101.36	13.86	-	12.94	-	-
	(10.50)	(41.35)	(23.68)	(-)	(25.19)	(-)	(-)
	-	100.66	13.85	-	12.95	-	-
	(10.52)	(41.29)	(23.68)	(-)	(25.19)	(-)	(-)
1E-1E (2)	-	88.35	13.98	-	13.49	-	-
	(-)	(54.28)	(12.76)	(-)	(19.00)	(-)	(-)

Contd......

Species	$n_{N2} \to \sigma^*_{C1-X3}$	$n_{N2} \to \pi^*_{C1-X3}$	$n_{X3} \to \sigma^*_{C1-H6}$	$n_{X3} \to \sigma^*_{O4-H5}$	$n_{X3} \to \sigma^*_{C1-N2}$	$n_{X3} \to \pi^*_{C1-N2}$	$n_{O4} \to \pi^*_{C1-N2}$
1E-1E (3)	- (12.30) -	74.41 (31.36) 58.11	13.50 (23.27) 12.92	- (-) -	14.04 (26.80) 18.83	- (-) -	- (-) -
1E-1E (4)	- (12.69) -	82.47 (32.15) 81.00	12.78 (23.95) -	- (-) -	16.77 (29.18) 18.13	- (-) -	- (-) -
1E-1E (5)	- (11.08) - -	- (29.02) 44.19 53.20	(17.99) 12.89 - -	- (-) - -	(29.42) 19.35 18.91	- - - -	- - - -
2Z	14.57 (13.50)	- -	- (0.84)	- -	39.47 (8.63)	6.45 (54.42)	- (13.22)
2Z-2Z (1)	14.12 14.62	- -	- -	- -	- -	26.22 38.78	19.73 17.26
2Z-2Z (2)	14.26 (13.12) 14.26 (13.11)	- (-) - (-)	- (-) - (-)	- (-) - (-)	- (-) - (-)	40.34 (55.12) 40.32 (55.37)	(12.51) 17.25 (12.56)
2E	0.86 (3.82)	- -	- (0.77)	- -	3.95 (6.12)	34.78 (48.37)	16.87 (13.80)
2E-2E (1)	(-) -	(-) -	(-) -	(-) -	(-) -	40.30 (51.67) 34.47	15.67 (11.09) 19.36
2E-2E (2)	(-) -	(-) -	(10.16) - (10.16)	(-) -	(-) -	(50.23) 33.97 (43.62) 23.93	(15.60) 18.96 (14.22) 8.12 (8.16)
1Z-1E (1)	- (-) -	- (65.95) 118.91	14.34 (25.00) 13.32	- (-) -	(48.90) 13.92 (26.22) 11.36	- (-) -	- (-) -

Contd......

Species	$n_{N2} \to \sigma^*_{C1-X3}$	$n_{N2} \to \pi^*_{C1-X3}$	$n_{X3} \to \sigma^*_{C1-H6}$	$n_{X3} \to \sigma^*_{O4-H5}$	$n_{X3} \to \sigma^*_{C1-N2}$	$n_{X3} \to \pi^*_{C1-N2}$	$n_{O4} \to \pi^*_{C1-N2}$
1Z-1E (2)	(-)	(52.63)	(22.93)	(-)	(22.79)	(-)	(-)
	11.24	61.65	13.96	-	5.16	-	-
	(-)	(84.58)	(24.50)	(-)	(27.44)	(-)	(-)
	(12.28)	60.23	12.74	-	19.04	-	-
	-	(24.94)	(25.04)	(-)	(31.85)	(-)	(-)
1Z-1E (3)	-	86.67	12.78	-	16.36	-	-
	(10.75)	(34.13)	(23.81)	(-)	(28.75)	(-)	(-)
	-	156.53	-	-	16.38	-	-
2E-1E (1)	(-)	(58.85)	(17.77)	(-)	(27.36)	(-)	(-)
	-	84.76	13.86	-	-	28.26	-
	(10.07)	(36.16)	(23.60)	(-)	(25.74)	(-)	(-)
	-	-	-	-	13.98	-	14.33 (11.14)
2E-2Z (1)	(-)	(-)	(-)	(-)	(-)	(29.84)	15.25
	-	-	-	-	-	41.46	-
	14.53	(-)	(-)	(-)	(17.57)	(20.95)	(-)
	(12.81)	-	-	-	-	39.69	21.17 (15.34)
	-	-	-	-	-	(57.12)	-
2Z-1E (1)	(10.12)	(-)	(-)	(-)	(-)	(-)	-
	14.03	93.26	13.88	-	13.09	37.09	18.16 (12.68)
	(13.14)	(39.98)	(23.36)	(-)	(24.43)	(48.96)	-
1Z-2Z (1)	-	(-)	(-)	(-)	(-)	-	-
	(15.96)	100.51	13.55	-	18.05	26.23	(-)
	13.70	(35.58)	(25.03)	(-)	(26.25)	(22.70)	-
1Z-2E (1)	(14.29)	(-)	(-)	(-)	(18.77)	-	-
	17.28	47.97	13.34	-	17.67	26.23	(-)
	(-)	(58.14)	(24.62)	(-)	(24.75)	27.12	10.06
	-	-	-	-	(55.56)	(-)	(-)

According to Desiraju, the H-bond is a complex conglomerate of at least four component interaction types (1) polarization (2) electrostatic (3) van der Waal (4) covalent. The electrostatic components are expected to be higher in **1Z-1Z (1)**, **1E-1E (1)** of FHA due to the presence of electronegative oxygen that participates in the H-bonding. But the stabilization energies for the dimerization for TFHA and FHA suggest the comparable bond strength that differ only by ±2 kcal/mol in thirteen out of twenty one optimized dimers.

The orbital interactions $n_{S3} \rightarrow \sigma^*_{O11-H12}$ and $n_{O3} \rightarrow \sigma^*_{O11-H12}$ explain the charge transfer interactions in **1Z-1Z (1)** of TFHA and FHA respectively and these impart covalent character to the H-bonds (Table 5.32). The analysis of $E^{(2)}$ values for the orbital interactions involving charge transfer interactions from lone pair of thiocarbonyl sulfur in TFHA are larger than that from carbonyl oxygen in FHA. Similar orbital interactions in most of the other dimers of TFHA with few exceptions are also higher in comparison to that in FHA as suggested by higher $E^{(2)}$ values and population of σ^*_{O-H} orbitals. Thus the covalent character of H-bond C=S...H-O is higher in comparison to C=O...H-O. Hence it can be concluded that the following factors contribute to cooperativity in the dimers of TFHA and FHA

1. Polarization of bonds
2. Conjugation effects
3. Charge transfer interactions responsible for providing covalent character to the hydrogen bonds.

Table 5.32: The second order delocalization energies $E^{(2)}$ in kcal/mol for the orbital interactions important for H-bond formation in dimeric forms of TFHA (FHA) and occupancies of acceptor orbitals at MP2/6-31+G* [L3] theoretical level.

Species	Donor Unit[1]		Acceptor Unit[2]	$E^{(2)}$	Donor Unit[2]		Acceptor Unit[1]	$E^{(2)}$	Occupancies Unit[2]		Unit[1]	
1Z-1Z (1)	$n_{X3(1)}$	→	σ*$_{O11-H12}$	4.33 (8.79)	$n_{X8(1)}$	→	σ*$_{O4-H5}$	4.37 (8.79)	σ*$_{O11-H12}$	0.056	σ*$_{O4-H5}$	0.057
	$n_{X3(2)}$	→	σ*$_{O11-H12}$	20.82 (10.83)	$n_{X8(2)}$	→	σ*$_{O4-H5}$	21.35 (10.84)		(0.032)		(0.032)
1Z-1Z (2)	$n_{O4(1)}$	→	σ*$_{O11-H12}$	2.89 (1.04)	$n_{X8(1)}$	→	σ*$_{N2-H7}$	5.19 (10.44)	σ*$_{O11-H12}$	0.030	σ*$_{N2-H7}$	0.068
	$n_{O4(2)}$	→	σ*$_{O11-H12}$	18.89 (20.40)	$n_{X8(2)}$	→	σ*$_{N2-H7}$	24.35 (9.92)		(0.030)		(0.037)
1Z-1Z (3)	$n_{X3(1)}$	→	σ*$_{C10-H14}$	2.06 (3.18)	$n_{X8(1)}$	→	σ*$_{O4-H5}$	4.47 (9.27)	σ*$_{C10-H14}$	0.042	σ*$_{O4-H5}$	0.056
	$n_{X3(2)}$	→	σ*$_{C10-H14}$	5.53 (1.69)	$n_{X8(2)}$	→	σ*$_{O4-H5}$	25.06 (17.59)		(0.038)		(0.036)
1Z-1Z (4)	$n_{O4(1)}$	→	σ*$_{C9-H13}$	2.04 (1.28)	$n_{X8(1)}$	→	σ*$_{N2-H7}$	4.08 (7.65)	σ*$_{C9-H13}$	0.032	σ*$_{N2-H7}$	0.052
	$n_{O4(2)}$	→	σ*$_{C9-H13}$	1.45 (0.69)	$n_{X8(2)}$	→	σ*$_{N2-H7}$	19.15 (10.60)		(0.038)		(0.033)
1E-1E (1)	$n_{X3(1)}$	→	σ*$_{N10-H14}$	4.17 (9.22)	$n_{X8(1)}$	→	σ*$_{N2-H7}$	4.17 (9.22)	σ*$_{N10-H14}$	0.073	σ*$_{N2-H7}$	0.073
	$n_{X3(2)}$	→	σ*$_{N10-H14}$	26.38 (11.65)	$n_{X8(2)}$	→	σ*$_{N2-H7}$	26.38 (11.64)		(0.040)		(0.040)
1E-1E (2)	$n_{S3(1)}$	→	σ*$_{O9-H8}$	3.62	n_{N10}	→	σ*$_{N2-H7}$	11.50	σ*$_{O9-H8}$	0.052	σ*$_{N2-H7}$	0.040
	$n_{S3(2)}$	→	σ*$_{O9-H8}$	24.18								
1E-1E (3)	$n_{X3(1)}$	→	σ*$_{O8-H12}$	3.23	$n_{O8(1)}$	→	σ*$_{N2-H7}$	1.68	σ*$_{O8-H12}$	0.051	σ*$_{N2-H7}$	0.035

Contd.....

248

Species	Donor Unit $^{(1)}$		Acceptor Unit $^{(2)}$	$E^{(2)}$	Donor Unit $^{(2)}$		Acceptor Unit $^{(1)}$	$E^{(2)}$	Occupancies Unit $^{(2)}$		Unit $^{(1)}$	
1E-1E (4)	$n_{X3(2)}$	→	σ^*_{O8-H12}	(6.05) 22.01 (10.98)	$n_{O8(2)}$	→	σ^*_{N2-H7}	(2.64) 9.79 (3.38)		(0.026)		(0.024)
	$n_{X3(1)}$	→	σ^*_{C9-H8}	1.15 (1.54)	$n_{X12(1)}$	→	σ^*_{N2-H7}	2.72 (6.56)	σ^*_{C9-H8}	0.042	σ^*_{N2-H7}	0.061
	$n_{X3(2)}$	→	σ^*_{C9-H8}	5.03 (1.58)	$n_{X12(2)}$	→	σ^*_{N2-H7}	21.51 (11.56)		(0.042)		(0.036)
1E-1E (5)	$n_{O4(1)}$	→	σ^*_{O9-H8}	3.32			-	-	σ^*_{O9-H8}	0.015	-	-
	$n_{O4(2)}$	→	σ^*_{O9-H8}	6.65			-	-				
2Z-2Z (1)	$n_{O4(1)}$	→	σ^*_{O9-H14}	4.74	n_{N8}	→	σ^*_{X3-H5}	6.12	σ^*_{O9-H14}	0.026	σ^*_{X3-H5}	0.018
	$n_{O4(2)}$	→	σ^*_{O9-H14}	14.09								
2Z-2Z (2)	n_{N2}	→	$\sigma^*_{C10-H13}$	3.14 (2.94)	n_{N8}	→	σ^*_{C1-H6}	3.12 (2.91)	$\sigma^*_{C10-H13}$	0.018 (0.020)	σ^*_{C1-H6}	0.018 (0.020)
2E-2E (1)	n_{N2}	→	$\sigma^*_{O11-H12}$	22.28 (23.01)	n_{N8}	→	σ^*_{X3-H5}	10.93 (28.94)	$\sigma^*_{O11-H12}$	0.035 (0.037)	σ^*_{X3-H5}	0.026 (0.051)
2E-2E (2)	n_{N2}	→	$\sigma^*_{C10-H13}$	3.94 (4.10)	$n_{O8(1)}$	→	σ^*_{O4-H7}	6.51 (5.36)	$\sigma^*_{C10-H13}$	0.026 (0.029)	σ^*_{O4-H7}	0.024 (0.024)
					$n_{O8(2)}$	→	σ^*_{O4-H7}	12.40 (13.46)				
1Z-1E (1)	$n_{X3(1)}$	→	$\sigma^*_{N10-H14}$	4.83 (10.74)	$n_{X8(1)}$	→	σ^*_{O4-H5}	4.49 (9.90)	$\sigma^*_{N10-H14}$	0.075	σ^*_{O4-H5}	0.080
	$n_{X3(2)}$	→	$\sigma^*_{N10-H14}$	26.14 (11.33)	$n_{X8(2)}$	→	σ^*_{O4-H5}	35.73 (19.99)		(0.043)		(0.040)
1Z-1E (2)	$n_{X3(1)}$	→	σ^*_{O8-H12}	4.36 (7.82)	$n_{O8(1)}$	→	σ^*_{O4-H5}	1.44 (2.09)	σ^*_{O8-H12}	0.051	σ^*_{O4-H5}	0.028

Contd.....

Species	Donor Unit(1)		Acceptor Unit(2)	E(2)	Donor Unit(2)		Acceptor Unit(1)	E(2)	Occupancies Unit(2)	Unit(1)
1Z-1E (3)	nX3(2)	→	σ*O8-H12	17.34 (8.90)	nO8(2)	→	σ*O4-H5	16.76 (12.58)	(0.028)	(0.023)
	nX3(1)	→	σ*C9-H8	1.44 (2.02)	nX10(1)	→	σ*N2-H7	4.08 (7.62)	σ*C9-H8 0.041	σ*N2-H7 0.058
	nX3(2)	→	σ*C9-H8	6.02 (2.09)	nX10(2)	→	σ*N2-H7	18.42 (10.86)	(0.039)	(0.037)
2E-1E (1)	nX3(1)	→	σ*O9-H8	3.26 (7.46)	nN10	→	σ*N2-H7	23.02 (17.19)	σ*O9-H8 0.047	σ*N2-H7 0.053
	nX3(2)	→	σ*O9-H8	21.80 (13.40)					(0.029)	(0.040)
2Z-2E (1)	nN2	→	σ*O11-H12	24.99 (27.84)	nN8	→	σ*X3-H5	12.60 (29.12)	σ*O11-H12 0.040 (0.045)	σ*X3-H5 0.029 (0.050)
2Z-1E (1)	nX3(1)	→	σ*O9-H8	3.71 (8.99)	nN10	→	σ*N2-H7	23.36 (17.71)	σ*O9-H8 0.057	σ*N2-H7 0.053
	nX3(2)	→	σ*O9-H8	27.02 (16.23)					(0.034)	(0.040)
1Z-2Z (1)	nX3(1)	→	σ*X9-H8	1.99 (10.90)	nO12(1)	→	σ*O4-H5	4.84 (3.16)	σ*X9-H8 0.028	σ*O4-H5 0.032
	nX3(2)	→	σ*X9-H8	7.58 (13.15)	nO12(2)	→	σ*O4-H5	18.43 (20.39)	(0.038)	(0.032)
1Z-2E (1)	nX3(1)	→	σ*X9-H8	2.64 (12.77)	nN11	→	σ*O4-H5	30.51 (31.08)	σ*X9-H8 0.029	σ*O4-H5 0.048
	nX3(2)	→	σ*X9-H8	4.54 (14.51)					(0.042)	(0.053)

5.2.5 SOLUTION PHASE STUDIES:

The effect of dielectric of the aqueous medium on the dimerization energies have also been evaluated by employing PCM model using [L5] theoretical level by calculating solvation energies. The solvation energy has two components-electrostatic and non electrostatic. The non electrostatic component comprises the cavitation, repulsion and dispersion energies. The cavitation energy is dependent on the volume and since volume does not vary much for the dimers under study and dispersion interactions are weak, thus solvation energies result mainly from the electrostatic interactions. The ΔG_{sol} values of all dimers are listed in Table 5.22. The negative values of solvation free energies indicate favorable interactions of dimers with the medium. ΔG_{sol} values for the dimers of TFHA range from 4.62 to 14.98 kcal/mol and for the dimers of FHA the values lies in the range 8.43 to 16.45 kcal/mol. Thus dimers of FHA interact with the medium relatively more strongly than the dimers of TFHA. ΔG_{sol} is higher for dimers involving **1Z** and **1E** in comparison to dimers of **2Z** and **2E**. The mixed dimers of **1Z, 1E** with **2Z** and **2E** also have large ΔG_{sol}.

$\Delta \Delta G_{Assoc}$ is the difference between free energy of solvation of the dimer and the isolated monomers. The positive values of $\Delta \Delta G_{Assoc}$ indicate that the isolated monomers interact with water more strongly than do the dimers. Due to the presence of intermolecular H-bonding between the monomeric units, some of the active sites become unavailable for the interaction with the dielectric of the medium and hence relatively weaker solvation of the dimer is observed. The analysis of conformations of the dimers indicates that the dimers having high favorable solvation energies have either H-bond acceptors or H-bond donors exposed to the medium. The stabilization energy associated with the dimerization process in medium can be considered to be sum of $\Delta \Delta G_{Assoc}$ and $(S.E.)_g$ and the values are listed in the table as $(S.E.)_{aq}$.

. Only one half of the dimers optimized, show favorable interaction energies in aqueous medium. Also it is seen that the dimers with favorable interaction energies with the aqueous medium have reduced $(S.E.)_{aq}$ as compared to interaction energies in the gas phase. The dimers of **2Z** and **2E** which had low interaction energies in gas phase indicate unfavorable interactions for the dimerization in the medium also. The negative $(S.E.)_{aq}$ for **1Z-1Z (1)** of TFHA as well as FHA suggests that the dimerization for **1Z** tautomeric form in aqueous medium remains favorable.

5.3 CONCLUSIONS:

Protonation and deprotonation enthalpies of different sites of FHA and TFHA are calculated and from the results of protonation and deprotonation enthalpies, the X3 (O, S) of C=X group is observed to be the most basic site while the N-H is the most acidic site in both FHA and TFHA. The stabilization energies associated with the dimerization of four selected tautomeric forms of TFHA and FHA are evaluated at MP2/6-311++G**//MP2/6-31+G* [L5] theoretical level. The dimerization involves the formation of two H-bonds between the monomeric units. The most stable dimer in both FHA and TFHA is **1Z-1Z (1)** involving C=X (X=O, S) as H-bond acceptor and O-H as H-bond donor and the stabilization energy is 3.38 kcal/mol (including counterpoise correction and distortion energy) higher in TFHA relative to FHA. For thirteen dimers the stabilization energies are comparable and difference lies within ±2 kcal/mol. The comparison of the stabilization energies with the proton affinities and deprotonation enthalpies for different sites in TFHA and FHA respectively indicates that the H-bonds of the dimer that involve the most basic site of TFHA and FHA are comparatively stronger than the other basic sites. However the dimer with a combination of strongest basic and strongest acidic site is not the most stable dimer.

The interaction energies for the dimerization are higher in value in comparison to H-bond energies for the similar functionalities present in chosen model molecules. The reasons for the additional stabilization (cooperativity) in H-bond have been explored and it has been inferred that cooperativity arises due to polarization, charge transfer interactions providing covalent character to H-bonds and extended conjugation involving H-bond donor and acceptor. The enthalpy changes accompanying dimerization are reduced considerably in aqueous phase as the monomeric units interact with solvent more strongly, thus disfavoring the dimerization in solution although **1Z-1Z (1)** still remains the most stable dimer for both the HAs.

5.4 REFERENCES:

1. B. Chan, J.E. Del Bene, J. Elguero, L. Radom, J. Phys. Chem. A 109 (2005) 5509.
2. P.C. Singh, G.N. Patwari, J. Phys. Chem. A 111 (2007) 3178.
3. J. Graton, M. Berthelot, C. Laurence, J. Chem. Soc. Perkin Trans. 2 (2001) 2130.
4. H.S. Frank, W.-Y. Wen, Discuss Faraday Soc. 24 (1957) 133.
5. R.D. Parra, J. Ohlssen, J. Phys. Chem. A 112 (2008) 3492.
6. L. Ojamae, K. Hermansson, J. Phys. Chem. 98 (1994) 4271.
7. G.R. Desiraju, T. Steiner, The Weak hydrogen Bond in Structural Chemistry and Biology, Oxford University Press: Oxford, 1999, Chapter 1.
8. S. Scheiner, Hydrogen Bonding: A theoretical perspective; Oxford University Press, 1997, Chapter 5.
9. H. Guo, M. Karplus, J. Phys. Chem. 98 (1994) 7104.
10. L. Rincon, R. Almeida, D.G. Aldea, H.D. Riega, J. Chem. Phys. 114 (2001) 5552.
11. W. Chen, M.S. Gordon, J. Phys. Chem. 100 (1996) 14316.
12. A.D. Kulkarni, R.K. Pathak, L.J. Bartolotti, J. Chem. Phys. 124 (2006) 214309.
13. F.H. Allen, J.A.K. Howard, V.J. Hoy, G.R. Desiraju, D.S. Reddy, C.C. Wilson, J. Am. Chem. Soc. 118 (1996) 4081.
14. L.R. MacGillivray, M.M. Siebke, J.L. Reid, Org. Lett. 3 (2001) 1257.
15. L. Turi, J.J. Dannenberg, J. Am. Chem. Soc. 116 (1994) 8714.
16. A. Masunov, J.J. Dannenberg, J. Phys. Chem. B 104 (2000) 806.
17. J.J. Dannenberg, AIP Proceedings 851 (2006) 102.
18. M. Saldyka, Z. Mielke, J. Phys. Chem. A 106 (2002) 3714.
19. L. Bauer, O. Exner, Angew. Chem. Internat. Edit. 6 (1974) 376.
20. H.R. Bravo, W. Lazo, J. Agric. Food Chem. 44 (1996) 1569.
21. A.E. Fazary, M.M. Khalil, A. Fahmy, T.A. Tantawy Medical Journal of Islamic Academy of Sciences 14 (2001) 107.
22. B. Garcia, S. Ibeas, F.J. Hoyuelos, J.M. Leal, F. Secco, M. Venturini, J. Org. Chem. 66 (2001) 7896.
23. A. Bagno, C. Comuzzi, G. Scorrano, J. Am. Chem. Soc. 116 (1994) 916.
24. D. Kaur, R. Kohli, Int. J. Quantum Chem. 108 (2008) 119.

25. D. Kaur, R. Kohli, R.P. Kaur, J. Mol. Struct. (Theochem) 864 (2008) 72.
26. K. Sakota, Y. Kageura, H. Sekiya, J. Chem. Phys. 129 (2008) 54303.
27. S. Tsuzuki, H. Houjou, Y. Nagawa, M. Goto, K. Hiratani, J. Am. Chem. Soc. 123 (2001) 4255.

HYDROGEN BONDING ABILITY AND ACID-BASE BEHAVIOUR OF FORMYLPHOSPHINOUS ACID: AN ISOSTERE OF HYDROXAMIC ACID

6.1 INTRODUCTION:

Phosphorus is essential element to all life [1,2]. The phosphate-ester backbone of DNA and RNA is provided by phosphorus. Besides, phosphorus is structural constituent of phosphoproteins, phospholipids, teeth bones etc. Phosphinous acids belong to hydrophosphoryl (HPCs) class of compounds [3]. Phosphinous acids are compounds of general formula H_2POH i.e. hydroxyphosphines. They are isosterically equivalent to well known hydroxylamine H_2NOH. Differences in the properties of these two classes of compounds stem from difference in properties of nitrogen and phosphorus like their different size, electronegativity and availability of d-orbitals for bonding. Substituting one of the hydrogen with formyl group (-HC=O) leads to formation of H(C=O)NHOH in case of hydroxylamine which is well characterized compound called formohydroxamic acid (FHA) while in case of phosphinous acids the compound of formula H(C=O)PHOH called formylphosphinous acid (FPA) is formed.

Experimental studies suggest that phosphinous acids (R_2POH) containing trivalent phosphorus are in tautomeric equilibrium with phosphine oxide $R_2(H)PO$ containing pentavalent phosphorus with the equilibrium almost completely shifted to the side of phosphine oxide **(Scheme 6.1)** [4,5]. Chatt and Heaton described that it is possible to stabilize phosphinous acid derivatives by coordination to transition metal complexes and thus shift the equilibrium to the side of phosphinous acids [6].

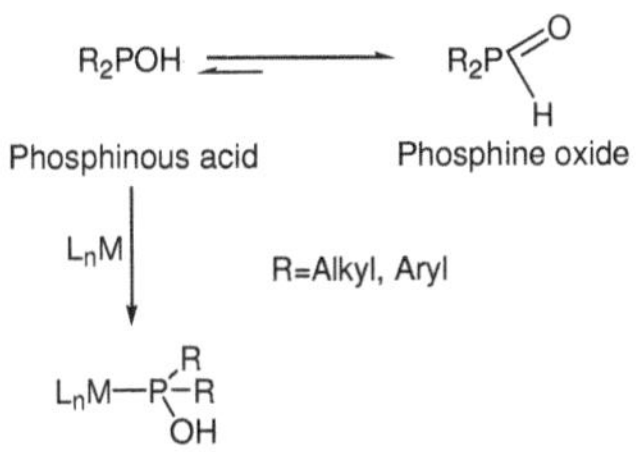

Scheme 6.1

A number of literature reports on FHA are available, but virtually nothing is known about its isostere FPA except a few reports on parent phosphinous acids. Only known example of thermally stable noncoordinated phosphinous acid is bis (trifluoromethyl) phosphinous acid $(CF_3)_2POH$ synthesized by Burg and Griffiths in 1960 [7]. It is extremely sensitive to hydrolysis evolving fluoroform on reaction with water [8]. Detailed knowledge is not available about it as its synthesis is risky [9].

In spite of their thermal instability there is interest in this class of compounds due to their numerous applications and unexplored chemistry. According to Dubrovina and Borner, transition metal complexes of phosphinous acids have catalytic activity in homogeneous catalysis [10]. Phosphorus containing compounds like phosphinous acid group, its derivatives can modify the surface of nanoparticles in vapor phase. FPA has been reported to be a radical cation in mass spectrometry study by Heydorn et al. [11]. The synthesis and study of FPA is thus challenging.

Theoretical studies suggest phosphinous acid to be more stable than phosphine oxide. Density functional treatments (B3LYP) and various model chemistries (e.g. G3 and G3MP2) reduce gas phase energy difference between phosphine oxide and phosphinous acid to less than 1 kcal/mol favoring acids, while oxide is slightly favored using CBS-Q method [12]. Theoretical and experimental studies reveal that electron withdrawing effect of organic groups attached to phosphorus strongly influences equlilibrium distribution between phosphine oxide and phosphinous acid tautomer [13,14]. Here an effort to the determination of properties of FPA is made using computational methods.

The present study has been carried out with the following aims: 1. Understanding reasons for instability of phosphinous acids. 2. Understanding tautomerism of phosphinous acid with secondary phosphine oxide as representative of trivalent and pentavalent phosphorous reagents. 3. Understanding chemistry of FPA relative to FHA.

6.2 RESULTS AND DISCUSSION:

6.2.1 RELATIVE STABILITIES AND ISOMERIC ANALYSIS:

Geometrical optimization of phosphinous acid and phosphine oxide has been carried out and it resulted in two conformations for the former denoted by **I** and **II** (Figure 6.1). The conformer **I** being 0.70 kcal/mol more stable than **II**. The phosphine oxide is observed to be 1.04 kcal/mol lower in stability relative to phosphinous acid conformer **I** at MP2/6-31+G* **[L3]** theoretical level. The nitrogen analog of phosphinous acid (NH_2OH) similar to phosphinous acid conformer **I** is magnificently stable (24.13 kcal/mol) than corresponding ammonia oxide form. The structural parameters of phosphine oxide, phosphinous acid **I** and **II** are reported in Table 6.1. The analysis of geometrical parameters indicates that the difference in energy of

PH$_2$OH and PH$_3$O arises because of presence of phosphoryl group in the latter and the sum of strengths of P=O bond and P-H bond is lower in comparison to sum of strength of P-O single bond and O-H bond in the phosphinous acid.

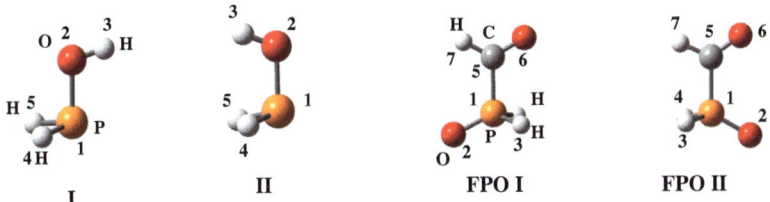

Figure 6.1: Isomeric forms of phosphinous acid (**I** and **II**) and formylphosphine oxide (**FPO I** and **FPO II**).

FHA being formyl derivative of hydroxylamine, the formyl derivatives of phosphinous acid (FPA) and phosphine oxide called formylphosphinous acid (FPA) and formylphosphine oxide (FPO) respectively have also been optimized. FPO optimizes to two conformations **FPO I** and **FPO II** which have anti and syn arrangement of C=O and P=O bonds respectively (Figure 6.1). **FPO I** is 2.79 kcal/mol more stable than **FPO II** at [L3] theoretical level that may have resulted from lone pair lone pair repulsions on oxygens (For parameters see Table 6.1).

The presence of formyl group at phosphorous of phosphinous acids enhances the stability of acid tautomer as the relative energy difference between the **1Z** of FPA and FPO is increased to 4.48 and 7.27 kcal/mol for **FPO I** and **FPO II** respectively. Reason of enhanced stability of FPA over its oxide form can be assigned to the onset of intramolecular H-bonding of C=O to OH in FPA (**1Z** in Figure 6.2b) along with the difference in bond strengths of P=O and P-H of FPO versus P-O and O-H of FPA as stated earlier.

The geometrical parameters of the tautomeric forms of FPA and FHA along with the transition states interconnecting them obtained at B3LYP/6-31+G* [L1] and MP2/6-31+G* [L3] theoretical levels are reported in Tables 6.2-6.5. The C-P bond length in **1Z** form of FPA is 1.860 Å at [L3] theoretical level which is very close to the experimental value of the average C-P bond length (1.84±0.006 Å). The electron diffraction studies by B. Hoge et al. reported P-O bond length as 1.661 (4) Å for (CF$_3$)$_2$POH [4]. In **1Z** form of FPA, the P-O bond distance is evaluated to be 1.662 Å at [L3] theoretical level. The C-O bond length in **1Z** of FPA is 0.004 Å shorter than that in FHA and in the enolic forms, it is longer than that in FHA. The sum of angles around P in **1Z** conformer of FPA is 290.5° at [L3] theoretical level which indicates the strong pyramidal character around phosphorus. The sum of angles deviates by 6.2° from the sum of angles around P in PH_3 (284.3°) at the same level. The sum of angles around N in FHA is 346.5° which shows deviation of 22.7° from that in NH_3 (323.8°). The comparison of geometrical parameters of **1Z** and **1E** of FPA reflects the largest difference in P2-O4 bond distance (0.027 Å) while the similar comparison in FHA highlights the difference in C1-N2 bond to be the largest (0.026 Å).

In the previous chapters, keto ↔ enol tautomerism involving carbonyl/thiocarbonyl group has been analyzed in formohydroxamic acid (FHA) and thioformohydroxamic acid (TFHA) respectively [15,16]. FPA isostere of FHA has been analysed for the relative stability of similar tautomeric forms as in FHA. The six isomeric forms of the molecule namely **1Z, 1E, 2Z, 2E, 2Z2, 2E2** along with one zwitterioic/dipolar form (**3**) have been optimized at B3LYP/6-31+G* [L1] and [L3] theoretical levels and are shown in Figure 6.2a and 6.2b.

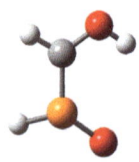

Figure 6.2a: Zwitter ion (3)

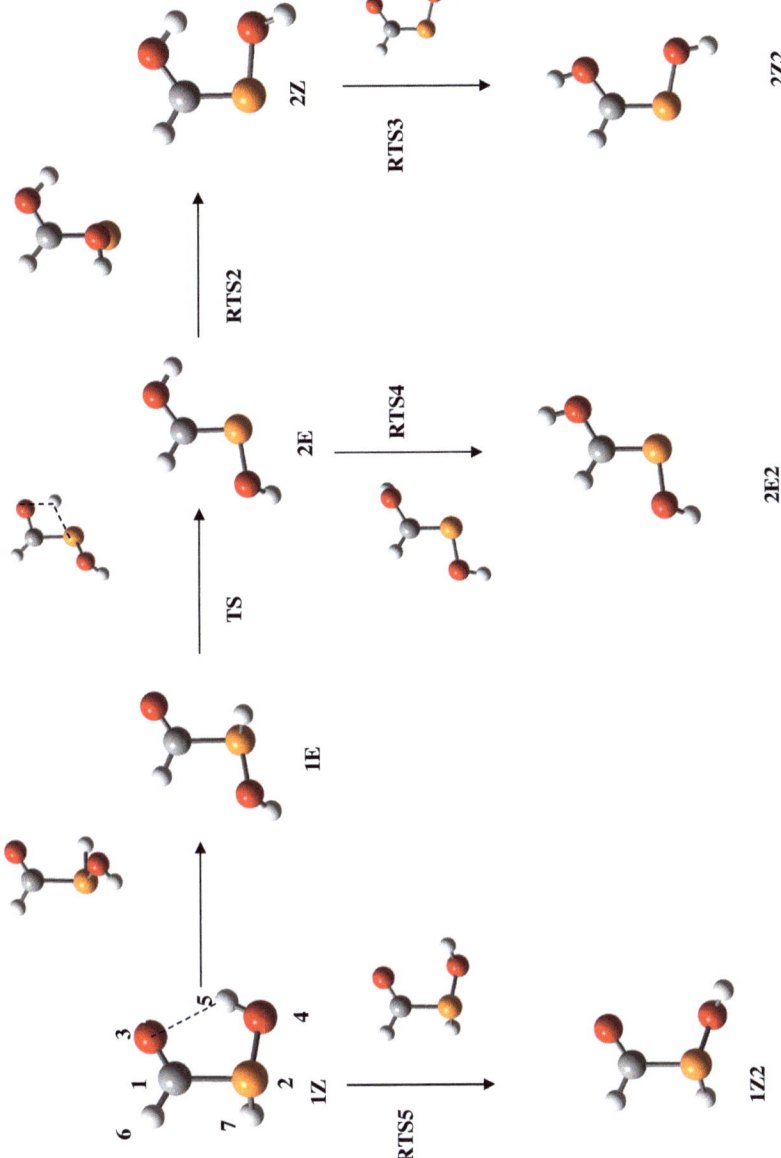

Figure 6.2b: The isomeric forms of formylphosphinous acid (FPA) and its various rotational transition states connecting them.

259

Table 6.1: Geometrical parameters of **PH₃O**, **PH₂OH (I and II)**, **HCOP(=O)H₂ (FPO I and FPO II)** at MP2/6-31+G* [L3] theoretical level. All bond distances are in angstrom (Å) and angles in degrees (°).

Parameters	PH₃O	Parameters	I	II	Parameters	FPO I	FPO II
O2-X1	1.505	O2-X1	1.694	1.683	O2-X1	1.505	1.500
H3-X1	1.410	H3-O2	0.972	0.974	H3-X1	1.412	1.416
H4-X1	1.410	H4-X1	1.415	1.422	H4-X1	1.412	1.416
H5-X1	1.410	H5-X1	1.415	1.422	C5-X1	1.869	1.866
H3-X1-O2	117.0	H3-O2-X1	109.7	114.9	O6-C5	1.227	1.221
H4-X1-O2	117.0	H4-X1-O2	98.0	100.8	H7-C5	1.106	1.115
H5-X1-O2	117.0	H5-X1-O2	98.0	100.8	H3-X1-O2	117.8	117.8
H4-X1-O2-H3	120.0	H4-X1-O2-H3	132.2	47.9	H4-X1-O2	117.8	117.8
H5-X1-O2-H3	-120.0	H5-X1-O2-H3	-132.4	-47.9	C5-X1-O2	116.1	119.2
					O6-C5-X1	119.7	124.3
					H7-C5-X1	117.9	114.0
					H4-X1-O2-H3	122.8	121.5
					H5-X1-O2-H3	241.4	-119.2
					O6-C5-X1-O2	180.0	0.1
					H7-C5-X1-O2	-0.0	-180.0

Table 6.2: Geometrical parameters of **1Z**, **1E** and the rotational transition state **RTS1** for FPA (FHA) at B3LYP/6-31+G* [**L1**] and MP2/6-31+G* [**L3**] theoretical levels. All bond distances are in angstrom (Å) and angles in degrees (°).

Parameters	1Z		1E		RTS1	
	L1	L3	L1	L3	L1	L3
X2-C1	1.876 (1.357)	1.860 (1.362)	1.880 (1.381)	1.859 (1.388)	1.928 (1.465)	1.894 (1.462)
O3-C1	1.217 (1.227)	1.232 (1.236)	1.211 (1.216)	1.227 (1.225)	1.207 (1.202)	1.226 (1.215)
O4-X2	1.656 (1.401)	1.662 (1.410)	1.685 (1.412)	1.689 (1.423)	1.686 (1.451)	1.687 (1.459)
H5-O4	0.981 (0.985)	0.985 (0.988)	0.969 (0.973)	0.973 (0.976)	0.969 (0.971)	0.974 (0.976)
H6-C1	1.110 (1.101)	1.107 (1.099)	1.114 (1.104)	1.110 (1.101)	1.113 (1.103)	1.109 (1.101)
H7-X2	1.439 (1.013)	1.429 (1.016)	1.422 (1.018)	1.414 (1.020)	1.423 (1.025)	1.415 (1.026)
O3-C1-X2	120.6 (122.0)	121.0 (122.0)	123.0 (123.0)	122.6 (122.7)	123.6 (124.9)	122.9 (124.5)
O4-X2-C1	97.5 (117.0)	97.6 (115.9)	98.3 (115.6)	97.9 (113.5)	98.1 (106.2)	97.6 (105.1)
H5-O4-X2	110.8 (102.1)	110.5 (101.8)	110.8 (104.5)	110.5 (103.6)	110.1 (102.1)	109.9 (101.4)
H6-C1-X2	118.2 (113.3)	118.2 (113.5)	114.9 (112.3)	115.6 (112.6)	115.9 (112.4)	116.8 (112.7)
H7-X2-C1	89.6 (122.4)	90.2 (120.6)	93.2 (116.2)	94.0 (114.7)	92.2 (107.0)	92.8 (106.9)
O4-X2-C1-O3	-7.1 (-11.0)	-5.6 (-13.2)	163.1 (159.6)	160.6 (157.7)	54.1 (55.4)	51.7 (53.9)
H5-O4-X2-C1	-7.5 (6.0)	-10.6 (7.7)	130.3 (117.5)	129.5 (119.9)	124.0 (123.6)	125.7 (125.6)

Contd.....

Parameters	1Z		1E		RTS1	
	L1	L3	L1	L3	L1	L3
H6-C1-X2-O4	174.5	175.5	-20.0	-22.3	-124.7	-126.8
	(172.1)	(170.9)	(-23.4)	(-25.9)	(-124.5)	(-125.8)
H7-X2-C1-O3	250.0	251.2	64.2	61.4	-43.2	-46.3
	(-154.1)	(-150.0)	(26.5)	(30.1)	(-54.1)	(-54.6)

Table 6.3: Geometrical parameters of **2Z**, **2E** and the rotational transition state **RTS2** for FPA (FHA) at B3LYP/6-31+G* [**L1**] and MP2/6-31+G* [**L3**] theoretical levels. All bond distances are in angstrom (Å) and angles in degrees (°).

Parameters	2Z		2E		RTS2	
	L1	L3	L1	L3	L1	L3
X2-C1	1.692	1.689	1.694	1.690	1.829	1.812
	(1.280)	(1.290)	(1.274)	(1.283)	(1.229)	(1.233)
O3-C1	1.349	1.357	1.359	1.369	1.300	1.301
	(1.344)	(1.348)	(1.350)	(1.355)	(1.397)	(1.405)
O4-X2	1.714	1.714	1.695	1.695	1.816	1.815
	(1.429)	(1.436)	(1.418)	(1.427)	(1.339)	(1.346)
H5-O3	0.979	0.985	0.973	0.978	0.994	0.998
	(0.977)	(0.982)	(0.975)	(0.980)	(0.973)	(0.977)
H6-C1	1.087	1.087	1.088	1.088	1.096	1.094
	(1.084)	(1.082)	(1.088)	(1.087)	(1.097)	(1.096)
H7-O4	0.968	0.974	0.970	0.975	0.972	0.977
	(0.969)	(0.974)	(0.970)	(0.974)	(0.979)	(0.981)
O3-C1-X2	126.0	125.6	126.7	126.6	122.0	122.1
	(126.8)	(126.7)	(121.1)	(120.8)	(127.2)	(127.3)
O4-X2-C1	94.8	94.7	98.8	98.2	73.5	74.2
	(108.6)	(107.7)	(110.6)	(109.3)	(179.5)	(179.9)
H5-O3-C1	108.0	106.8	111.2	110.5	106.3	105.5
	(107.8)	(107.2)	(108.3)	(107.7)	(108.8)	(107.9)

Contd.....

Parameters	2Z		2E		RTS2	
	L1	L3	L1	L3	L1	L3
H6-C1-X2	121.2	121.3	122.1	121.8	128.0	127.9
	(118.6)	(118.4)	(125.2)	(125.2)	(126.0)	(126.4)
H7-O4-X2	112.3	112.0	110.7	110.6	108.3	108.2
	(103.0)	(102.3)	(102.5)	(101.8)	(107.3)	(106.5)
O4-X2-C1-O3	-0.0	-1.2	174.5	174.3	81.7	79.3
	(0.0)	(0.0)	(180.0)	(180.0)	(89.5)	(91.6)
H5-O3-C1-X2	-0.0	-2.0	2.2	2.5	8.2	7.5
	(0.0)	(-0.0)	(0.0)	(-0.0)	(0.1)	(-0.1)
H6-C1-X2-O4	180.0	176.6	-2.5	-2.5	-89.7	-92.7
	(180.0)	(180.0)	(-0.0)	(0.0)	(-90.5)	(-88.4)
H7-O4-X2-C1	179.7	156.0	140.7	137.9	99.0	101.2
	(180.1)	(179.9)	(180.0)	(180.0)	(-89.5)	(-91.6)

Table 6.4: Geometrical parameters of **2Z**, **2E2** and the rotational transition state **RTS3** and **RTS4** for FPA (FHA) at B3LYP/6-31+G* [**L1**] and MP2/6-31+G* [**L3**] theoretical levels. All bond distances are in angstrom (Å) and angles in degrees (°).

Parameters	2Z2		2E2		RTS3		RTS4	
	L1	L3	L1	L3	L1	L3	L1	L3
X2-C1	1.683	1.681	1.689	1.687	1.677	1.677	1.679	1.679
	(1.274)	(1.284)	(1.272)	(1.281)	(1.275)	(1.285)	(1.272)	(1.282)
O3-C1	1.360	1.369	1.366	1.375	1.375	1.384	1.386	1.396
	(1.356)	(1.359)	(1.358)	(1.361)	(1.367)	(1.371)	(1.368)	(1.373)
O4-X2	1.685	1.684	1.700	1.701	1.678	1.677	1.686	1.686
	(1.409)	(1.417)	(1.418)	(1.430)	(1.400)	(1.409)	(1.409)	(1.420)
H5-O3	0.969	0.974	0.968	0.973	0.971	0.975	0.970	0.973
	(0.969)	(0.973)	(0.968)	(0.973)	(0.971)	(0.975)	(0.971)	(0.974)
H6-C1	1.091	1.092	1.091	1.092	1.094	1.094	1.092	1.092

Contd......

Parameters	2Z2		2E2		RTS3		RTS4	
	L1	L3	L1	L3	L1	L3	L1	L3
H7-O4	1.087	1.086	1.092	1.091	1.088	1.087	1.093	1.091
	(0.969)	(0.974)	(0.969)	(0.974)	(0.970)	(0.974)	(0.969)	(0.974)
	(0.970)	(0.974)	(0.970)	(0.974)	(0.970)	(0.975)	(0.970)	(0.975)
O3-C1-X2	127.2	126.2	121.7	121.1	127.7	126.7	121.2	120.7
	(124.7)	(124.2)	(118.6)	(117.9)	(125.7)	(125.3)	(119.2)	(118.6)
O4-X2-C1	101.7	100.9	97.5	96.9	102.4	101.8	99.7	99.0
	(112.0)	(110.6)	(109.6)	(108.1)	(112.7)	(111.5)	(110.4)	(109.0)
H5-O3-C1	110.5	109.5	110.4	109.5	111.1	110.3	110.9	110.1
	(109.8)	(109.1)	(109.4)	(108.9)	(111.4)	(110.7)	(110.9)	(110.2)
H6-C1-X2	115.7	116.1	121.1	120.9	116.1	116.4	123.3	123.3
	(116.8)	(116.7)	(123.2)	(123.2)	(116.5)	(116.3)	(123.2)	(123.3)
H7-O4-X2	110.2	110.0	110.0	110.1	110.5	110.2	110.5	110.3
	(102.2)	(101.5)	(102.3)	(101.5)	(102.5)	(101.6)	(102.7)	(101.9)
H4-O2-C1-O3	-0.0	-0.0	175.9	173.6	-6.6	-6.9	177.9	178.1
	(-0.0)	(-0.0)	(180.0)	(180.0)	(-5.3)	(-5.4)	(176.4)	(176.5)
H5-O3-C1-X2	180.0	180.0	180.6	174.3	99.2	100.4	82.3	80.0
	(180.0)	(180.0)	(180.0)	(180.0)	(100.5)	(101.7)	(94.6)	(95.2)
H6-C1-X2-O4	180.0	180.0	-2.1	-2.9	179.2	179.3	3.6	4.2
	(180.0)	(180.0)	(-0.0)	(-0.0)	(180.0)	(180.1)	(0.8)	(1.0)
H7-O4-X2-C1	180.0	180.0	156.3	148.6	176.8	176.5	181.7	181.9
	(180.0)	(180.0)	(180.0)	(180.0)	(179.1)	(179.4)	(181.3)	(181.5)

Table 6.5: Geometrical parameters of **dipolar3** and tautomerization transition state (**PTS**) for FPA (FHA) at B3LYP/6-31+G* (**L1**) and MP2/6-31+G* [**L3**] theoretical levels. All bond distances are in angstrom (Å) and angles in degrees (°).

Parameters	Dipolar 3		R=H	PTS	
	L1	L3		L1	L3
X2-C1	1.703	1.681	X2-C1	1.774	1.749
	(1.303)	(1.311)		(1.309)	(1.314)
O3-C1	1.346	1.364	O3-C1	1.282	1.300
	(1.326)	(1.333)		(1.287)	(1.293)
O4-X2	1.510	1.511	O4-X2	1.683	1.686
	(1.311)	(1.303)		(1.405)	(1.411)
H5-O3	0.976	0.979	H5-O4	0.971	0.975
	(0.995)	(0.999)		(0.973)	(0.978)
H6-C1	1.088	1.088	H6-C1	1.091	1.090
	(1.081)	(1.080)		(1.088)	(1.086)
H7-X2	1.416	1.406	H7-X2	1.757	1.746
	(1.022)	(1.024)		(1.340)	(1.334)
O3-C1-X2	125.1	126.1	O3-C1-X2	105.1	105.4
	(117.4)	(116.8)		(106.5)	(106.5)
O4-X2-C1	122.6	126.6	O4-X2-C1	109.6	109.0
	(120.9)	(121.1)		(122.4)	(122.0)
H5-O3-C1	110.1	109.7	H5-O4-C1	111.1	110.6
	(103.7)	(103.1)		(104.4)	(103.7)
H6-C1-X2	119.0	118.1	H6-C1-X2	134.5	134.1
	(123.3)	(123.4)		(128.2)	(128.1)
H7-X2-C1	103.4	105.8	H7-X2-C1	61.9	62.6
	(120.7)	(120.6)		(76.6)	(77.0)
O4-X2-C1-O3	-23.9	-19.8	O4-X2-C1-O3	137.7	139.3
	(-0.0)	(0.0)		(165.7)	(164.5)
H5-O3-C1-X2	24.6	24.1	H5-O4-C1-X2	131.9	133.3
	(0.0)	(-0.0)		(132.6)	(131.1)

Contd.....

Parameters	Dipolar 3		R=H	PTS	
	L1	L3		L1	L3
H6-C1-X2-O4	177.1 (180.0)	178.8 (180.0)	H6-C1-X2-O4	-56.5 (-14.8)	-56.0 (-16.6)
H7-X2-C1-O3	197.3 (180.0)	190.1 (180.0)	H7-X2-C1-O3	10.5 (3.2)	11.1 (3.8)

The $(CF_3)_2POH$ being the known example of isolated phosphinous acid. The influence of CH_3 and CF_3 substituents at carbonyl carbon of the FPA has been analysed. The resulting variations in C-X (X=N, P in FHA and FPA respectively) and C-O bond lengths are recorded in Table 6.6. The CH_3 substitution results in elongation of both C-X and C-O bonds in all isomers of both FPA and FHA. The contraction of C-X bond in **1Z** and **1E** conformer of FHA is observed with substitution of CF_3 group. Negligible to small elongation of C-X bond is indicated in other isomeric forms of CF_3 substituted FHA. The elongation of C-X bond is reflected in all the isomeric forms of CF_3 substituted FPA. The contraction of C-O with a few exceptions is observed for CF_3 substituted FPA and FHA. These variations suggest that both the substituents affect C-X bond strength comparatively more than the C-O bond. Full set of geometrical parameters for CH_3 and CF_3 substituted FHA and FPA are listed in Tables 6.7-6.14.

Table 6.6: Variation of important bond distances in angstrom (Å) on CH_3 and CF_3 substitution in FPA (FHA) at MP2/6-31+G* **[L3]** theoretical level.

Species	R=CH$_3$		R=CF$_3$	
	C-X	C-O	C-X	C-O
1Z	0.019	0.002	0.013	-0.005
	(0.010)	(0.004)	(-0.015)	(0.000)
1E	0.016	0.006	0.017	-0.003
	(0.012)	(0.005)	(-0.010)	(-0.002)
2Z	0.008	0.004	0.010	-0.003
	(0.004)	(0.007)	(0.000)	(-0.005)
2E	0.007	0.012	0.008	0.008
	(0.005)	(0.011)	(0.004)	(-0.002)
2Z2	0.008	0.003	0.013	-0.003
	(0.004)	(0.006)	(0.001)	(-0.007)
2E2	0.004	0.015	0.005	0.008
	(0.006)	(0.012)	(0.006)	(-0.006)

Table 6.7: Geometrical parameters of **1Z**, **1E** and the rotational transition state **RTS1** for CH$_3$ substituted FPA (FHA) at B3LYP/6-31+G* [**L1**] and MP2/6-31+G* [**L3**] theoretical levels. All bond distances are in angstrom (Å) and angles in degrees (°).

Parameters	1Z		1E		RTS1	
	L1	L3	L1	L3	L1	L3
X2-C1	1.898	1.879	1.896	1.875	1.934	1.903
	(1.367)	(1.372)	(1.394)	(1.400)	(1.471)	(1.466)
O3-C1	1.221	1.234	1.218	1.233	1.216	1.233
	(1.233)	(1.240)	(1.222)	(1.230)	(1.213)	(1.255)
O4-X2	1.656	1.661	1.680	1.684	1.683	1.686
	(1.403)	(1.413)	(1.416)	(1.426)	(1.440)	(1.450)
C5-C1	1.509	1.504	1.513	1.507	1.512	1.506
	(1.510)	(1.504)	(1.511)	(1.503)	(1.498)	(1.494)
H6-C5	1.098	1.096	1.093	1.092	1.093	1.092
	(1.095)	(1.093)	(1.095)	(1.093)	(1.091)	(1.090)
H7-C5	1.096	1.094	1.098	1.095	1.098	1.096
	(1.095)	(1.093)	(1.091)	(1.090)	(1.097)	(1.095)
H8-C5	1.094	1.092	1.097	1.096	1.098	1.096
	(1.095)	(1.093)	(1.096)	(1.094)	(1.097)	(1.095)
H9-O4	0.983	0.987	0.971	0.975	0.973	0.977
	(0.987)	(0.990)	(0.972)	(0.976)	(0.980)	(0.984)
H10-X2	1.439	1.430	1.426	1.418	1.428	1.420
	(1.013)	(1.016)	(1.018)	(1.020)	(1.025)	(1.027)
O3-C1-X2	117.7	118.4	116.8	117.3	120.8	120.6
	(119.5)	(119.8)	(118.8)	(118.7)	(120.8)	(120.8)
O4-X2-C1	96.9	97.3	105.2	104.1	97.9	97.1
	(116.1)	(115.1)	(117.0)	(115.0)	(106.2)	(105.1)

Contd.....

Parameters	1Z		1E		RTS1	
	L1	L3	L1	L3	L1	L3
C5-C1-X2	119.2	118.8	119.8	119.5	116.2	116.6
	(116.2)	(115.9)	(116.7)	(116.4)	(113.8)	(113.5)
H6-C5-C1	108.5	108.6	109.6	109.0	110.6	110.1
	(113.0)	(112.5)	(110.4)	(110.2)	(111.0)	(110.6)
H7-C5-C1	111.8	111.4	110.3	111.0	110.0	109.8
	(109.0)	(108.8)	(108.4)	(108.0)	(108.8)	(108.4)
H8-C5-C1	110.2	109.7	110.5	109.8	109.5	109.5
	(108.8)	(108.6)	(111.2)	(110.8)	(109.5)	(109.3)
H9-O4-X2	109.7	109.3	115.8	114.8	113.9	113.1
	(101.6)	(101.3)	(104.4)	(103.4)	(106.5)	(105.8)
H10-X2-C1	89.7	90.0	92.7	93.3	92.1	92.6
	(121.7)	(119.8)	(113.8)	(112.3)	(107.3)	(107.3)
O4-X2-C1-O3	3.4	2.6	199.2	204.9	-80.2	-80.6
	(11.3)	(13.3)	(157.0)	(154.7)	(-65.2)	(-65.1)
C5-C1-X2-O4	180.7	180.0	24.3	31.2	97.6	96.4
	(188.3)	(189.2)	(-26.0)	(-29.1)	(113.4)	(113.3)
H6-C5-C1-X2	-85.4	-86.6	-190.4	-183.8	-173.9	-172.4
	(1.3)	(2.0)	(66.2)	(66.4)	(-178.4)	(-178.3)
H7-C5-C1-X2	32.9	32.1	-70.0	-62.4	-51.9	-50.7
	(123.0)	(123.9)	(186.6)	(187.0)	(-56.3)	(-56.4)
H8-C5-C1-X2	155.9	154.6	47.8	55.5	64.8	66.6
	(240.3)	(241.2)	(307.4)	(307.6)	(59.4)	(59.6)
H9-O4-X2-C1	7.9	10.4	60.0	61.3	51.6	52.2
	(-6.4)	(-7.6)	(119.8)	(122.3)	(52.2)	(52.8)
H10-X2-C1-O3	105.8	105.5	-58.6	-52.4	20.6	20.6
	(150.2)	(146.5)	(26.3)	(29.1)	(47.5)	(46.6)

Table 6.8: Geometrical parameters of **2Z**, **2E** and the rotational transition state **RTS2** for CH$_3$ substituted FPA (FHA) at B3LYP/6-31+G* [**L1**] and MP2/6-31+G* [**L3**] theoretical levels. All bond distances are in angstrom (Å) and angles in degrees (°).

Parameters	2Z		2E		RTS2	
	L1	L3	L1	L3	L1	L3
X2-C1	1.705	1.697	1.703	1.697	1.879	1.857
	(1.286)	(1.294)	(1.280)	(1.288)	(1.233)	(1.236)
O3-C1	1.350	1.361	1.370	1.381	1.295	1.298
	(1.351)	(1.355)	(1.362)	(1.366)	(1.410)	(1.418)
O4-X2	1.725	1.720	1.699	1.697	1.765	1.772
	(1.434)	(1.440)	(1.421)	(1.429)	(1.342)	(1.350)
C5-C1	1.495	1.492	1.493	1.490	1.483	1.480
	(1.493)	(1.488)	(1.494)	(1.488)	(1.515)	(1.514)
H6-C5	1.095	1.095	1.091	1.091	1.097	1.094
	(1.092)	(1.091)	(1.090)	(1.089)	(1.096)	(1.093)
H7-C5	1.098	1.095	1.098	1.096	1.094	1.093
	(1.096)	(1.094)	(1.096)	(1.094)	(1.096)	(1.091)
H8-C5	1.097	1.095	1.098	1.096	1.099	1.097
	(1.096)	(1.094)	(1.096)	(1.094)	(1.091)	(1.093)
H9-O3	0.982	0.987	0.974	0.979	1.002	1.005
	(0.977)	(0.982)	(0.975)	(0.980)	(0.972)	(0.978)
H10-O4	0.969	0.974	0.970	0.975	0.969	0.975
	(0.969)	(0.974)	(0.969)	(0.974)	(0.978)	(0.981)
O3-C1-X2	122.9	123.3	120.7	120.9	116.8	117.2
	(124.4)	(124.5)	(117.2)	(117.0)	(124.5)	(124.8)
O4-X2-C1	95.1	95.1	100.6	100.0	87.2	84.8
	(108.7)	(108.1)	(112.1)	(111.0)	(176.1)	(178.9)

Contd......

Parameters	2Z		2E		RTS2	
	L1	L3	L1	L3	L1	L3
C5-C1-X2	123.5 (120.6)	123.5 (120.6)	127.9 (129.2)	128.1 (129.5)	129.0 (127.4)	128.9 (126.4)
H6-C5-C1	111.3 (109.9)	111.0 (109.6)	110.9 (110.7)	110.6 (110.5)	108.4 (110.2)	108.1 (109.3)
H7-C5-C1	110.4 (110.1)	110.1 (109.9)	110.1 (109.7)	110.0 (109.3)	113.7 (110.4)	113.1 (112.3)
H8-C5-C1	110.1 (110.1)	109.8 (109.9)	109.7 (109.7)	109.2 (109.3)	107.7 (109.9)	107.5 (109.3)
H9-O3-C1	107.4 (107.6)	106.7 (107.1)	111.3 (107.8)	110.7 (107.2)	104.5 (108.1)	103.9 (107.1)
H10-O4-X2	111.9 (102.9)	111.6 (102.1)	110.3 (102.1)	109.8 (101.3)	109.0 (107.3)	109.0 (106.5)
O4-X2-C1-O3	1.6 (-0.0)	1.6 (-0.0)	175.8 (180.0)	175.4 (180.1)	85.6 (91.0)	85.0 (90.3)
C5-C1-X2-O4	183.5 (180.0)	184.0 (180.0)	-1.8 (-0.0)	-1.9 (0.0)	-91.9 (-90.6)	-90.9 (-90.0)
H6-C5-C1-X2	-2.1 (-0.0)	-2.6 (0.0)	-5.0 (0.2)	-6.9 (0.5)	51.6 (122.5)	49.6 (59.5)
H7-C5-C1-X2	118.5 (120.6)	117.9 (120.6)	115.7 (121.3)	113.9 (121.6)	175.1 (240.7)	172.9 (180.6)
H8-C5-C1-X2	237.3 (239.3)	236.9 (239.4)	233.8 (239.2)	232.1 (239.5)	298.1 (1.6)	295.6 (-58.4)
H9-O3-C1-X2	1.6 (-0.0)	2.1 (-0.0)	1.4 (0.0)	1.1 (0.0)	3.5 (-1.3)	4.1 (-0.4)
H10-O4-X2-C1	214.5 (180.0)	213.9 (179.9)	146.9 (180.0)	146.8 (180.1)	105.0 (-91.1)	106.9 (-90.4)

Table 6.9: Geometrical parameters of **2Z2**, **2E2** and the rotational transition state **RTS3** and **RTS4** for CH$_3$ substituted FPA (FHA) at B3LYP/6-31+G* **[L1]** and MP2/6-31+G* **[L3]** theoretical levels. All bond distances are in angstrom (Å) and angles in degrees (°).

Parameters	2Z2		2E2		RTS3		RTS4	
	L1	L3	L1	L3	L1	L3	L1	L3
X2-C1	1.693	1.689	1.697	1.691	1.687	1.686	1.685	1.685
	(1.279)	(1.288)	(1.278)	(1.287)	(1.279)	(1.288)	(1.277)	(1.287)
O3-C1	1.363	1.372	1.377	1.390	1.382	1.391	1.400	1.409
	(1.362)	(1.365)	(1.368)	(1.373)	(1.376)	(1.380)	(1.382)	(1.386)
O4-X2	1.689	1.687	1.705	1.702	1.683	1.681	1.694	1.693
	(1.412)	(1.421)	(1.421)	(1.430)	(1.406)	(1.414)	(1.414)	(1.424)
C5-C1	1.503	1.500	1.497	1.494	1.503	1.500	1.494	1.491
	(1.500)	(1.495)	(1.501)	(1.496)	(1.499)	(1.494)	(1.498)	(1.493)
H6-C5	1.095	1.095	1.092	1.092	1.096	1.096	1.093	1.093
	(1.092)	(1.091)	(1.090)	(1.089)	(1.092)	(1.091)	(1.091)	(1.090)
H7-C5	1.100	1.097	1.100	1.097	1.098	1.095	1.099	1.096
	(1.098)	(1.096)	(1.098)	(1.096)	(1.097)	(1.094)	(1.097)	(1.095)
H8-C5	1.100	1.098	1.101	1.099	1.100	1.097	1.100	1.097
	(1.098)	(1.096)	(1.098)	(1.096)	(1.098)	(1.096)	(1.098)	(1.094)
H9-O3	0.971	0.976	0.971	0.976	0.972	0.975	0.970	0.974
	(0.969)	(0.974)	(0.969)	(0.973)	(0.971)	(0.975)	(0.971)	(0.975)
H10-O4	0.969	0.974	0.969	0.974	0.970	0.975	0.969	0.974
	(0.970)	(0.975)	(0.969)	(0.974)	(0.970)	(0.975)	(0.970)	(0.975)
O3-C1-X2	123.1	122.6	115.8	114.9	123.4	123.0	115.7	115.4
	(122.0)	(121.7)	(114.6)	(114.0)	(123.0)	(122.9)	(115.3)	(114.9)
O4-X2-C1	101.3	100.7	99.1	98.7	101.8	101.4	100.9	100.2
	(111.8)	(110.7)	(111.3)	(110.1)	(112.5)	(111.5)	(111.8)	(110.4)

Contd.....

Parameters	2Z2		2E2		RTS3		RTS4	
	L1	L3	L1	L3	L1	L3	L1	L3
C5-C1-X2	119.9 (119.4)	120.3 (119.3)	127.7 (127.9)	128.4 (128.2)	120.8 (119.7)	121.0 (119.5)	129.3 (128.3)	129.4 (128.1)
H6-C5-C1	111.5 (109.8)	111.1 (109.3)	111.0 (110.6)	110.6 (110.4)	111.9 (110.6)	111.5 (110.2)	111.3 (111.5)	111.0 (111.0)
H7-C5-C1	110.9 (110.9)	110.6 (110.7)	110.5 (110.3)	110.3 (110.0)	110.0 (109.9)	109.6 (109.5)	109.5 (109.2)	109.2 (109.3)
H8-C5-C1	110.9 (110.9)	110.8 (110.7)	110.6 (110.3)	110.2 (110.0)	111.0 (110.6)	110.8 (110.4)	110.7 (110.1)	110.4 (109.6)
H9-O3-C1	110.3 (109.8)	109.2 (109.0)	109.9 (109.4)	108.7 (108.8)	111.1 (111.2)	110.2 (110.4)	110.6 (110.5)	109.8 (109.8)
H10-O4-X2	109.8 (102.0)	109.8 (101.2)	109.4 (101.8)	109.5 (101.0)	109.9 (102.1)	109.8 (101.3)	110.1 (102.1)	109.9 (101.3)
O4-X2-C1-O3	0.0 (-0.0)	-2.0 (0.0)	176.2 (180.1)	171.8 (180.0)	-6.5 (-5.2)	-7.0 (-5.4)	178.5 (176.3)	178.0 (175.2)
C5-C1-X2-O4	180.1 (180.0)	178.1 (180.0)	-2.0 (0.0)	-3.5 (0.0)	178.7 (179.6)	179.0 (180.0)	4.2 (0.6)	5.0 (1.1)
H6-C5-C1-X2	-0.1 (0.0)	0.4 (-0.0)	-2.6 (0.1)	-7.3 (0.2)	-0.6 (-1.3)	-1.2 (-1.9)	-0.6 (-2.3)	-1.6 (-16.6)
H7-C5-C1-X2	119.9 (120.0)	120.3 (119.9)	117.9 (120.4)	113.3 (120.5)	120.2 (119.7)	119.4 (119.0)	120.3 (118.9)	119.1 (104.0)
H8-C5-C1-X2	239.9 (240.0)	240.6 (240.1)	237.2 (239.8)	232.6 (239.9)	238.7 (238.0)	238.4 (237.4)	238.1 (236.3)	237.2 (221.6)
H9-O3-C1-X2	180.0 (180.0)	173.3 (180.0)	163.8 (180.0)	152.2 (180.0)	100.5 (101.9)	102.2 (102.7)	83.4 (97.6)	83.7 (97.3)
H10-O4-X2-C1	180.3 (180.1)	169.5 (180.0)	178.5 (180.1)	164.8 (180.1)	175.8 (179.1)	175.6 (179.4)	181.4 (181.3)	180.5 (179.2)

Table 6.10: Geometrical parameters of **PTS** for CH_3 substituted FPA (FHA) at B3LYP/6-31+G* **[L1]** and MP2/6-31+G* **[L3]** theoretical levels. All bond distances are in angstrom (Å) and angles in degrees (°).

Parameters	L1	L3
X2-C1	1.789	1.762
	(1.317)	(1.321)
O3-C1	1.288	1.303
	(1.295)	(1.300)
O4-X2	1.690	1.693
	(1.409)	(1.414)
C5-C1	1.488	1.485
	(1.486)	(1.482)
H6-C5	1.093	1.092
	(1.095)	(1.093)
H7-C5	1.096	1.094
	(1.095)	(1.093)
H8-C5	1.099	1.097
	(1.093)	(1.091)
H9-O4	0.971	0.975
	(0.973)	(0.978)
H10-X2	1.746	1.738
	(1.326)	(1.321)
O3-C1-X2	103.9	104.4
	(104.8)	(105.0)
O4-X2-C1	107.9	107.2
	(121.9)	(121.3)
C5-C1-X2	134.5	134.2
	(129.5)	(129.3)
H6-C5-C1	111.0	110.5
	(109.5)	(109.4)

Contd.....

Parameters	L1	L3
H7-C5-C1	110.6 (109.3)	110.3 (109.0)
H8-C5-C1	109.0 (110.8)	109.0 (110.3)
H9-O4-X2	110.8 (104.3)	110.3 (103.6)
H10-X2-C1	61.7 (77.3)	62.4 (77.7)
O4-X2-C1-O3	138.0 (166.1)	140.3 (164.5)
C5-C1-X2-O4	-52.6 (-14.1)	-51.4 (-16.4)
H6-C5-C1-X2	21.8 (117.7)	20.2 (118.4)
H7-C5-C1-X2	143.9 (235.3)	141.8 (236.3)
H8-C5-C1-X2	262.1 (356.9)	260.3 (357.7)
H9-O4-X2-C1	132.8 (131.5)	134.1 (131.0)
H10-X2-C1-O3	11.3 (3.5)	12.1 (4.3)

Table 6.11: Geometrical parameters of **1Z**, **1E** and the rotational transition state **RTS1** for CF$_3$ substituted FPA (FHA) at B3LYP/6-31+G* [**L1**] and MP2/6-31+G* [**L3**] theoretical levels. All bond distances are in angstrom (Å) and angles in degrees (°).

R=CF$_3$	1Z		1E		RTS1	
	L1	L3	L1	L3	L1	L3
X2-C1	1.889	1.873	1.895	1.876	1.932	1.903
	(1.344)	(1.347)	(1.373)	(1.378)	(1.446)	(1.443)
O3-C1	1.211	1.227	1.208	1.224	1.207	1.225
	(1.227)	(1.236)	(1.213)	(1.223)	(1.203)	(1.217)
O4-X2	1.652	1.658	1.661	1.666	1.672	1.675
	(1.389)	(1.398)	(1.404)	(1.415)	(1.435)	(1.445)
C5-C1	1.551	1.538	1.558	1.544	1.557	1.542
	(1.542)	(1.531)	(1.558)	(1.545)	(1.548)	(1.536)
F6-C5	1.356	1.360	1.348	1.351	1.354	1.357
	(1.362)	(1.364)	(1.348)	(1.350)	(1.339)	(1.342)
F7-C5	1.349	1.352	1.334	1.337	1.337	1.340
	(1.342)	(1.346)	(1.357)	(1.359)	(1.344)	(1.346)
F8-C5	1.338	1.341	1.364	1.367	1.351	1.355
	(1.341)	(1.343)	(1.336)	(1.341)	(1.351)	(1.355)
H9-O4	0.979	0.983	0.972	0.976	0.972	0.977
	(0.985)	(0.989)	(0.974)	(0.978)	(0.980)	(0.984)
H10-X2	1.434	1.426	1.428	1.420	1.427	1.420
	(1.010)	(1.014)	(1.017)	(1.020)	(1.024)	(1.026)
O3-C1-X2	121.3	122.2	120.1	120.6	124.2	124.4
	(122.8)	(123.3)	(122.9)	(123.2)	(125.1)	(125.3)
O4-X2-C1	96.7	96.9	105.9	104.6	98.9	98.1
	(117.9)	(116.8)	(119.2)	(116.8)	(106.8)	(105.7)
C5-C1-X2	118.6	118.2	120.1	119.6	116.3	116.4
	(115.1)	(114.7)	(115.9)	(115.4)	(112.9)	(112.4)
F6-C5-C1	109.8	109.6	110.4	110.4	108.8	108.8

Contd.....

R=CF$_3$	1Z		1E		RTS1	
	L1	L3	L1	L3	L1	L3
F7-C5-C1	111.4 (109.8)	111.3 (109.7)	110.9 (112.6)	110.6 (112.8)	111.3 (112.6)	111.5 (112.9)
F8-C5-C1	110.7 (112.1)	110.6 (112.2)	111.6 (109.1)	111.5 (108.7)	110.7 (110.2)	110.5 (109.9)
H9-O4-X2	110.5 (112.6)	110.4 (112.2)	109.9 (116.9)	109.9 (116.1)	109.1 (115.4)	108.9 (114.6)
H10-X2-C1	102.6 (89.8)	102.3 (90.0)	104.9 (90.4)	103.9 (90.9)	107.3 (89.8)	106.4 (90.1)
O4-X2-C1-O3	126.8 (-3.5)	125.0 (-2.7)	115.4 (190.8)	114.1 (193.8)	108.7 (-83.4)	108.4 (-82.8)
C5-C1-X2-O4	(-5.9) 179.9	(-9.6) 180.6	201.9 (14.6)	205.0 (17.0)	(-67.4) 97.2	(-66.8) 96.6
F6-C5-C1-X2	176.4 (27.7)	175.0 (-27.7)	25.6 (56.5)	29.8 (52.2)	112.3 (72.4)	112.8 (72.2)
F7-C5-C1-X2	(-7.0) 90.6	(-13.1) 90.6	(-76.6) 178.4	(-78.9) 174.3	(-177.2) 192.8	(-176.2) 192.8
F8-C5-C1-X2	112.9 (211.6)	106.8 (211.7)	44.1 (298.2)	41.9 (294.1)	(-56.2) 314.2	(-55.1) 314.2
H9-O4-X2-C1	233.2 (-13.3)	227.1 (-17.5)	163.7 (49.7)	161.3 (51.7)	63.2 (53.2)	64.2 (53.4)
H10-X2-C1-O3	(4.0) 253.6 (193.4)	(6.9) 254.0 (200.9)	250.8 (-66.6) (-20.5)	248.5 (-63.3) (-23.8)	56.6 (17.3) (48.0)	56.9 (18.3) (47.2)

Table 6.12: Geometrical parameters of **2Z**, **2E** and the rotational transition state **RTS2** for CF₃ substituted FPA (FHA) at B3LYP/6-31+G* [L1] and MP2/6-31+G* [L3] theoretical levels. All bond distances are in angstrom (Å) and angles in degrees (°).

Parameters	2Z		2E		RTS2	
	L1	L3	L1	L3	L1	L3
X2-C1	1.697	1.699	1.695	1.698	1.819	1.788
	(1.279)	(1.290)	(1.276)	(1.287)	(1.224)	(1.228)
O3-C1	1.347	1.354	1.370	1.377	1.306	1.312
	(1.340)	(1.343)	(1.349)	(1.353)	(1.381)	(1.387)
O4-X2	1.703	1.702	1.670	1.671	1.835	1.842
	(1.414)	(1.422)	(1.401)	(1.411)	(1.331)	(1.338)
C5-C1	1.503	1.495	1.513	1.504	1.542	1.534
	(1.516)	(1.507)	(1.530)	(1.520)	(1.541)	(1.534)
F6-C5	1.357	1.361	1.341	1.345	1.347	1.352
	(1.340)	(1.343)	(1.348)	(1.351)	(1.351)	(1.354)
F7-C5	1.352	1.354	1.358	1.360	1.343	1.345
	(1.352)	(1.354)	(1.345)	(1.347)	(1.352)	(1.353)
F8-C5	1.352	1.354	1.355	1.357	1.350	1.352
	(1.352)	(1.354)	(1.348)	(1.351)	(1.351)	(1.353)
H9-O3	0.979	0.985	0.973	0.979	0.987	0.992
	(0.977)	(0.983)	(0.975)	(0.981)	(0.973)	(0.978)
H10-O4	0.969	0.974	0.970	0.976	0.974	0.980
	(0.970)	(0.975)	(0.970)	(0.975)	(0.980)	(0.983)
O3-C1-X2	125.5	125.9	120.3	120.6	123.6	125.2
	(127.2)	(127.7)	(120.2)	(120.3)	(128.3)	(128.9)
O4-X2-C1	93.7	93.2	104.7	103.8	68.9	69.3
	(108.5)	(107.4)	(113.3)	(111.7)	(178.4)	(178.8)
C5-C1-X2	122.0	121.8	131.4	131.4	125.8	125.0
	(119.4)	(119.0)	(126.6)	(126.1)	(123.8)	(124.1)

Contd.....

Parameters	2Z		2E		RTS2	
	L1	L3	L1	L3	L1	L3
F6-C5-C1	110.4 (112.0)	110.4 (112.0)	112.9 (110.9)	113.0 (110.8)	111.0 (110.0)	110.3 (110.0)
F7-C5-C1	112.0 (110.4)	112.0 (110.3)	110.7 (111.2)	110.7 (111.1)	113.8 (111.8)	114.1 (111.8)
F8-C5-C1	112.0 (110.4)	112.0 (110.3)	110.7 (110.9)	110.4 (110.8)	107.5 (112.1)	107.7 (111.8)
H9-O3-C1	108.0 (107.9)	107.2 (107.3)	111.4 (107.8)	110.7 (107.1)	106.6 (108.1)	106.0 (107.3)
H10-O4-X2	112.7 (103.4)	112.4 (102.5)	109.7 (102.8)	109.7 (102.0)	107.8 (106.5)	107.0 (105.6)
O4-X2-C1-O3	0.0 (0.0)	0.0 (-0.0)	179.1 (179.9)	175.6 (180.0)	81.5 (10.1)	77.3 (10.2)
C5-C1-X2-O4	180.0 (180.0)	180.0 (179.9)	0.1 (-0.0)	-2.5 (0.0)	-96.4 (190.3)	-102.0 (190.1)
F6-C5-C1-X2	-0.0 (-0.0)	0.0 (0.2)	-5.4 (60.5)	-9.7 (60.1)	45.8 (-2.3)	52.9 (0.2)
F7-C5-C1-X2	119.9 (120.7)	119.9 (121.0)	115.4 (180.4)	111.4 (180.1)	167.6 (117.0)	174.5 (119.7)
F8-C5-C1-X2	240.0 (239.3)	240.2 (239.5)	233.7 (300.4)	229.7 (300.0)	287.4 (238.0)	294.7 (240.7)
H9-O3-C1-X2	-0.0 (0.0)	0.0 (-0.1)	1.8 (-0.0)	-3.4 (0.0)	7.9 (-0.3)	6.5 (-0.1)
H10-O4-X2-C1	180.0 (180.0)	180.0 (179.9)	179.0 (180.0)	166.0 (180.0)	95.6 (170.1)	94.3 (169.9)

279

Table 6.13: Geometrical parameters of **2Z2**, **2E2** and the rotational transition state **RTS3** and **RTS4** for CF$_3$ substituted FPA (FHA) at B3LYP/6-31+G* [**L1**] and MP2/6-31+G* [**L3**] theoretical levels. All bond distances are in angstrom (Å) and angles in degrees (°).

Parameters	2Z2		2E2		RTS3		RTS4	
	L1	L3	L1	L3	L1	L3	L1	L3
X2-C1	1.692 (1.274)	1.694 (1.285)	1.690 (1.275)	1.692 (1.287)	1.687 (1.275)	1.690 (1.285)	1.686 (1.275)	1.691 (1.288)
O3-C1	1.358 (1.349)	1.366 (1.352)	1.374 (1.352)	1.383 (1.355)	1.370 (1.359)	1.377 (1.362)	1.387 (1.362)	1.392 (1.366)
O4-X2	1.671 (1.392)	1.672 (1.403)	1.673 (1.401)	1.674 (1.414)	1.666 (1.385)	1.667 (1.396)	1.666 (1.390)	1.667 (1.402)
C5-C1	1.502 (1.517)	1.494 (1.505)	1.512 (1.533)	1.503 (1.524)	1.505 (1.517)	1.496 (1.508)	1.509 (1.535)	1.500 (1.525)
F6-C5	1.347 (1.333)	1.351 (1.334)	1.340 (1.342)	1.343 (1.344)	1.358 (1.340)	1.361 (1.342)	1.343 (1.349)	1.347 (1.351)
F7-C5	1.360 (1.355)	1.360 (1.364)	1.355 (1.370)	1.357 (1.371)	1.353 (1.351)	1.354 (1.353)	1.355 (1.348)	1.357 (1.350)
F8-C5	1.368 (1.368)	1.370 (1.364)	1.374 (1.341)	1.374 (1.344)	1.362 (1.362)	1.364 (1.364)	1.362 (1.351)	1.363 (1.354)
H9-O3	0.973 (0.972)	0.979 (0.977)	0.973 (0.971)	0.978 (0.976)	0.972 (0.973)	0.977 (0.977)	0.970 (0.972)	0.974 (0.975)
H10-O4	0.970 (0.971)	0.975 (0.976)	0.970 (0.970)	0.976 (0.975)	0.970 (0.972)	0.976 (0.976)	0.970 (0.971)	0.976 (0.976)
O3-C1-X2	124.8 (125.2)	124.6 (125.0)	115.8 (117.7)	115.2 (117.3)	125.7 (126.5)	125.7 (126.8)	117.3 (118.4)	117.8 (118.1)
O4-X2-C1	99.5 (112.0)	98.8 (110.4)	101.7 (112.1)	101.1 (110.4)	99.7 (112.4)	99.2 (111.1)	104.8 (113.2)	103.9 (111.6)
C5-C1-X2	119.6 (117.4)	119.6 (117.7)	129.5 (125.2)	130.0 (124.8)	120.1 (118.2)	120.1 (117.8)	132.1 (125.3)	132.1 (124.8)

Contd......

Parameters	2Z2 L1	2Z2 L3	2E2 L1	2E2 L3	RTS3 L1	RTS3 L3	RTS4 L1	RTS4 L3
F6-C5-C1	112.8 (113.7)	112.7 (113.9)	113.7 (112.3)	113.7 (112.2)	111.7 (113.4)	111.7 (113.4)	113.2 (111.6)	113.1 (111.5)
F7-C5-C1	112.2 (111.1)	112.3 (109.7)	112.7 (109.2)	112.5 (109.1)	112.6 (110.9)	112.4 (110.8)	110.9 (110.6)	110.4 (110.6)
F8-C5-C1	110.6 (109.0)	110.3 (109.7)	109.0 (112.4)	108.8 (112.2)	111.2 (109.2)	111.0 (109.0)	110.7 (111.5)	110.7 (111.2)
H9-O3-C1	111.0 (110.7)	109.9 (110.5)	109.8 (109.9)	109.0 (109.4)	111.2 (111.7)	110.2 (110.8)	111.1 (111.6)	110.2 (110.8)
H10-O4-X2	110.6 (102.7)	110.2 (101.8)	110.5 (102.6)	110.4 (101.7)	110.7 (102.8)	110.3 (101.9)	110.2 (102.9)	109.9 (102.1)
O4-X2-C1-O3	-1.6 (-1.5)	-2.9 (0.0)	170.2 (179.9)	168.2 (180.0)	-6.6 (-5.3)	-7.0 (-5.6)	181.4 (175.7)	182.9 (175.7)
C5-C1-X2-O4	181.5 (182.5)	181.4 (180.0)	-4.7 (-0.1)	-5.0 (0.0)	178.5 (179.4)	179.0 (179.7)	6.3 (1.0)	8.1 (1.3)
F6-C5-C1-X2	-10.7 (-22.6)	-11.2 (0.2)	-34.4 (61.7)	-31.1 (61.4)	-3.6 (-5.0)	-4.4 (-6.0)	3.3 (64.4)	3.9 (62.4)
F7-C5-C1-X2	111.1 (99.9)	110.6 (122.1)	88.7 (180.3)	92.0 (180.0)	117.4 (117.0)	116.6 (116.1)	124.5 (184.3)	124.8 (182.4)
F8-C5-C1-X2	228.5 (216.6)	228.2 (238.3)	206.3 (298.8)	209.6 (298.5)	236.6 (234.7)	235.8 (233.7)	242.5 (303.6)	243.0 (301.6)
H9-O3-C1-X2	166.7 (169.5)	160.6 (180.0)	147.6 (179.9)	140.9 (180.0)	96.4 (96.1)	98.0 (97.6)	63.0 (88.4)	57.6 (89.3)
H10-O4-X2-C1	180.3 (180.5)	179.0 (180.0)	175.3 (180.0)	170.9 (180.0)	177.5 (179.3)	177.7 (179.5)	179.5 (181.2)	183.7 (181.3)

Table 6.14: Geometrical parameters of **dipolar 3** and **PTS** of CF$_3$ substituted FPA (FHA) at B3LYP/6-31+G* **[L1]** and MP2/6-31+G* **[L3]** theoretical levels. All bond distances are in angstrom (Å) and angles in degrees (°).

Parameters	Dipolar form		Parameters	PTS	
	L1	L3		L1	L3
X2-C1	1.692	1.687	X2-C1	1.771	1.743
	(1.306)	(1.320)		(1.309)	(1.313)
O3-C1	1.353	1.364	O3-C1	1.282	1.303
	(1.326)	(1.334)		(1.284)	(1.290)
O4-X2	1.504	1.508	O4-X2	1.668	1.669
	(1.298)	(1.286)		(1.397)	(1.403)
C5-C1	1.504	1.494	C5-C1	1.517	1.506
	(1.499)	(1.487)		(1.522)	(1.512)
F6-C5	1.358	1.363	F6-C5	1.349	1.352
	(1.358)	(1.362)		(1.348)	(1.350)
F7-C5	1.351	1.354	F7-C5	1.349	1.353
	(1.347)	(1.350)		(1.348)	(1.352)
F8-C5	1.350	1.352	F8-C5	1.349	1.351
	(1.347)	(1.350)		(1.339)	(1.342)
H9-O3	0.976	0.981	H9-O4	0.972	0.976
	(0.994)	(0.998)		(0.974)	(0.979)
H10-X2	1.406	1.400	H10-X2	1.742	1.722
	(1.023)	(1.027)		(1.346)	(1.343)
O3-C1-X2	122.3	122.8	O3-C1-X2	105.1	105.6
	(118.1)	(118.0)		(107.1)	(107.5)
O4-X2-C1	121.1	122.6	O4-X2-C1	111.0	110.8
	(120.3)	(120.8)		(123.0)	(122.2)
C5-C1-X2	123.4	123.2	C5-C1-X2	134.5	134.1
	(123.1)	(122.9)		(128.5)	(127.5)
F6-C5-C1	110.6	110.7	H6-C5-C1	110.8	110.8
	(109.7)	(109.6)		(110.0)	(109.9)

Contd.....

Parameters	Dipolar form		Parameters	PTS	
	L1	L3		L1	L3
F7-C5-C1	112.7 (111.7)	112.5 (111.9)	H7-C5-C1	110.7 (110.6)	110.6 (110.4)
F8-C5-C1	111.0 (111.7)	111.1 (111.9)	H8-C5-C1	111.0 (111.0)	111.1 (111.2)
H9-O3-C1	110.2 (104.0)	109.6 (103.7)	H9-O4-X2	112.1 (104.6)	111.5 (104.0)
H10-X2-C1	109.8 (120.8)	110.6 (120.4)	H10-X2-C1	62.1 (75.9)	63.3 (76.1)
O4-X2-C1-O3	-17.5 (0.0)	-16.1 (-0.0)	O4-X2-C1-O3	138.3 (194.5)	140.2 (195.6)
C5-C1-X2-O4	179.4 (180.0)	181.1 (180.1)	C5-C1-X2-O4	-57.1 (14.5)	-57.5 (16.0)
F6-C5-C1-X2	-13.3 (0.1)	-13.9 (-0.3)	F6-C5-C1-X2	-102.2 (-73.3)	-99.5 (-72.2)
F7-C5-C1-X2	107.1 (119.8)	106.3 (119.4)	F7-C5-C1-X2	17.9 (46.2)	20.6 (47.2)
F8-C5-C1-X2	227.2 (240.3)	226.5 (240.1)	F8-C5-C1-X2	137.6 (166.5)	140.4 (167.5)
H9-O3-C1-X2	24.0 (-0.0)	25.3 (0.0)	H9-O4-X2-C1	128.7 (227.2)	131.1 (230.3)
H10-X2-C1-O3	194.4 (180.0)	190.5 (180.0)	H10-X2-C1-O3	10.6 (-2.9)	11.5 (-3.4)

The relative energies of all the isomeric forms are listed in Table 6.15. As can be seen from the table, the order of stability for the representative isomeric forms in case of FPA is **1Z > 2Z >1E > 2E > 3** at **[L3]** theoretical level which is similar to that determined for FHA. The order remains the same with the presence of CH_3 at carbonyl carbon of FPA. The presence of electron withdrawing CF_3 at the carbonyl carbon however decreases the energy difference between the isomeric forms and the order is changed to **2Z > 1Z >1E > 2E > 3**. The **2Z** form of $(CF_3)_2$ (C=O) PHOH is only 1.21 kcal/mol more stable than **1Z** form. The relative energy difference between **2Z** and **2Z2** forms is 8.07 kcal/mol in case of FPA that increases to 9.54 kcal/mol in CH_3 substituted FPA and decreases to 6.83 kcal/mol in CF_3 substituted FPA. The relative energy difference between **2E** and **2E2** forms is 1.27 kcal/mol that increases in the presence of CH_3 to 1.67 kcal/mol and to 2.69 kcal/mol in presence of CF_3.

Table 6.15: Relative energies (in kcal/mol) of isomeric forms of FPA at B3LYP/6-31+G* **[L1]** and MP2/6-31+G* **[L3]** theoretical levels. ZPE uncorrected values at **L3**.

Species	R=H		R=CH$_3$		R=CF$_3$	
	L1	L3	L1	L3	L1	L3
1Z	0.00	0.00	0.00	0.00	0.00	0.00
1E	1.60	1.20	4.06	3.55	2.57	2.00
2Z	2.95	0.77	5.87	3.45	1.77	-1.21
2E	10.29	8.64	13.16	11.05	12.87	11.12
2Z2	10.25	8.84	14.65	12.99	7.98	5.62
2E2	11.66	9.91	15.03	12.72	11.05	8.43
3	24.70	23.07	-	-	25.44	22.72

The **1Z, 1E** and **2Z, 2E** isomeric pairs are interrelated through rotation around C-X bond while **2Z, 2Z2** and **2E, 2E2** are rotamers of C-O bond (Figure 6.2) and **1Z, 1Z2** are rotamers of X-O bond. The **1Z2** conformations, i.e. P-O rotamer of **1Z** for FPA and CH_3 and CF_3 substituted FPA have also been optimized and their geometrical parameters are recorded in Table 6.16. For unsubstituted FPA, the structural parameters of rotational transition states **RTS1, RTS2, RTS3** and **RTS4** for interconversion among **1Z → 1E, 2Z → 2E, 2Z → 2Z2** and **2E → 2E2** isomeric pairs respectively are given in the Tables 6.2-6.4. For CH_3 substituted FPA, the parameters for **RTS1, RTS2, RTS3 and RTS4** are listed in Tables 6.7-6.9 and for similar CF_3 substituted rotational transition states are given in Tables 6.11-6.13. The geometrical parameters for **1Z** → **1Z2** (**RTS5**) interconversions are recorded in Table 6.16. Geometrical parameters for tautomerization transition state **1E** → **2E** (**PTS**) are recorded in Tables 6.5, 6.10 for unsubstituted and CH_3 substituted FPA and 6.14 for CF_3 substituted FPA. The activation barriers for the interconversions among isomeric forms are listed in the Table 6.17. The Figure 6.2 depicts the probable pathways for gas phase interconversions amongst the various isomeric forms of FPA. In chapter 2, it was observed that the rotation around C-N of **1Z** of FHA is restricted because of presence of partial π bond character arising from the conjugative interactions between the lone pair of electrons present on N and the C-O π bond. The rotational transition state for **1Z** → **1E** transition (**RTS1**) in case of FPA indicates activation barrier of 7.41 kcal/mol which is nearly half than evaluated in case of FHA (15.45 kcal/mol) for similar transition at [**L3**] theoretical level. In case of FHA the barrier was assigned as arising due to rupture of intramolecular H-bond and breaking of partial C-X π bond but in case of FPA, the strong pyramidal character does not favor the electron delocalization, hence formation of **RTS** involves breaking of intramolecular H-bond. Though intramolecular H-bond donor and acceptor are same in FHA and FPA but the long C-P bond distance disfavors the intramolecular H-bond strength. For **1Z** → **1E** interconversion, the presence of CH_3 at C1 decreases the activation barrier by 1 kcal/mol and CF_3 decreases the barrier by 3 kcal/mol in FPA. The rotational transition state for **1Z** → **1Z2** transition (**RTS5**) indicates the activation barrier of 5.58 kcal/mol. The energy difference of 2.13 kcal/mol (between **1Z** and **1Z2** of FPA) is mainly due to intramolecular H-bond. The energy barrier for **2Z** → **2E** transition through **RTS2** is magnificently high 61.76 kcal/mol which is anticipated due to the presence of C-X π bond. But this barrier is 5.95 kcal/mol higher than that in case of FHA. The rotation around C-O bond of **2Z** in FPA requires energy barrier of 9.93 kcal/mol which is 0.69 kcal/mol lower than the similar rotation in case of FHA. The **2E** → **2E2** transition requires activation barrier of 2.96 kcal/mol in comparison to 6.92 kcal/mol in FHA at [**L3**] theoretical level. **2Z** → **2Z2** has larger barrier over **2E** → **2E2** because during this conversion, stabilization of **2Z** by intramolecular H-bond is lost in **2Z2** while none of the two forms **2E** and **2E2** is stabilized by H-bond.

Table 6.16: Geometrical parameters of **1Z2** conformation of FPA and **RTS5**, **1Z2** of CH$_3$ and CF$_3$ substituted FPA at MP2/6-31+G* [L3] theoretical level. All bond distances are in angstrom (Å) and angles in degrees (°).

Parameters	1Z2	RTS5	Parameters	CH$_3$ substituted FPA	Parameters	CF$_3$ substituted FPA
P2-C1	1.857	1.857	P2-C1	1.874	P2-C1	1.874
O3-C1	1.226	1.230	O3-C1	1.230	O3-C1	1.223
O4-P2	1.673	1.677	O4-P2	1.676	O4-P2	1.670
H5-O4	0.973	0.977	C5-C1	1.511	C5-C1	1.543
H6-C1	1.113	1.110	H6-C5	1.096	F6-C5	1.363
H7-P2	1.415	1.416	H7-C5	1.096	F7-C5	1.341
O3-C1-P2	124.2	121.5	H8-C5	1.093	F8-C5	1.353
O4-P2-C1	101.5	100.1	H9-O4	0.974	H9-O4	0.974
H5-O4-P2	111.1	111.2	H10-P2	1.415	H10-P2	1.413
H6-C1-P2	114.7	117.4	O3-C1-P2	120.0	O3-C1-P2	123.1
H7-P2-C1	93.7	94.8	O4-P2-C1	101.1	O4-P2-C1	98.7
O4-P2-C1-O3	−30.9	−29.1	C5-C1-P2	117.2	C5-C1-P2	117.4
H5-O4-P2-C1	111.4	59.8	H6-C5-C1	108.8	F6-C5-C1	109.5
H6-C1-P2-O4	153.7	156.6	H7-C5-C1	111.6	F7-C5-C1	112.7
H7-P2-C1-O3	230.7	233.6	H8-C5-C1	109.5	F8-C5-C1	109.8
			H9-O4-P2	111.0	H9-O4-P2	111.1
			H10-P2-C1	95.1	H10-P2-C1	94.6
			O4-P2-C1-O3	−34.2	O4-P2-C1-O3	−34.6
			C5-C1-P2-O4	148.9	C5-C1-P2-O4	148.9
			H6-C5-C1-P2	−90.5	F6-C5-C1-P2	21.8
			H7-C5-C1-P2	28.6	F7-C5-C1-P2	142.5
			H8-C5-C1-P2	150.6	F8-C5-C1-P2	263.8
			H9-O4-P2-C1	102.3	H9-O4-P2-C1	107.6
			H10-P2-C1-O3	228.2	H10-P2-C1-O3	227.3

Table 6.17: Activation barriers (kcal/mol) of various isomeric forms of FPA and (FHA) at B3LYP/6-31+G* [**L1**] and MP2/6-31+G* [**L3**] theoretical levels. (Frequencies of HF/6-31+G* are used at **L3**)

Transition state	R=H		R=CH$_3$		R=CF$_3$	
	L1	L3	L1	L3	L1	L3
RTS1	8.19	7.41	7.03	6.41	4.96	4.41
	(17.63)	(15.45)	(16.75)	(15.10)	(20.32)	(18.54)
RTS2	56.12	61.76	51.18	57.11	60.60	66.79
	(49.90)	(55.81)	(52.53)	(58.69)	(45.31)	(51.01)
RTS3	9.63	9.93	10.99	11.24	7.31	7.51
	(9.66)	(10.52)	(10.82)	(11.59)	(7.34)	(8.11)
RTS4	3.47	2.96	3.58	2.95	1.22	-
	(6.75)	(6.92)	(7.53)	(7.66)	(5.48)	(5.74)
PTS	50.85	52.06	50.38	51.59	50.02	51.11
	(44.39)	(44.41)	(41.81)	(41.90)	(47.50)	(47.52)

The 1, 3 prototropic shift in **1E** isomer leading to **2E** form involves four membered ring transition state (**PTS**) as shown in Figure 6.2. The angles C1-X2-H7 and X2-H7-O3, H7-O3-C1 deviate from 90° desirable for four membered ring, thereby making the ring highly strained. The dihedral H7-X2-C1-O3 of 11.1° in case of FPA also suggests the distortion of the ring. The activation energy in case of FPA is 52.06 kcal/mol while it is 44.41 kcal/mol in case of FHA. The 1, 3 prototropic shift barriers are larger in FPA over FHA for all the substitutions (R=H, CH$_3$, CF$_3$) since four membered ring in FPA is more strained as ∠H7-X2-C1 is 12-15° lower than similar transition states in FHA.

6.2.2 INTRAMOLECULAR H-BONDING:

The presence of intramolecular hydrogen bonding (H-bonding) in **1Z** isomeric form of FHA has been assigned to be the cause of higher stability of the form in comparison to other forms. In order to explore the presence of intramolecular H-bonding in the tautomeric forms of FPA and substituted FPA, the important atomic separations between the H-bond donors and acceptors and angles at bridging hydrogen are scrutinized. The atomic separations and angles at bridging hydrogen that indicate the presence of intramolecular H-bonding in these, are recorded in Table 6.18. The intramolecular H-bond is indicated in **1Z** and **2Z** isomeric forms of FPA involving H5...O3 and H5...O4 hydrogen bonds respectively. The Δr values (difference of H-bond contact distances and van der Waals radii) are indicative of strength of intramolecular H-bond. The smaller Δr for **1Z** of FPA (0.442Å) in comparison to **1Z** of FHA (0.554Å) show relatively weaker strength of intramolecular H-bond in FPA. The presence of CH_3 substituent at the carbonyl carbon strengthens the intramolecular H-bond of **1Z** and **2Z** forms in both FHA and FPA as indicated by decrease in H...O distance and increase in O-H...O angle. However the similar considerations in CF_3 substituted FPA and FHA suggests the weakening of the H-bond. The longer H-bond distances and slight decrease in linearity of angles at bridging hydrogen suggest that intramolecular H-bond is weakened in the presence of substituent CF_3.

6.2.3 ELECTRON DELOCALIZATIONS IN ISOMERIC FORMS OF FPA:

The second order delocalization energies ($E^{(2)}$) for the important orbital interactions in the isomeric forms of FPA obtained at **[L3]** theoretical level are recorded in Table 6.19. The $E^{(2)}$ values for the similar orbital interactions in case of FHA are also included in the table for comparison. As can be seen from the table the $n_{X2} \rightarrow \pi^*_{C1-O3}$ orbital interactions (with $E^{(2)}$=5.57 kcal/mol) in **1Z** conformer that impart partial π bond character to C-X bond are much weaker in FPA in comparison to that in FHA. The similar conclusion has also been drawn from the lower rotational barrier for **1Z** → **1E** interconversion in FPA than FHA. The $E^{(2)}$ values for $n_{O3} \rightarrow \sigma^*_{C1-X2}$ orbital interaction are also lower in **1Z** of FPA relative to that in FHA.

The $E^{(2)}$ values for the orbital interactions $n_{X2} \rightarrow \pi^*_{C1-O3}$ and $n_{O3} \rightarrow \sigma^*_{C1-X2}$ are of comparative magnitude in the **1E** and **1Z** conformers of FPA. The $n_{O3} \rightarrow \sigma^*_{C1-X2}$ orbital interactions in **2Z** and **2E** tautomers impart C-O bond a partial π character. This is reflected in activation barrier of **2Z** → **2Z2** and **2E** → **2E2** interconversions. The presence of CF_3 substituent on FHA favors electron delocalizations as reflected by enhanced $E^{(2)}$ value for $n_{X2} \rightarrow \pi^*_{C1-O3}$ orbital interactions, however no significant variation in $E^{(2)}$ value results with CF_3 substituted FPA. All the $E^{(2)}$ values in case of FPA are lower in comparison to $E^{(2)}$ values for corresponding orbital interactions in FHA. Thus the electron delocalizations are weaker in CF_3 substituted FPA.

The $E^{(2)}$ values for orbital interactions $n_{O3} \rightarrow \sigma^*_{O4-H5}$ in **1Z** conformer from both lone pairs of electrons at oxygen are 3.31 and 1.09 kcal/mol in FPA. The values are reflective of contribution of charge transfer interactions that attribute the covalent character to intramolecular H-bond. The values are 2.05 and 0.15 kcal/mol lower than the values in case of FHA suggesting weaker covalent component of H-bond in FPA. Similarly $E^{(2)}$ value of 1.46 and 5.20 kcal/mol (for both lone pairs) for $n_{O4} \rightarrow \sigma^*_{O3-H5}$ in **2Z** supports the intramolecular H-bond in the isomer of FPA thereby strengthening the bond. Value for similar interaction in **2Z** of FHA is 5.06 kcal/mol.

Table 6.18: Important intramolecular H-bonding parameters for **1Z** of FPA (FHA) at MP2/6-31+G* [**L3**] theoretical level. All bond distances are in angstrom (Å) and angles in degrees (°).

Parameters	1Z							2Z					
	R=H	Δr	R=CH$_3$	Δr	R=CF$_3$	Δr		R=H	Δr	R=CH$_3$	Δr	R=CF$_3$	Δr
H...O	2.158	0.442	2.081	0.519	2.217	0.383		2.029	0.571	1.973	0.627	2.005	0.595
	(2.046)	(0.554)	(1.980)	(0.620)	(2.076)	(0.524)		(2.020)	(0.581)	(1.968)	(0.632)	(2.037)	(0.563)
O-H...O	119.8		122.1		116.1			122.4		124.7		121.6	
	(117.8)		(119.3)		(116.7)			(111.5)		(113.1)		(110.6)	

Table 6.19: Second order delocalization energies (E$^{(2)}$ in kcal/mol) of the important orbital interactions from NBO analysis of neutral FPA (FHA) at MP2/6-31+G* [**L3**] theoretical levels.

Species		n$_{X2}$→σ*$_{C1-O3}$	n$_{X2}$→π*$_{C1-O3}$	n$_{O3}$→σ*$_{C1-X2}$	n$_{O3}$→σ*$_{O4-H5}$	n$_{O3}$→π*$_{C1-X2}$	n$_{O4}$→σ*$_{O3-H5}$	n$_{O4}$→π*$_{C1-X2}$	n$_{O4}$→σ*$_{C1-X2}$
FPA (FHA)	1Z	-	5.57	20.73	1.09+3.31	-	-	-	-
			(32.68)	(29.25)	(1.24+5.36)	(-)	(-)	(-)	(-)
	1E	(14.98)	5.85	22.59	-	(-)	(-)	(-)	6.70
		(12.11)	(20.90)	(32.25)	(-)	(-)	(-)	(-)	(-)
	2Z	7.75	(-)	6.30+5.01	-	34.31	5.20+1.46	3.82	4.75
		(13.50)	(-)	(8.63)	(-)	(54.42)	(5.06)	(13.22)	(-)
	2E	-	(-)	11.03+4.41	-	18.03	-	9.49	-
		(3.82)	(-)	(6.12)	(-)	(48.37)	(-)	(13.80)	(-)
	2Z2	10.01	-	-	-	42.40	-	15.01	-
		(14.88)	(-)	(1.62)	(-)	(48.02)	(-)	(18.00)	(-)
	2E2	-	-	-	-	35.22	-	11.08	-
		(3.66)	(-)	(-)	(-)	(45.08)	(-)	(-)	(14.29)
	3	-	-	6.39	-	36.21	-	46.16	15.31+5.62
		(-)	(-)	(-)	(-)	(62.21)	(11.83)	(51.67)	(5.30)

Contd.....

Species		$n_{X2}\to\sigma^*_{Cl-O3}$	$n_{X2}\to\pi^*_{Cl-O3}$	$n_{O3}\to\sigma^*_{Cl-X2}$	$n_{O3}\to\sigma^*_{O4-H5}$	$n_{O3}\to\pi^*_{Cl-X2}$	$n_{O4}\to\sigma^*_{O3-H5}$	$n_{O4}\to\pi^*_{Cl-X2}$	$n_{O4}\to\sigma^*_{Cl-X2}$
CH$_3$-FPA (FHA)	1Z	3.92 (17.58)	4.43 (29.53)	22.24 (29.74)	4.40+1.52 (6.78)	- (-)	- (-)	- (-)	9.04 (-)
	1E	- (5.62)	8.33 (33.68)	22.56 (32.94)	- (-)	54.39 (55.42)	4.47+4.85 (6.31)	3.77 (11.33)	- (-)
	2Z	9.33 (15.78)	- (-)	7.74 (9.83)	- (-)	42.84 (47.25)	- (-)	7.45 (13.16)	3.29 (-)
	2E	- (-)	- (-)	6.11 (6.41)	- (-)	43.74 (49.10)	- (-)	13.03 (15.69)	- (-)
	2Z2	10.41 (16.48)	- (-)	- (-)	- (-)	15.19 (44.17)	- (-)	7.02 (13.95)	- (-)
	2E2	- (-)	- (-)	8.64 (-)	- (-)	- (65.38)	- (16.09)	- (41.17)	- (-)
	3	- (-)	- (-)	- (-)	- (-)	- (-)	- (-)	- (-)	- (6.93)
CF$_3$-FPA (FHA)	1Z	4.30 (9.60)	5.50 (68.24)	19.68 (29.36)	2.32 (4.62)	- (-)	- (-)	- (-)	- (4.60)
	1E	- (4.72)	8.11 (49.85)	19.28 (30.94)	- (-)	- (-)	- (-)	- (-)	9.90 (5.15)
	2Z	9.01 (16.38)	- (-)	8.36 (10.92)	- (-)	48.92 (56.34)	6.61 (4.90)	11.66 (15.03)	- (-)
	2E	- (-)	- (-)	6.17 (7.46)	- (-)	37.22 (50.12)	- (-)	16.22 (18.12)	- (-)
	2Z2	9.93 (17.16)	- (-)	- (-)	- (-)	33.92 (49.85)	- (-)	17.12 (20.38)	- (-)
	2E2	- (-)	- (-)	- (-)	- (-)	17.13 (47.64)	- (-)	16.57 (18.75)	- (-)
	3	- (-)	- (-)	6.74 (8.46)	- (-)	33.96 (60.79)	- (9.50)	54.31 (66.06)	18.49 (4.84+9.75)

6.2.4 DEPROTONATION ENTHALPIES AND STABILITIES OF ANIONS:

With the probable isomerism in FPA and more than one deprotonation site, several isomeric forms for the anions resulting from deprotonation of FPA are expected. Considering three probable deprotonation sites (P-H, O3-H and O4-H), seven anionic isomers of FPA have been optimized at [L1] and [L3] theoretical levels. The P-H deprotonated anion of **1E** i.e. **1EP⁻** and O-H deprotonated anion of **2E** (**2EO3⁻**) are identical.

The optimized anions are shown in Figure 6.3 and their structural data is compiled in Tables 6.20-6.21. Table 6.22 records the relative stabilities of all the anionic species. The stability of anionic species of FPA decreases in the order **1ZP⁻ > 1ZO⁻ > 2ZO4⁻ > 1EO⁻ > 1EP⁻ = 2EO3⁻ > 2EO4⁻** at [L3] theoretical level. The order differs from that in case of FHA. As can be seen from the table, the anions of FPA do not show large variations in relative energies as the relative energies range between 4.81-14.85 kcal/mol. In comparison, relative energies for anions of FHA range between 8.75-27.97 kcal/mol. The formation of the most stable **1ZP⁻** species with P-H deprotonation is accompanied by C-P bond contraction while the C-O and O-P are lengthened. The total planarity of the anionic species favors the orientation for stronger conjugative interactions. The C-P bond contraction and C-O, O-P bond elongation accompany the **1EP⁻** formation. Significant variation in geometrical parameters of anions are indicative of the fact that the bond strengths are altered either through conjugative interactions or through the polarization of the bonds. O3-H deprotonation of **2Z** lead to formation of **2ZO3⁻** in FHA but similar deprotonation in case of FPA leads to formation of **1ZP⁻**. O4-H deprotonation of FPA leads to **2ZO4⁻** while similar deprotonation in FHA results in forming **1ZN⁻**.

The enthalpy changes accompanying the deprotonation of isomeric forms of FPA and FHA are listed in Table 6.23. The values range 328.99-337.92 kcal/mol in FPA while the range in case of FHA is 335.34-357.72 kcal/mol at [L3] theoretical level. The process of heterolytic dissociation is relatively easier in FPA than FHA. The most stable anion in FPA is **1ZP⁻** generated with gas phase acidity of 332.63 kcal/mol while the gas phase acidity associated with **1ZN⁻** in FHA is 335.34 kcal/mol at [L3] theoretical level. Thus X-site (X=N, P) in both FHA and FPA is the most acidic site. The deprotonation from hydroxyl group requires gas phase enthalpy change of 337.01 kcal/mol in FPA and 350.02 kcal/mol in FHA. Thus X-H bond is more acidic than O-H by 4.38 kcal/mol and 14.68 kcal/mol in FPA and FHA respectively.

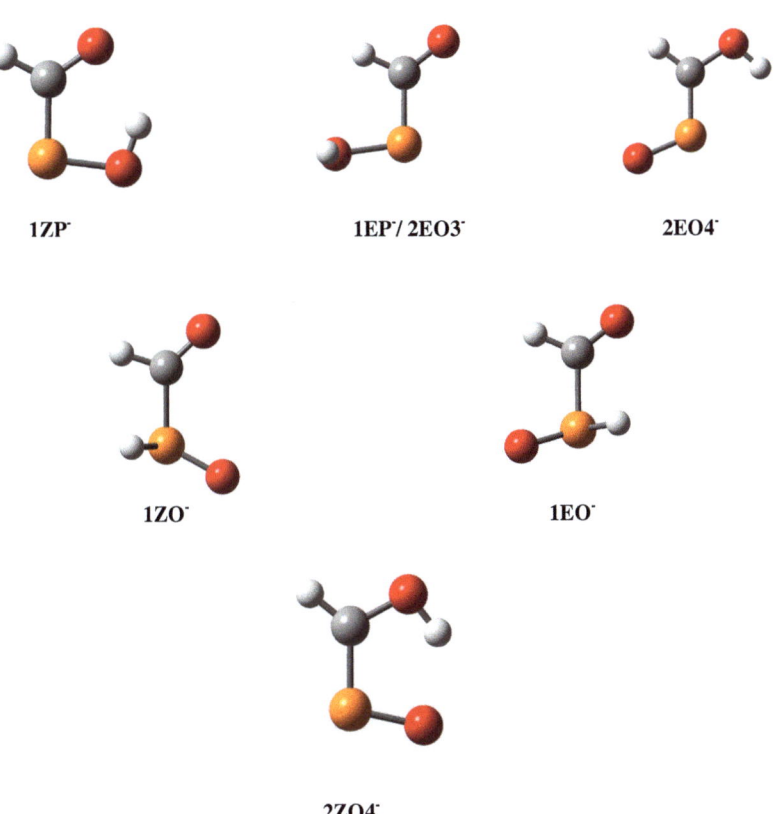

Figure 6.3: Anions of formylphosphinous acid (FPA) optimized at MP2/6-31+G* [**L3**] theoretical level.

Table 6.20: Geometrical parameters of deprotonated FPA (FHA) species at B3LYP/6-31+G* **[L1]** and MP2/6-31+G* **[L3]** theoretical levels. All bond distances are in angstrom (Å) and angles in degrees (°).

Parameters	1ZX⁻		1EX⁻		Parameters	2EO4⁻	
	L1	L3	L1	L3		L1	L3
X2-C1	1.758	1.738	1.783	1.764	X2-C1	1.711	1.713
	(1.318)	(1.326)	(1.324)	(1.330)		(1.281)	(1.294)
O3-C1	1.270	1.289	1.252	1.267	O3-C1	1.420	1.425
	(1.280)	(1.291)	(1.264)	(1.273)		(1.410)	(1.409)
O4-X2	1.729	1.723	1.770	1.769	O4-X2	1.556	1.564
	(1.455)	(1.457)	(1.504)	(1.502)		(1.333)	(1.344)
H5-O4	0.985	0.993	0.967	0.971	H5-O3	0.974	0.978
	(0.987)	(0.995)	(0.967)	(0.972)		(0.977)	(0.981)
H6-C1	1.116	1.112	1.125	1.121	H6-C1	1.090	1.090
	(1.108)	(1.104)	(1.119)	(1.115)		(1.091)	(1.090)
O3-C1-X2	125.3	124.7	128.5	128.2	O3-C1-X2	122.4	122.4
	(128.1)	(127.5)	(126.1)	(125.9)		(119.7)	(119.2)
O4-X2-C1	96.7	96.5	98.9	98.6	O4-X2-C1	113.7	113.2
	(108.8)	(108.1)	(107.1)	(106.7)		(120.6)	(118.7)
H5-O4-X2	103.3	102.4	107.7	107.8	H5-O3-X1	106.3	105.6
	(102.0)	(100.9)	(99.5)	(99.3)		(102.5)	(101.9)
H6-C1-X2	115.7	116.1	112.7	113.0	H6-C1-X2	127.5	127.0
	(111.8)	(112.1)	(114.6)	(114.6)		(127.0)	(126.9)
O4-X2-C1-O3	0.0	0.0	180.5	180.2	O4-X2-C1-O3	180.0	180.0
	(0.1)	(0.0)	(181.9)	(182.2)		(179.9)	(180.0)
H5-O4-X2-C1	0.0	0.0	268.3	268.0	H5-O3-C1-X2	0.0	0.0
	(-0.4)	(-0.0)	(208.3)	(219.8)		(0.0)	(-0.0)
H6-C1-X2-O4	180.0	180.0	-0.4	-0.9	H6-C1-X2-O4	0.0	0.0
	(180.0)	(180.0)	(1.6)	(1.7)		(0.0)	(-0.0)

294

Table 6.21: Geometrical parameters of deprotonated FPA (FHA) in the gas phase at B3LYP/6-31+G* [L1] and MP2/6-31+G* [L3] theoretical levels. All bond distances are in angstrom (Å) and angles in degrees (°).

Parameters	1ZO⁻		1EO⁻		Parameters	2ZO4⁻	
	L1	L3	L1	L3		L1	L3
X2-C1	1.855 (1.342)	1.840 (1.343)	1.885 (1.330)	1.870 (1.330)	X2-C1	1.714 (1.260)	1.715
O3-C1	1.230 (1.250)	1.241 (1.259)	1.231 (1.266)	1.242 (1.274)	O3-C1	1.388 (1.360)	1.395
O4-X2	1.545 (1.337)	1.554 (1.338)	1.557 (1.358)	1.570 (1.361)	O4-X2	1.583 (1.341)	1.590
H5-C1	1.136 (1.119)	1.132 (1.116)	1.128 (1.106)	1.122 (1.103)	H5-O3	0.995 (0.967)	1.001
H6-X2	1.469 (1.026)	1.460 (1.025)	1.467 (1.023)	1.455 (1.023)	H6-C1	1.088 (1.074)	1.089
O3-C1-X2	132.9 (129.7)	132.4 (129.4)	127.8 (126.0)	127.6 (125.7)	O3-C1-X2	117.8 (121.8)	117.6
O4-X2-C1	117.7 (129.8)	116.8 (129.6)	108.5 (127.6)	106.9 (127.1)	O4-X2-C1	102.2 (113.4)	102.0
H5-C1-X2	107.1 (109.0)	107.8 (109.1)	113.1 (111.8)	113.4 (111.9)	H5-O3-C1	101.3 (100.0)	100.5
H6-X2-C1	86.1 (113.9)	85.4 (114.3)	89.2 (115.6)	89.0 (116.1)	H6-C1-X2	128.5 (122.8)	128.2
O4-X2-C1-O3	27.6 (-0.0)	27.2 (-0.0)	169.5 (180.0)	166.3 (180.0)	O4-X2-C1-O3	-0.1 (-0.0)	0.0
H5-C1-X2-O4	194.5 (180.0)	195.6 (180.0)	-19.4 (-0.0)	-20.5 (-0.0)	H5-O3-C1-X2	-0.1 (-0.0)	0.1
H6-X2-C1-O3	139.1 (180.0)	138.6 (180.0)	59.2 (0.0)	56.2 (0.0)	H6-C1-X2-O4	179.9 (180.0)	180.0

Table 6.22: Relative energies (kcal/mol) of anionic forms of FPA (FHA) at B3LYP/6-31+G* **[L1]** and MP2/6-31+G* **[L3]** theoretical levels.

Deprotonated Species	L1	L3
1ZX⁻	0.00	0.00
	(0.00)	(0.00)
1ZO⁻	2.02	4.81
	(10.10)	(14.69)
1EX⁻	5.12	6.82
	(7.84)	(9.61)
1EO⁻	3.26	5.56
	(6.41)	(10.24)
2ZO3⁻	-	-
	(7.58)	(8.75)
2EO4⁻	17.47	14.85
	(26.03)	(27.97)
2ZO4⁻	8.77	5.32
	(-)	(-)

Table 6.23: Deprotonation enthalpies (kcal/mol) of FPA (FHA) at B3LYP/6-31+G* **[L1]** and MP2/6-31+G* **[L3]** theoretical levels.

Deprotonated Species	L1	L3
1ZX⁻	336.32	332.63
	(340.35)	(335.34)
1ZO⁻	338.49	337.01
	(350.33)	(350.02)
1EX⁻	339.97	337.92
	(347.20)	(344.66)
1EO⁻	338.13	336.73
	(346.10)	(345.80)
2ZO3⁻	331.51	-
	(344.06)	(342.43)
2EO4⁻	341.90	338.08
	(359.29)	(357.72)
2EO3⁻	-	328.99
	(340.88)	(339.24)
2ZO4⁻	340.36	334.75
	(-)	(-)

6.2.5 ATOMIC CHARGE ANALYSIS OF THE ANIONS:

The atomic charges (NPA) for the molecules as well as for the anions of FPA and FHA are reported in Table 6.24. Nitrogen being more electronegative in comparison to phosphorus, there is accumulation of charge density at N in neutral as well as anionic species of FHA. Phosphorus on the other hand, in all neutral and anionic species of FPA, is positively charged. The carbonyl group of FHA is highly polarized in all the species while that in FPA is relatively less polarized. The oxygen of hydroxyl group attached to phosphorus in FPA carries more than one unit of negative charge in all the neutral isomeric species. As a result, the P-O bond of FPA is highly polar with strong ionic character. The anion formation in FPA results in increase in charge density at O4 but the increase is less than 0.22 units in all the FPA deprotonations. The oxygen of hydroxyl group is comparatively less electron dense in all the neutral species of FHA and the increase in electron density is less than 0.04 units upon anion formation. Charges indicate that the electrostatic component of C1-X2 and C1-O3 bonds is relatively stronger in all the species of FHA in comparison to that in FPA. As can be seen from the table, the additional unit of negative charge is dispersed over all the atoms in the anionic species. With the P-H deprotonation of **1Z** and **1E** of FPA, the variation in charge density indicates that P holds additional 0.548 units of negative charge while similar variation at N in case of FHA is much smaller (0.006). Inspite of O-H deprotonation, the variation in charge density at the O4 is small in both FHA and FPA (0.187 and 0.196 in **1ZO⁻** of FHA and FPA respectively). Thus, the deprotonation enthalpies in these molecules are reflective of ability of anions to stabilize the negative charge. Phosphorus, being larger in size and polarizability, has comparatively higher ability to hold the additional charge thereby stabilizing the anions and resulting in lower deprotonation enthalpies.

Table 6.24: Atomic charges (NPA) for FPA (FHA) anions at MP2/6-31+G* [L3] theoretical level.

Species	C1	X2	O3	O4	H5	H6	H7
1Z	0.149	0.949	-0.643	-1.074	0.541	0.172	-0.093
	(0.636)	(-0.426)	(-0.732)	(-0.616)	(0.535)	(0.173)	(0.430)
1ZX⁻	-0.052	0.401	-0.884	-1.127	0.524	0.139	-
	(0.493)	(-0.432)	(-0.941)	(-0.744)	(0.507)	(0.118)	(-)
1ZO⁻	0.119	0.958	-0.727	-1.261	-	0.098	-0.187
	(0.577)	(-0.353)	(-0.849)	(-0.812)	(-)	(0.086)	(0.351)
1E	0.163	0.886	-0.620	-1.069	0.529	0.169	-0.058
	(0.661)	(-0.454)	(-0.685)	(-0.626)	(0.517)	(0.166)	(0.422)
1EX⁻	0.043	0.272	-0.828	-1.092	0.484	0.121	-
	(0.564)	(-0.486)	(-0.896)	(-0.752)	(0.476)	(0.094)	(-)
1EO⁻	0.159	0.912	-0.731	-1.285	-	0.116	-0.171
	(0.594)	(-0.370)	(-0.892)	(-0.827)	(-)	(0.128)	(0.366)
2Z	0.213	0.785	-0.780	-1.103	0.547	0.233	0.531
	(0.478)	(-0.282)	(-0.783)	(-0.703)	(0.547)	(0.214)	(0.528)
2ZO3⁻	-	-	-	-	-	-	-
	(0.551)	(-0.501)	(-0.890)	(-0.721)	(-)	(0.093)	(0.468)
2ZO4⁻	-0.446	0.881	-0.872	-1.299	0.548	0.188	-
	(-)	(-)	(-)	(-)	(-)	(-)	(-)
2E	-0.222	0.779	-0.767	-1.063	0.514	0.240	0.520
	(0.488)	(-0.312)	(-0.772)	(-0.650)	(0.524)	(0.206)	(0.516)
2EO3⁻	0.043	0.272	-0.828	-1.092	-	0.121	0.484
	(0.564)	(-0.486)	(-0.896)	(-0.752)	(-)	(0.094)	(0.476)
2EO4⁻	-0.519	0.933	-0.829	-1.264	0.485	0.194	-
	(0.253)	(-0.224)	(-0.842)	(-0.831)	(0.495)	(0.150)	(-)

6.2.6 ELECTRON DELOCALIZATIONS IN THE ANIONS:

Analysis of molecular orbitals obtained by the NBO analysis (Table 6.25) indicated the formation of π_{C-X} bond in place of π_{C-O} in **1ZX⁻**, **1ZO⁻**, **1EX⁻** and **1EO⁻** of FHA but in case of FPA π_{C-X} formation occurs only in **1ZX⁻**. The $n_{X2} \rightarrow \pi^*_{C1-O3}$ orbital interactions vanish in the **1ZX⁻** anion because of the absence of π_{C-O} bond. The additional stability that arises in **1ZX⁻** anion both in case of FPA and FHA is due to much stronger $E^{(2)}$ value for $n_{O3} \rightarrow \pi^*_{C-X}$ orbital interaction. The $E^{(2)}$ values for $n_{X2} \rightarrow \sigma^*_{C1-O3}$, $n_{X2} \rightarrow \pi^*_{C1-O3}$ and $n_{O3} \rightarrow \sigma^*_{C1-X2}$ orbital interactions in **1ZO⁻** indicate stabilization of the anion through conjugative interactions but the values are comparatively weaker relative to the $n_{O3} \rightarrow \pi^*_{C1-X2}$ interactions in **1ZX⁻** anion in FPA. The variations in $E^{(2)}$ values for other orbital interaction as the result of anion formation are small with a few exceptions only.

In case of FPA, the $n_{O3} \rightarrow \sigma^*_{O4-H5}$ orbital interactions with $E^{(2)}$ value of 8.40 kcal/mol in **1ZP⁻** anion reflect the charge transfer interactions favoring the presence of intramolecular H-bond. The $E^{(2)}$ value is 4.00 kcal/mol higher than the $E^{(2)}$ value in neutral **1Z** species. The stronger intramolecular H-bond is also suggested by $E^{(2)}$ value of 14.09 kcal/mol for $n_{O4} \rightarrow \sigma^*_{O3-H5}$ in **2ZO4⁻** anion which is 8.89 kcal/mol higher than $E^{(2)}$ value for **2Z** species of FPA.

Thus it can be concluded that the lower deprotonation enthalpies in FPA than FHA are the result of better dispersal of additional unit of negative charge in addition to the differences in their conjugative interactions and intramolecular H-bonds strength.

Table 6.25: Second order delocalization energies ($E^{(2)}$) in kcal/mol for important orbital interactions in neutral and anionic species of FPA and (FHA) at MP2/6-31+G* [L3] theoretical level.

Species	$n_{X2} \rightarrow \sigma^*_{C1-O3}$	$n_{X2} \rightarrow \pi^*_{C1-O3}$	$n_{O3} \rightarrow \sigma^*_{C1-X2}$	$n_{O3} \rightarrow \sigma^*_{O4-H5}$	$n_{O3} \rightarrow \pi^*_{C1-X2}$	$n_{O4} \rightarrow \sigma^*_{O3-H5}$	$n_{O4} \rightarrow \pi^*_{C1-X2}$	$n_{O4} \rightarrow \sigma^*_{C1-X2}$
1Z	-	5.57	20.73	1.09+3.31	-	(-)	(-)	(-)
	(14.98)	(32.68)	(29.25)	(1.24+5.36)				
1ZX	7.26	-	13.02	8.40	154.95	-	6.08	-
	(12.55)	(-)	(18.75)	(11.21)	(119.51)		(9.04)	
1ZO⁻	6.98	15.74	16.71	-	-	-	-	8.79+22.01
	(-)	(-)	(-)		(27.57+168.14)	(-)	(26.34)	(10.48)
1E	-	5.85	22.59	-	-	-	-	6.70
	(12.11)	(20.90)	(32.25)					(-)
1EX	-	107.85	15.49	-	-	-	-	5.94
1EO⁻	(-)	(-)	(20.91+11.19)	-	(8.39+91.27)	(-)	(-)	(-)
		9.82	16.50					20.75
	(-)	(-)	(24.29)		(147.78)	(-)	(20.21)	(-)
2Z	7.75	-	6.30+5.01	-	34.31	5.20	-	4.75
	(13.50)	(-)	(8.63)		(54.42)	(5.06)	(13.22)	(-)
2ZO3⁻	-	-	(23.40)	-	(151.70)	(-)	(6.74)	(-)
	(13.13)	(-)						
2ZO4⁻	7.49	-	-	-	29.78	14.09	40.03	13.76
	(-)	(-)			(-)	(-)	(-)	(-)
2E	-	-	11.03+4.41	-	18.03	-	9.49	-
	(3.82)	(-)	(6.12)		(48.37)	(-)	(13.80)	
2EO3⁻	-	107.85	15.49	-	(8.39+91.27)	(-)	-	5.94
	(-)	(-)	(20.91+11.19)					-
2EO4⁻	-	-	5.05	-	19.91	(-)	51.29	19.43
	(-)	(-)	(24.92)		(-)		(59.42)	(7.92)

6.2.7 PROTON AFFINITIES:

Protonation of all the probable sites with lone pairs of electrons have been carried out (Figure 6.4). It resulted in optimization of ten protonated species, out of which two pair of protonated species are isoenergetic with similar structural parameters. The protonation at O3 in **1Z** and at X in **2Z** tautomers resulted in the same protonated species which is labeled as **1ZO3H$^+$A/2ZXH$^+$**. Similarly O3-protonation of **1E** converged to same structure as obtained by X-protonation of **2E** and labeled as **1EO3H$^+$A/2EXH$^+$**. The corresponding proton affinity (PA) values are listed in Table 6.26 and the geometrical parameters of protonated species are recorded in Tables 6.27-6.29. As can be seen from the Table 6.26, the P is the most basic site in all the isomeric forms of FPA. It is the O3 site in **1Z** and **1E** conformers of FHA that has higher PA than N site while reverse is the case in **2Z** and **2E** conformations. The **1ZO3H$^+$A/2ZXH$^+$, 1ZPH$^+$** and **2ZO3H$^+$** protonated species have intramolecular H-bonds intact. The important parameters associated with intramolecular H-bonds are shown in Figure 6.4.

The PA values reflect that P in **1Z** and **1E** forms of FPA has higher PA in comparison to that of N in respective forms of FHA. But for **2Z** and **2E** forms, the PA of the P is lower in comparison to N in the respective FHA conformers. The X-protonation of **1Z** conformer of FPA has PA 8.48 kcal/mol higher than similar protonation in case of FHA. The O3-protonation of FHA in **1ZO3H$^+$A** orientation has PA value 16.30 kcal/mol higher than similar protonation in case of FPA while the O3-protonation in **1ZO3H$^+$B** orientation, the difference is reduced to 6.50 kcal/mol. The basicity of O3 site of **1Z** and **1E** conformations of FPA is lower than the respective values in case of FHA. The results suggest that basicity of all the sites in FPA and FHA is also determined by nature of neighbouring atoms and bonds in addition to the nature of basic centers. Relative energies of the protonated species of FPA and FHA also suggest the same.

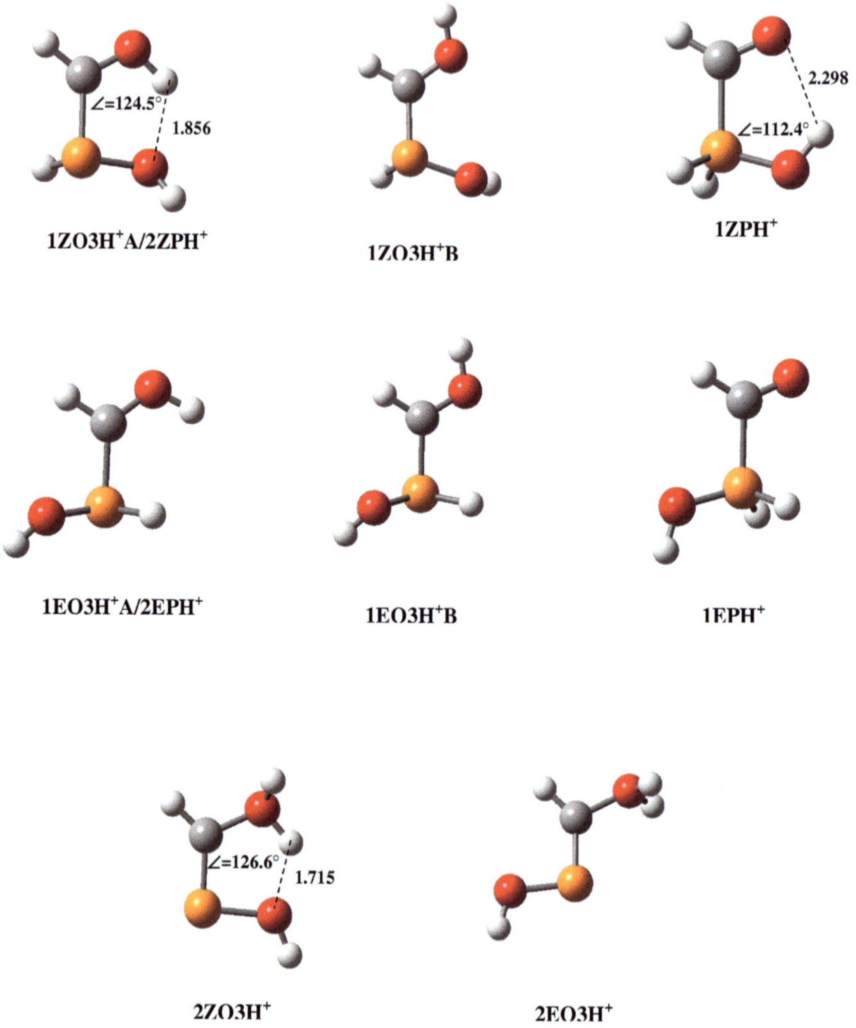

1ZO3H⁺A/2ZPH⁺   **1ZO3H⁺B**   **1ZPH⁺**

1EO3H⁺A/2EPH⁺   **1EO3H⁺B**   **1EPH⁺**

2ZO3H⁺   **2EO3H⁺**

Figure 6.4: Protonated species of FPA

Table 6.26: Proton affinities (kcal/mol) of FPA and FHA at MP2/6-31+G* [L3] theoretical level.

Protonated Species	FPA	FHA
1ZO3H$^+$A	173.36	189.66
1ZO3H$^+$B	180.81	187.31
1ZXH$^+$	184.84	176.36
1EO3H$^+$A	175.13	187.40
1EO3H$^+$B	172.95	183.37
1EXH$^+$	185.57	174.88
2ZXH$^+$	183.56	191.05
2ZO3H$^+$	171.65	155.10
2EXH$^+$	179.60	189.07
2EO3H$^+$	167.85	157.38

The atomic charges of the protonated FPA and FHA species are listed in Table 6.30. The atomic charge on P undergoes magnificent variation as the result of protonation at different sites in nearly all the isomeric forms of FPA. The similar variations on N in case of FHA are indicated to be relatively much smaller. The electronegativity of N in FHA does not allow to get further polarized. The P protonated **1Z, 1E, 2Z and 2E** species indicate that positive charge at P increases by 0.506, 0.591, 0.251 and 0.310 respectively. As a result the C1 and O3 also undergo variations in their charge densities. Hence it suggests that bond polarities of P protonated species are affected significantly. The O3 protonation on the other hand does not show considerable charge variation at O3 but atomic charges at C1 and P are altered.

Table 6.27: Geometrical parameters of O3-protonated **1Z** and **1E** of FPA (FHA) in the gas phase at B3LYP/6-31+G* [**L1**] and MP2/6-31+G* [**L3**] theoretical level. All bond distances are in angstrom (Å) and angles in degrees (°).

Parameters	1ZO3H⁺A		1ZO3H⁺B		1EO3H⁺A		1EO3H⁺B	
	L1	L3	L1	L3	L1	L3	L1	L3
X2-C1	1.827	1.829	1.791	1.771	1.812	1.807	1.795	1.786
	(1.307)	(1.306)	(1.294)	(1.294)	(1.305)	(1.304)	(1.300)	(1.299)
O3-C1	1.269	1.270	1.285	1.291	1.282	1.283	1.285	1.287
	(1.286)	(1.290)	(1.307)	(1.313)	(1.293)	(1.298)	(1.297)	(1.302)
O4-X2	1.663	1.676	1.609	1.614	1.630	1.638	1.632	1.640
	(1.379)	(1.384)	(1.362)	(1.364)	(1.378)	(1.384)	(1.374)	(1.381)
H5-O4	0.975	0.980	0.975	0.980	0.975	0.979	0.975	0.979
	(0.981)	(0.986)	(0.982)	(0.987)	(0.981)	(0.985)	(0.981)	(0.985)
H6-C1	1.093	1.091	1.097	1.096	1.095	1.094	1.096	1.096
	(1.088)	(1.087)	(1.088)	(1.087)	(1.088)	(1.087)	(1.090)	(1.089)
H7-X2	1.426	1.419	1.421	1.413	1.419	1.413	1.416	1.410
	(1.021)	(1.023)	(1.019)	(1.022)	(1.023)	(1.025)	(1.024)	(1.026)
H8⁺-O3	1.000	1.012	0.982	0.988	0.983	0.990	0.982	0.988
	(0.985)	(0.990)	(0.977)	(0.981)	(0.980)	(0.984)	(0.977)	(0.982)
O3-C1-X2	120.8	120.1	121.5	121.0	128.3	128.2	120.0	120.0
	(124.0)	(123.8)	(117.5)	(116.6)	(126.1)	(126.1)	(118.4)	(117.9)
O4-X2-C1	92.2	90.7	108.5	108.6	100.7	99.2	101.2	100.1
	(117.6)	(116.9)	(125.7)	(125.0)	(118.0)	(117.5)	(118.4)	(118.0)
H5-O4-X2	115.1	114.2	120.9	119.5	114.3	113.4	114.2	113.3
	(107.3)	(106.6)	(108.1)	(107.5)	(106.9)	(106.3)	(106.9)	(106.2)
H6-C1-X2	124.3	125.1	118.2	118.5	118.8	119.0	120.0	120.1
	(119.0)	(119.2)	(118.9)	(119.3)	(118.1)	(118.1)	(118.3)	(118.4)
H7-X2-C1	91.5	91.0	95.6	97.0	96.6	95.9	97.7	97.5
	(123.0)	(123.5)	(123.3)	(123.7)	(125.0)	(125.7)	(122.7)	(123.1)
H8⁺-O3-C1	110.7	108.9	115.4	114.3	116.6	115.8	115.4	114.6

Contd.....

Parameters	1ZO3H$^+$A		1ZO3H$^+$B		1EO3H$^+$A		1EO3H$^+$B	
	L1	L3	L1	L3	L1	L3	L1	L3
O4-X2-C1-O3	113.3	112.3	114.4	113.7	117.3	116.6	113.9	113.0
	(113.3)	(112.3)	(114.4)	(113.7)	(117.3)	(116.6)	(113.9)	(113.0)
H5-O4-X2-C1	-14.0	-12.3	-20.6	-21.5	160.8	161.0	156.8	156.6
	(7.9)	(7.8)	(-0.0)	(-0.0)	(171.0)	(172.0)	(170.2)	(171.0)
H6-C1-X2-O4	-215.8	-217.8	-70.1	-69.6	127.4	126.5	130.3	129.4
	(-124.0)	(-121.9)	(0.0)	(0.0)	(114.4)	(111.2)	(119.6)	(116.1)
H7-X2-C1-O3	171.8	172.3	173.2	172.9	-31.5	-30.3	-33.8	-33.9
	(187.2)	(186.7)	(180.0)	(180.0)	(-9.2)	(-8.4)	(-9.5)	(-8.9)
H8$^+$-O3-C1-X2	243.6	246.3	226.7	224.4	56.3	57.4	50.7	50.8
	(-190.0)	(-190.2)	(-180.0)	(-180.0)	(6.8)	(7.2)	(5.8)	(6.2)
	7.6	6.4	194.2	194.6	-12.0	-10.6	168.5	169.0
	(-3.6)	(-4.3)	(180.0)	(180.0)	(-1.1)	(-1.2)	(178.6)	(178.4)

Table 6.28: Geometrical parameters of protonated species of FPA (FHA) in the gas phase at B3LYP/6-31+G* **[L1]** and MP2/6-31+G* **[L3]** theoretical levels. All bond distances are in angstrom (Å) and angles in degrees (°).

Parameters	2ZO3H$^+$		2EO3H$^+$	
	L1	L3	L1	L3
X2-C1	1.677	1.681	1.680	1.681
	(1.253)	(1.268)	(1.253)	(1.265)
O3-C1	1.467	1.465	1.490	1.488
	(1.493)	(1.478)	(1.481)	(1.472)
O4-X2	1.686	1.690	1.627	1.633
	(1.373)	(1.390)	(1.352)	(1.366)
H5-O3	0.984	0.990	0.981	0.987
	(0.996)	(1.004)	(0.992)	(0.998)
H6-C1	1.082	1.083	1.084	1.084
	(1.081)	(1.082)	(1.084)	(1.085)

Contd.....

Parameters	2ZO3H⁺		2EO3H⁺	
	L1	L3	L1	L3
H7-O4	0.974	0.979	0.974	0.980
	(0.977)	(0.981)	(0.977)	(0.981)
H8⁺-O3	1.013	1.019	0.981	0.987
	(0.986)	(0.991)	(0.987)	(0.993)
O3-C1-X2	117.1	117.3	117.0	117.7
	(117.9)	(118.0)	(111.5)	(111.1)
O4-X2-C1	93.2	92.7	98.3	97.5
	(112.7)	(110.6)	(115.3)	(113.1)
H5-O3-C1	114.3	113.1	118.7	116.5
	(108.8)	(107.6)	(110.0)	(109.1)
H6-C1-X2	130.7	130.0	133.9	132.8
	(128.8)	(127.8)	(136.4)	(135.9)
H7-O4-X2	117.2	116.6	114.5	113.8
	(106.0)	(105.1)	(105.1)	(104.2)
H8⁺-O3-C1	106.2	105.6	118.7	116.6
	(115.8)	(114.8)	(114.7)	(113.9)
O4-X2-C1-O3	3.4	3.7	179.9	180.1
	(-2.9)	(-3.1)	(176.1)	(176.0)
H5-O3-C1-X2	-126.4	-123.9	-71.3	-69.2
	(5.4)	(5.5)	(-4.4)	(-4.1)
H6-C1-X2-O4	176.7	177.0	-0.0	0.2
	(181.9)	(181.8)	(1.5)	(1.5)
H7-O4-N2-C1	181.6	183.2	180.0	179.9
	(179.9)	(179.6)	(180.1)	(179.8)
H8⁺-O3-C1-X2	-3.6	-3.5	72.2	65.7
	(131.9)	(129.5)	(120.4)	(118.5)

Table 6.29: Geometrical parameters of protonated species of FPA (FHA) in the gas phase at B3LYP/6-31+G* [L1] and MP2/6-31+G* [L3] theoretical levels. All bond distances are in angstrom (Å) and angles in degrees (°).

Parameters	1ZXH$^+$		1EXH$^+$		Parameters	2ZXH$^+$		2EXH$^+$	
	L1	L3	L1	L3		L1	L3	L1	L3
X2-C1	1.927	1.891	1.927	1.885	X2-C1	1.827	1.829	1.818	1.805
	(1.580)	(1.570)	(1.595)	(1.580)		(1.307)	(1.306)	(1.305)	(1.304)
O3-C1	1.195	1.216	1.190	1.211	O3-C1	1.269	1.270	1.284	1.286
	(1.177)	(1.191)	(1.173)	(1.187)		(1.286)	(1.290)	(1.293)	(1.298)
O4-X2	1.581	1.584	1.588	1.591	O4-X2	1.663	1.676	1.617	1.625
	(1.398)	(1.404)	(1.406)	(1.412)		(1.379)	(1.384)	(1.378)	(1.384)
H5-O4	0.983	0.989	0.977	0.982	H5-O3	1.000	1.012	0.983	0.989
	(0.989)	(0.992)	(0.981)	(0.985)		(0.985)	(0.990)	(0.980)	(0.984)
H6-C1	1.101	1.099	1.104	1.101	H6-C1	1.093	1.091	1.095	1.094
	(1.097)	(1.096)	(1.099)	(1.098)		(1.088)	(1.086)	(1.088)	(1.087)
H7-X2	1.404	1.398	1.407	1.400	H7-O4	0.975	0.980	0.975	0.980
	(1.034)	(1.036)	(1.035)	(1.037)		(0.981)	(0.986)	(0.981)	(0.985)
H8$^+$-X2	1.404	1.398	1.407	1.400	H8$^+$-X2	1.427	1.419	1.422	1.416
	(1.034)	(1.036)	(1.033)	(1.035)		(1.021)	(1.023)	(1.023)	(1.025)
O3-C1-X2	113.6	113.7	116.2	115.5	O3-C1-X2	120.8	120.1	128.3	128.3
	(116.2)	(116.5)	(117.7)	(117.8)		(124.0)	(123.8)	(126.1)	(126.1)
O4-X2-C1	105.4	105.2	107.4	107.0	O4-X2-C1	92.1	90.7	104.3	103.6
	(112.6)	(112.2)	(113.6)	(112.9)		(117.6)	(116.9)	(118.0)	(117.5)
H5-O4-X2	116.2	115.1	122.3	121.8	H5-O3-C1	110.7	108.9	116.5	115.7
	(105.7)	(105.4)	(107.7)	(107.4)		(113.3)	(112.3)	(117.3)	(116.6)
H6-C1-X2	119.8	120.6	115.9	117.4	H6-C1-X2	124.3	125.1	118.3	118.4
	(112.6)	(112.6)	(109.5)	(109.5)		(119.0)	(119.2)	(118.1)	(118.1)
H7-X2-C1	112.3	111.9	108.7	108.4	H7-O4-X2	115.2	114.2	120.3	119.1
	(110.4)	(110.9)	(109.8)	(110.4)		(107.3)	(106.6)	(106.9)	(106.3)
H8$^+$-X2-C1	112.4	111.9	108.8	108.5	H8$^+$-X2-C1	91.5	91.0	97.5	97.9
	(110.4)	(110.9)	(109.2)	(109.7)		(123.0)	(123.5)	(125.0)	(125.7)

Contd.....

Parameters	1ZXH$^+$		1EXH$^+$		Parameters	2ZXH$^+$		2EXH$^+$	
	L1	L3	L1	L3		L1	L3	L1	L3
O4-X2-C1-O3	-0.0 (0.0)	-0.1 (-0.2)	180.1 (206.7)	180.9 (202.3)	O4-X2-C1-O3	14.2 (7.9)	12.3 (7.8)	188.3 (189.0)	190.1 (188.0)
H5-O4-X2-C1	0.0 (-0.0)	0.1 (0.5)	180.0 (77.9)	180.0 (80.1)	H5-O3-C1-X2	-7.7 (-3.6)	-6.4 (-4.3)	15.6 (1.1)	15.0 (1.2)
H6-C1-X2-O4	180.0 (180.0)	179.9 (179.8)	0.1 (26.4)	1.0 (22.3)	H6-C1-X2-O4	188.5 (187.2)	187.8 (186.7)	25.1 (9.2)	26.6 (8.4)
H7-X2-C1-O3	119.4 (120.3)	119.2 (119.6)	-57.7 (-26.0)	-57.1 (-30.6)	H7-O4-X2-C1	215.2 (236.0)	217.8 (238.1)	79.4 (245.6)	78.3 (248.8)
H8$^+$-X2-C1-O3	240.6 (239.7)	240.5 (240.0)	57.8 (91.3)	59.0 (87.3)	H8$^+$-X2-C1-O3	116.6 (170.0)	113.7 (169.8)	-61.3 (-6.8)	-59.5 (-7.3)

Table 6.30: Atomic charges (NPA) in neutral and protonated FPA (FHA) at MP2/6-31+G* [L3] theoretical level.

Species	C1	X2	O3	O4	H5	H6	H7	H8
1Z	0.149 (0.636)	0.949 (-0.426)	-0.643 (-0.732)	-1.074 (-0.616)	0.541 (0.535)	0.172 (0.173)	-0.093 (0.430)	- (-)
1ZXH⁺	0.170 (0.657)	1.455 (-0.437)	-0.498 (-0.494)	-1.039 (-0.538)	0.599 (0.562)	0.243 (0.248)	0.035 (0.501)	0.035 (0.501)
1ZO3H⁺A	0.265 (0.706)	1.036 (-0.348)	-0.643 (-0.685)	-1.092 (-0.572)	0.578 (0.557)	0.266 (0.273)	-0.010 (0.490)	0.599 (0.580)
1ZO3H⁺B	0.187 (0.632)	1.136 (-0.272)	-0.668 (-0.738)	-1.035 (-0.527)	0.562 (0.553)	0.252 (0.267)	-0.012 (0.499)	0.577 (0.586)
1E	0.163 (0.661)	0.886 (-0.454)	-0.620 (-0.685)	-1.069 (-0.626)	0.529 (0.517)	0.169 (0.166)	-0.058 (0.422)	-
1EXH⁺	0.175 (0.652)	1.447 (-0.440)	-0.474 (-0.468)	-1.049 (-0.540)	0.594 (0.549)	0.243 (0.244)	0.031 (0.494)	0.031 (0.510)
1EO3H⁺A	0.263 (0.711)	1.034 (-0.349)	-0.651 (-0.688)	-1.044 (-0.539)	0.569 (0.548)	0.269 (0.278)	-0.012 (0.473)	0.573 (0.566)
1EO3H⁺B	0.251 (0.696)	1.056 (-0.325)	-0.667 (-0.705)	-1.047 (-0.543)	0.567 (0.548)	0.252 (0.264)	0.006 (0.485)	0.581 (0.581)
2Z	-0.213 (0.478)	0.785 (-0.282)	-0.780 (-0.783)	-1.103 (-0.703)	0.547 (0.547)	0.233 (0.214)	0.531 (0.528)	- (-)
2ZXH⁺	0.265 (0.706)	1.036 (-0.348)	-0.643 (-0.685)	-1.092 (-0.572)	0.599 (0.580)	0.266 (0.273)	0.578 (0.557)	-0.010 (0.490)
2ZO3H⁺	-0.478 (0.370)	1.244 (-0.066)	-0.753 (-0.766)	-1.126 (-0.658)	0.605 (0.636)	0.290 (0.286)	0.585 (0.579)	0.634 (0.619)
2E	-0.223 (0.488)	0.779 (-0.312)	-0.767 (-0.772)	-1.063 (-0.650)	0.514 (0.524)	0.240 (0.206)	0.520 (0.516)	- (-)
2EXH⁺	0.228 (0.711)	1.089 (-0.349)	-0.653 (-0.688)	-1.038 (-0.539)	0.572 (0.566)	0.268 (0.278)	0.561 (0.548)	-0.026 (0.473)
2EO3H⁺	-0.540 (0.362)	1.262 (-0.120)	-0.760 (-0.744)	-1.048 (-0.566)	0.613 (0.620)	0.294 (0.278)	0.566 (0.553)	0.613 (0.616)

6.2.8 NBO ANALYSIS OF PROTONATED SPECIES:

The variations in bond strengths arising from the change in electron delocalizations explain the increase or decrease in molecular stability as the result of protonation. The second order delocalization energy $E^{(2)}$ values associated with the orbital interactions in the protonated species are recorded in the Table 6.31. The orbital interactions $n_{O3} \rightarrow \sigma^*_{C1-X2}$ increase in **1ZXH**$^+$ protonated species as reflected by increase in $E^{(2)}$ value by 12.21 kcal/mol in case of FPA while the similar increase in $E^{(2)}$ value in case of FHA is 30.36 kcal/mol. This also points out that higher PA of X site of FPA in comparison to that in FHA is the result of larger size of P and its higher ability to hold the additional charge in comparison to N in FHA.

The **1EXH**$^+$ protonated species resulting from X protonation of **1E** conformer also shows increase in $E^{(2)}$ value from 22.59 kcal/mol to 35.46 kcal/mol for $n_{O3} \rightarrow \sigma^*_{C1-X2}$ orbital interactions. Both **1ZXH**$^+$ and **1EXH**$^+$ species undergo nearly similar variations in $E^{(2)}$ values and it is important to note that comparable PA values are associated with both in case of FPA. The **2ZXH**$^+$ and **2EXH**$^+$ species indicate decrease in overall electron delocalizations as the $n_{O3} \rightarrow \pi^*_{C1-X2}$ and $n_{X2} \rightarrow \sigma^*_{C1-O3}$ orbital interactions of **2Z** are absent in X protonated species. But the high PA values associated with the formation of these species suggest variations in bond polarizations through σ electrons playing important role. The O3 protonation of **1Z**, **1E**, **2Z** and **2E** also suggest the overall decrease in electron delocalization and suggesting that stabilizations associated with these protonated species have contribution from the ability of P to get polarized and dispersal of the additional positive charge on most of the atoms.

The PA values in case of FHA suggest that O3 protonation is preferred over X protonation in **1Z** and **1E** conformers while X protonation is favored over O3 in case of **2Z** and **2E** of FHA which are rationalized in terms of enhancement in sum of $E^{(2)}$ values associated with several orbital interactions. The formation of C=N π bond occurs in **1ZO3H**$^+$**A** and **1ZO3H**$^+$**B** species of FHA but no such C-P π bond formation is indicated in FPA and stabilization of protonated species result from increase in bond polarization.

It can be finally concluded that the higher protonation and deprotonation enthalpies of FPA relative to those in FHA are the result of ability of P to get polarized and hold additional charge.

Table 6.31: NBO analysis of FPA (FHA) neutral and protonated at MP2/6-31+G* [L3] theoretical level.

Species	$n_{O3}\to\sigma^*_{C1\text{-}X2}$	$n_{O3}\to\pi^*_{C1\text{-}X2}$	$n_{O4}\to\sigma^*_{C1\text{-}X2}$	$n_{O4}\to\pi^*_{C1\text{-}X2}$	$n_{X2}\to\pi^*_{C1\text{-}O3}$	$n_{X2}\to\sigma^*_{C1\text{-}O3}$	$n_{O3}\to\sigma^*_{O4\text{-}H5}$	$n_{O4}\to\sigma^*_{O3\text{-}H5}$
1Z	20.73 (29.25)	-	-	-	-	-	1.09+3.31 (1.24+5.36) 1.77+0.85 (1.86+4.44)	(-) (-)
1ZXH⁺	32.94 (59.61)	-	4.73 (-)	-	5.57 (32.68)	(14.98)	-	(-)
1ZO3H⁺A	6.71 (6.47)	(3.17)	7.03 (4.16)	-	10.34 (-)	-	3.86 (1.33+2.01)	(-)
1ZO3H⁺B	-	(76.36)	14.15 (7.30) 6.70	(18.82)	-	-	-	(-)
1E	22.59 (32.25)	-	(-)	-	5.85 (20.90)	(12.11)	(2.00)	(-)
1EXH⁺	35.46 (64.76)	-	5.88 (5.94)	-	-	-	-	(-)
1EO3H⁺A	6.77 (8.87+10.96)	-	9.63 (6.41)	-	8.57 (-)	8.58 (-)	-	(-)
1EO3H⁺B	-	(55.22)	8.59	(-)	14.71 (-)	7.01 (-)	-	(-)
2Z	(19.17)	(35.07) 34.31 (54.52)	(-) 4.75	(7.32)	-	7.75 (13.50)	-	1.46+5.20 (5.06)
2ZXH⁺	6.30 (8.63)	-	(-)	(13.22)	10.34 (-)	-	-	3.86+12.69 (1.33+2.01)
2ZO3H⁺	6.71 (6.47)	(-) 7.65 (9.56)	7.03 (4.16)	(-) 8.51 (15.09)	-	6.94 (20.59)	-	23.48 (9.72)
2E	(-) 11.03+4.41 (6.12)	18.03 (48.37)	(-)	9.49 (13.80)	(-)	(3.82)	-	(-)
2EXH⁺	6.79 (8.87+10.96)	(-) (55.23)	(-) 14.39 (6.41)	-	11.87 (-)	10.71 (-)	-	(-)
2EO3H⁺	(5.80)	(-)	(9.82)	23.84 (10.82)	(-)	(11.87)	(-)	(-)

6.2.9 HYDROGEN BONDING OF FPA WITH WATER:

Exploring the H-bonding ability of FPA isostere of FHA, thirteen aggregates of FPA with water have been optimized. These aggregates are shown in Figure 6.5 and geometrical parameters are recorded in Tables 6.32-6.35. The analysis of interatomic distances and respective angles between H-bond donors and acceptors has been carried out in order to locate the H-bonds. The important intermolecular H-bond distances, angles that bridging hydrogen makes along with stabilization energies are listed in Table 6.36 and the important geometrical parameters for intramolecular H-bonding are recorded in Table 6.37. Since some of the resulting optimized FPA-H_2O adducts differ in geometric orientation to that in FHA-H_2O adducts, therefore comparison of H-bonding strength is made only for similar structures in the FHA and FPA. **1ZW1** adduct of FPA and the similar adduct of FHA both indicate the presence of two homonuclear O-H...O H-bonds, however the stabilization arising out of the adduct formation is 0.99 kcal/mol lower in case of **1ZW1** of FPA. The H-bond donor and acceptor groups in both the cases are same but O4-H5...O8 bond is relatively stronger and O8-H9...O3 is relatively weaker in **1ZW1** of FPA in comparison to that in case of FHA. This is supported by the H-bond distances and angles in the two cases. The difference in stabilization energy of adducts of FPA and FHA with water is less than 1.00 kcal/mol in **1ZW1, 1EW2, 1EW3, 2ZW3, 2EW1, 2EW2** and **2EW3**.

In none of the adducts, the P-H bond acts as H-bond donor however P acts as H-bond acceptor in **2EW1**. It is interesting to note that in **1ZW2** adduct of FPA with water only very weak H-bond is present involving C1-H6....O8 interaction, but the stabilization energy of 4.52 kcal/mol indicates the presence of other interactions also. The favorable electrostatic interactions between P2 and O8 at a distance of 2.896 Å to one another with atomic charges of 0.934 and -1.021 respectively are responsible for the additional stability of the adduct. The charge transfer interactions imparting covalent character to H-bonds are listed in Table 6.38. Occupancies of acceptor orbitals are also included in the table that verify the shift of charge density from H-bond donor to acceptor antibonding orbital. The values in case of adducts of FPA with water are comparatively smaller relative to the values in adducts of FHA with water with only few exceptions. The intramolecular H-bond present in **1Z** conformer remains intact in the **1ZW2** aggregate, supported by important parameters for the H-bonding and $n_{O3} \rightarrow \sigma^*_{O4-H5}$ orbital interactions. The **1ZW3** with two H-bonds between **1Z** of FPA and water has stabilization nearly of the same order as in case of **1ZW2**.

Here lone pairs on oxygen participate in a bifurcated fashion of H-bonding i.e. simultaneously interacting with two H-bond acceptor atoms with intramolecular component on one side and intermolecular on the other side (see Figure 6.5). The intramolecular H-bond in case of **1ZW3** is slightly weakened as suggested by the geometrical parameters. The aggregate **1EW1** of FPA is 2.49 kcal/mol less stabilized in comparison to the similar adduct in FHA due to the absence of X2-H7...O8 H-bond in the former. The O4-H5...O8 H-bonded **1EW2** adduct is stabilized with stabilization energy of 7.07 kcal/mol and 7.25 kcal/mol in case of FPA and FHA respectively. The **2ZW2** adduct of FPA is stabilized by 9.61 kcal/mol of energy relative to 7.32 kcal/mol in case of FHA with two conventional H-bonds in each. The **2ZW1** and **2ZW3** aggregates of FPA suggest the presence of intramolecular H-bond from the analysis of relevant parameters and the presence of orbital interactions $n_{O4} \rightarrow \sigma^*_{O3-H5}$ while the intramolecular H-bond in **2ZW2** is ruptured and replaced by intermolecular association. The **2EW1** adduct with two H-bonds and phosphorus as H-bond acceptor is only 1.00 kcal/mol less stabilized than the similar aggregate of FHA. One of the two H-bonds in **2EW1** involve O3-H5...O9 conventional H-bond in both FPA and FHA. The strength of this bond is anticipated to be approximately of the same order in FPA and FHA keeping in view the H-bond distance and angle. The second H-bond in FPA is comparatively weaker as reflected by the O9-H8...X2 angle.

Second order stabilization energies $E^{(2)}$ associated with various orbital interactions are recorded in Table 6.39. As discussed earlier the conjugative interactions are comparatively weaker in isomeric forms of FPA relative to those in FHA. The $E^{(2)}$ values of the orbital interactions present in the conformers undergo insignificant change as the result of aggregate formation with water. The orbital interactions specially $n_{X2} \rightarrow \pi^*_{C1-O3}$ in **1Z** conformer of FHA displayed magnificent variation as the aggregation results.

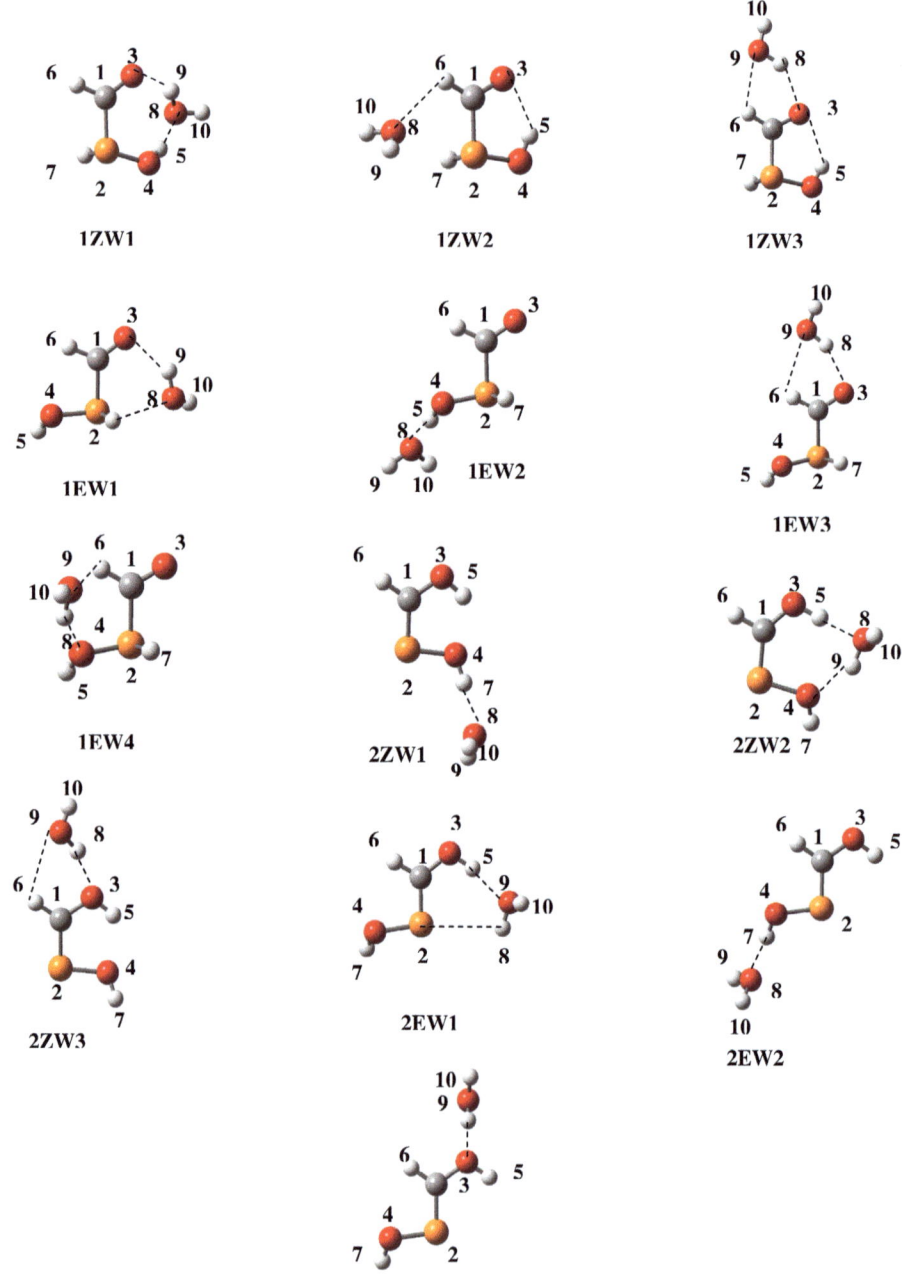

Figure 6.5: Hydrogen bonded complexes of FPA with water

Table 6.32: Geometrical parameters of the adducts of FPA (FHA) with water at MP2/6-31+G* [L3] theoretical level. All bond distances are in angstrom (Å) and bond angles in degrees (°).

Parameters	1ZW1	Parameters	1ZW2	Parameters	1ZW3	Parameters	1EW1
X2-C1	1.853 (1.357)	X2-C1	1.866 (1.359)	X2-C1	1.855 (1.350)	X2-C1	1.873 (1.369)
O3-C1	1.233 (1.236)	O3-C1	1.234 (1.238)	O3-C1	1.237 (1.244)	O3-C1	1.231 (1.235)
O4-X2	1.644 (1.399)	O4-X2	1.669 (1.416)	O4-X2	1.662 (1.405)	O4-X2	1.702 (1.417)
H5-O4	0.993 (0.993)	H5-O4	0.985 (0.990)	H5-O4	0.984 (0.988)	H5-O4	0.973 (0.976)
H6-C1	1.112 (1.103)	H6-C1	1.101 (1.099)	H6-C1	1.104 (1.097)	H6-C1	1.110 (1.100)
H7-X2	1.428 (1.014)	H7-X2	1.425 (1.022)	H7-X2	1.428 (1.014)	H7-X2	1.410 (1.025)
O8-H5	1.783 (1.792)	O8-H7	2.800 (2.004)	H8-O3	2.052 (1.978)	O8-H7	2.600 (2.014)
H9-O8	0.979 (0.983)	H9-O8	0.973 (0.976)	O9-H8	0.976 (0.979)	H9-O8	0.980 (0.981)
H10-O8	0.972 (0.971)	H10-O8	0.973 (0.971)	H10-O9	0.970 (0.970)	H10-O8	0.972 (0.971)
O3-C1-X2	126.7 (125.6)	O3-C1-X2	120.2 (122.3)	O3-C1-X2	120.6 (121.3)	O3-C1-X2	127.4 (123.4)
O4-X2-C1	104.7 (120.8)	O4-X2-C1	96.7 (114.7)	O4-X2-C1	98.0 (117.1)	O4-X2-C1	93.6 (114.4)
H5-O4-X2	115.4 (104.6)	H5-O4-X2	110.1 (101.7)	H5-O4-X2	110.7 (101.8)	H5-O4-X2	110.0 (103.8)
H6-C1-X2	113.7 (111.4)	H6-C1-X2	118.3 (113.4)	H6-C1-X2	119.1 (114.8)	H6-C1-X2	112.2 (113.1)

Contd.....

Parameters	1ZW1	Parameters	1ZW2	Parameters	1ZW3	Parameters	1EW1
H7-X2-C1	90.3 (120.0)	H7-X2-C1	91.5 (122.1)	H7-X2-C1	90.1 (122.5)	H7-X2-C1	95.7 (117.2)
O8-H5-O4	165.8 (155.5)	O8-H7-X2	79.3 (132.8)	H8-O3-C1	99.6 (98.3)	O8-H7-X2	82.1 (139.1)
H9-O8-H5	95.9 (91.9)	H9-O8-H7	116.8 (88.8)	O9-H8-O3	140.5 (146.2)	H9-O8-H7	98.0 (87.2)
H10-O8-H9	106.8 (106.6)	H10-O8-H7	106.0 (139.9)	H10-O9-H8	106.3 (106.2)	H10-O8-H9	106.4 (106.9)
O4-X2-C1-O3	-11.4 (-19.6)	O4-X2-C1-O3	-4.1 (12.4)	O4-X2-C1-O3	-6.3 (-11.3)	O4-X2-C1-O3	168.3 (161.3)
H5-O4-X2-C1	-39.2 (68.3)	H5-O4-X2-C1	-10.1 (-6.4)	H5-O4-X2-C1	-11.3 (6.0)	H5-O4-X2-C1	489.9 (118.3)
H6-C1-X2-O4	172.1 (165.7)	H6-C1-X2-O4	177.5 (188.7)	H6-C1-X2-O4	174.8 (172.0)	H6-C1-X2-O4	-14.8 (-21.8)
H7-X2-C1-O3	243.6 (195.0)	H7-X2-C1-O3	253.2 (146.8)	H7-X2-C1-O3	250.5 (205.1)	H7-X2-C1-O3	70.1 (26.2)
O8-H5-O4-X2	56.2 (-69.2)	O8-H7-X2-C1	285.8 (230.4)	H8-O3-C1-X2	-183.3 (-180.4)	O8-H7-X2-O4	189.2 (207.4)
H9-O8-H5-O3	17.7 (-8.3)	H9-O8-H7-X2	-59.5 (-5.8)	O9-H8-O3-C1	-6.6 (-3.5)	H9-O8-H7-O3	12.1 (-5.1)
H10-O8-H9-O3	200.4 (248.7)	H10-O8-H7-X2	183.4 (239.7)	H10-O9-H8-O3	180.2 (158.6)	H10-O8-H9-O3	128.6 (153.3)

Table 6.33: Geometrical parameters of the adducts of FPA (FHA) with water at MP2/6-31+G* [L3] theoretical level. All bond distances are in angstrom (Å) and bond angles in degrees (°).

Parameters	1EW2	Parameters	1EW3	Parameters	1EW4	Parameters	2ZW1
X2-C1	1.859	X2-C1	1.854	X2-C1	1.864	N2-C1	1.690
	(1.383)		(1.376)		(1.394)		(1.290)
O3-C1	1.228	O3-C1	1.232	O3-C1	1.227	O3-C1	1.360
	(1.227)		(1.232)		(1.225)		(1.347)
O4-X2	1.672	O4-X2	1.685	O4-X2	1.704	O4-X2	1.696
	(1.418)		(1.417)		(1.426)		(1.427)
H5-O4	0.984	H5-O4	0.973	H5-O4	0.974	H5-O3	0.986
	(0.985)		(0.976)		(0.977)		(0.983)
H6-C1	1.111	H6-C1	1.108	H6-C1	1.106	H6-C1	1.087
	(1.101)		(1.099)		(1.099)		(1.083)
H7-X2	1.415	H7-X2	1.414	H7-X2	1.413	H7-O4	0.985
	(1.019)		(1.019)		(1.020)		(0.986)
O8-H5	1.840	H8-O3	2.004	H8-O4	1.952	O8-H7	1.835
	(1.841)		(1.969)		(2.008)		(1.870)
H9-O8	0.972	O9-H8	0.977	O9-H8	0.977	H9-O8	0.972
	(0.972)		(0.979)		(0.975)		(0.972)
H10-O8	0.973	H10-O9	0.970	H10-O9	0.971	H10-O8	0.972
	(0.973)		(0.970)		(0.971)		(0.977)
O3-C1-X2	123.1	O3-C1-X2	122.0	O3-C1-X2	120.4	O3-C1-X2	124.7
	(123.0)		(122.2)		(121.5)		(126.0)
O4-X2-C1	99.1	O4-N2-C1	98.4	O4-N2-C1	99.4	O4-X2-C1	95.3
	(114.4)		(114.6)		(114.3)		(108.4)
H5-O4-X2	109.5	H5-O4-X2	110.6	H5-O4-X2	110.7	H5-O3-C1	106.2
	(102.5)		(103.8)		(103.7)		(106.5)
H6-C1-X2	115.3	H6-C1-X2	116.8	H6-C1-X2	117.2	H6-C1-X2	122.0
	(112.4)		(113.7)		(113.5)		(119.0)

Contd......

Parameters	1EW2	Parameters	1EW3	Parameters	1EW4	Parameters	2ZW1
H7-X2-C1	93.9 (114.8)	H7-X2-C1	94.0 (116.1)	H7-X2-C1	94.2 (114.6)	H7-O4-X2	111.6 (101.3)
O8-H5-O4	174.1 (163.2)	H8-O3-C1	102.0 (100.4)	H8-O4-X2	113.0 (117.7)	O8-H7-O4	174.2 (147.1)
H9-O8-H5	120.3 (124.9)	O9-H8-O3	149.0 (150.8)	O9-H8-O4	145.2 (152.3)	H9-O8-H7	120.8 (124.2)
H10-O8-H5	119.5 (108.1)	H10-O9-H8	106.1 (106.1)	H10-O9-H8	106.6 (106.5)	H10-O8-H7	120.7 (87.1)
O4-X2-C1-O3	161.1 (159.1)	O4-X2-C1-O3	159.0 (159.2)	O4-X2-C1-O3	163.1 (157.5)	O4-X2-C1-O3	-1.0 (-0.2)
H5-O4-X2-C1	485.1 (116.4)	H5-O4-X2-C1	487.2 (116.9)	H5-O4-X2-C1	486.2 (116.9)	H5-O3-C1-X2	-1.7 (-0.7)
H6-C1-X2-O4	-21.9 (-24.5)	H6-C1-X2-O4	-24.7 (-24.4)	H6-C1-N2-O4	-20.2 (-25.9)	H6-C1-X2-O4	177.0 (179.8)
H7-X2-C1-O3	61.0 (29.6)	H7-X2-C1-O3	59.7 (27.7)	H7-X2-C1-O3	64.9 (30.1)	H7-O4-X2-C1	158.8 (177.1)
O8-H5-O4-X2	-3.3 (-7.1)	H8-O3-C1-X2	-190.9 (-190.0)	H8-O4-X2-C1	-38.4 (-6.3)	O8-H7-O4-X2	-0.2 (-1.8)
H9-O8-H5-O4	161.9 (134.9)	O9-H8-O3-C1	-6.0 (-2.1)	O9-H8-O4-H6	-24.8 (31.4)	H9-O8-H7-O4	76.9 (110.6)
H10-O8-H5-O4	27.6 (9.2)	H10-O9-H8-O3	164.7 (156.1)	H10-O9-H8-O4	-149.7 (-169.9)	H10-O8-H7-X2	-59.9 (1.6)

Table 6.34: Geometrical parameters of the adducts of FPA (FHA) with water at MP2/6-31+G* [L3] theoretical levels. All bond distances are in angstrom (Å) and bond angles in degrees (°).

Parameters	2ZW1	Parameters	2ZW2	Parameters	2ZW3
N2-C1	1.690 (1.290)	X2-C1	1.689 (1.291)	X2-C1	1.686 (1.288)
O3-C1	1.360 (1.347)	O3-C1	1.345 (1.338)	O3-C1	1.367 (1.357)
O4-X2	1.696 (1.427)	O4-X2	1.724 (1.441)	O4-X2	1.712 (1.436)
H5-O3	0.986 (0.983)	H5-O3	0.995 (0.994)	H5-O3	0.986 (0.982)
H6-C1	1.087 (1.083)	H6-C1	1.092 (1.085)	H6-C1	1.087 (1.083)
H7-O4	0.985 (0.986)	H7-O4	0.975 (0.974)	H7-O4	0.974 (0.974)
O8-H7	1.835 (1.870)	O8-H5	1.763 (1.787)	H8-O3	1.979 (2.106)
H9-O8	0.972 (0.972)	H9-O8	0.981 (0.976)	O9-H8	0.976 (0.974)
H10-O8	0.972 (0.977)	H10-O8	0.972 (0.972)	H10-O9	0.970 (0.971)
O3-C1-X2	124.7 (126.0)	O3-C1-X2	135.6 (130.9)	O3-C1-X2	124.8 (125.7)
O4-X2-C1	95.3 (108.4)	O4-X2-C1	102.2 (110.6)	O4-X2-C1	94.6 (107.9)
H5-O3-C1	106.2 (106.5)	H5-O3-C1	113.2 (113.6)	H5-O3-C1	107.0 (107.3)
H6-C1-X2	122.0 (119.0)	H6-C1-X2	112.9 (115.4)	H6-C1-X2	122.2 (119.8)
H7-O4-X2	111.6	H7-O4-X2	109.5	H7-O4-X2	112.2

Contd......

Parameters	2ZW1	Parameters	2ZW2	Parameters	2ZW3
O8-H7-O4	101.3 174.2 (147.1)	O8-H5-O3	(100.8) 173.4 (171.4)	H8-O3-C1	(102.3) 106.3 (101.7)
H9-O8-H7	120.8 (124.2)	H9-O8-H5	91.3 (90.9)	O9-H8-O3	157.0 (133.1)
H10-O8-H7	120.7 (87.1)	H10-O8-H5	116.8 (117.4)	H10-O9-H8	105.7 (106.2)
O4-X2-C1-O3	-1.0 (-0.2)	O4-X2-C1-O3	0.2 (-0.1)	O4-N2-C1-O3	-1.2 (-0.0)
H5-O3-C1-X2	-1.7 (-0.7)	H5-O3-C1-X2	-3.4 (-2.0)	H5-O3-C1-X2	-2.3 (-0.9)
H6-C1-X2-O4	177.0 (179.8)	H6-C1-X2-O4	181.9 (179.9)	H6-C1-X2-O4	177.9 (180.1)
H7-O4-X2-C1	158.8 (177.1)	H7-O4-X2-C1	206.5 (187.2)	H7-O4-X2-C1	160.0 (179.0)
O8-H7-O4-X2	-0.2 (-1.8)	O8-H5-O3-C1	271.5 (225.9)	H8-O3-C1-X2	208.2 (186.4)
H9-O8-H7-O4	76.9 (110.6)	H9-O8-H5-O4	-4.6 (-8.2)	O9-H8-O3-C1	-12.1 (-0.0)
H10-O8-H7-X2	-59.9 (1.6)	H10-O8-H5-O3	-20.3 (19.6)	H10-O9-H8-O3	229.6 (203.9)

Table 6.35: Geometrical parameters of the adducts of FPA (FHA) with water at MP2/6-31+G* [L3] theoretical levels. All bond distances are in angstrom (Å) and bond angles in degrees (°).

Parameters	2EW1	Parameters	2EW2	Parameters	2EW3
X2-C1	1.697 (1.290)	X2-C1	1.690 (1.283)	X2-C1	1.687 (1.281)
O3-C1	1.354 (1.341)	O3-C1	1.374 (1.356)	O3-C1	1.379 (1.364)
O4-X2	1.700 (1.433)	O4-X2	1.678 (1.420)	O4-X2	1.690 (1.425)
H5-O3	0.989 (0.994)	H5-O3	0.978 (0.979)	H5-O3	0.979 (0.980)
H6-C1	1.090 (1.088)	H6-C1	1.088 (1.086)	H6-C1	1.088 (1.086)
H7-O4	0.975 (0.974)	H7-O4	0.984 (0.984)	H7-O4	0.975 (0.974)
H8-X2	2.982 (2.124)	O8-H7	1.842 (1.902)	H8-O3	1.975 (2.098)
O9-H8	0.974 (0.981)	H9-O8	0.972 (0.972)	O9-H8	0.975 (0.974)
H10-O9	0.973 (0.972)	H10-O8	0.972 (0.976)	H10-O9	0.970 (0.971)
O3-C1-X2	127.6 (122.5)	O3-C1-X2	126.2 (121.2)	O3-C1-X2	125.9 (119.5)
O4-X2-C1	98.5 (108.7)	O4-X2-C1	99.3 (110.1)	O4-N2-C1	98.1 (109.5)
H5-O3-C1	110.0 (109.1)	H5-O3-C1	110.1 (107.9)	H5-O3-C1	110.7 (107.8)
H6-C1-X2	123.0 (120.0)	H6-C1-X2	122.2 (124.8)	H6-C1-X2	122.5 (126.7)

Contd......

Parameters	2EW1	Parameters	2EW2	Parameters	2EW3
H7-O4-X2	110.7 (102.1)	H7-O4-X2	110.6 (101.2)	H7-O4-X2	110.7 (101.8)
H8-X2-C1	92.9 (108.3)	O8-H7-O4	178.6 (148.0)	H8-O3-C1	116.8 (102.4)
O9-H8-X2	103.1 (130.1)	H9-O8-H7	121.1 (123.4)	O9-H8-O3	178.6 (134.8)
H10-O9-H8	105.5 (106.7)	H10-O8-H7	120.2 (86.5)	H10-O9-H8	105.5 (106.2)
O4-X2-C1-O3	174.8 (178.4)	O4-X2-C1-O3	174.6 (179.9)	O4-X2-C1-O3	174.2 (180.1)
H5-O3-C1-X2	1.0 (-0.5)	H5-O3-C1-X2	2.0 (0.7)	H5-O3-C1-X2	1.8 (-1.1)
H6-C1-X2-O4	-2.4 (-1.3)	H6-C1-X2-O4	-2.5 (-0.0)	H6-C1-X2-O4	-3.1 (0.3)
H7-O4-X2-C1	124.9 (166.7)	H7-O4-X2-C1	142.2 (177.8)	H7-O4-X2-C1	142.7 (180.8)
H8-X2-C1-O3	-2.0 (2.6)	O8-H7-O4-X2	2.0 (0.8)	H8-O3-C1-X2	134.2 (189.0)
O9-H8-X2-H5	11.4 (-3.9)	H9-O8-H7-O4	-111.9 (108.6)	O9-H8-O3-C1	-21.9 (-1.2)
H10-O9-H8-O3	124.4 (123.9)	H10-O8-H7-X2	-246.3 (2.2)	H10-O9-H8-O3	142.6 (207.7)

Table 6.36: Important intermolecular H-bonding parameters (hydrogen bond distances[a], hydrogen bond angles[b]), charges on hydrogen bond acceptor and hydrogen along with stabilization energies S.E.[c] and distortion energies E_{Dis}^{d} of various hydrogen bonded adducts of FPA (FHA) with water at MP2/6-31+G* [L3] theoretical level.

Species	Hydrogen bond distances[a]		Hydrogen bond Angles[b]	Atomic charges		S.E.[c]	E_{Dis}^{d}
1ZW1	H9...O3	2.037 (1.964)	O8-H9...O3	136.4 (145.0)	$q_H(q_O)$	(0.541(-0.677) (0.546(-0.758))	9.89 (10.88)	1.89 (1.72)
	H5...O8	1.783 (1.792)	O4-H5...O8	165.8 (155.5)		0.563(-1.034) (0.560(-1.043))		
1ZW2	H6...O8	2.477	C1-H6...O8	109.3	$q_H(q_O)$	0.517(-1.087)	4.52	0.11
1ZW3	H8...O3	2.052 (1.978)	O9-H8...O3	140.5 (146.2)	$q_H(q_O)$	0.520(-0.681) (0.527(-0.777))	4.56 (5.89)	0.08 (0.21)
	H6...O9	2.535 (2.588)	C1-H6...O9	108.6 (104.3)		0.197(-1.030) (0.200(-1.037))		
1EW1	H7...O8	-	X2-H7...O8	-	$q_H(q_O)$	-	5.89 (8.38)	0.84 (0.45)
	H9...O3	(2.014) 1.968 (2.011)	O8-H9...O3	(139.1) 135.9 (140.6)	$q_H(q_O)$	(0.463(-1.044)) (0.533(-0.660) (0.537(-0.741))		
1EW2	H5...O8	1.840 (1.841)	O4-H5...O8	174.1 (163.2)	$q_H(q_O)$	0.559(-1.019) (0.552(-1.018))	7.07 (7.26)	0.23 (0.25)
1EW3	H8...O3	2.004 (1.969)	O9-H8...O3	149.0 (150.8)	$q_H(q_O)$	0.521(-0.658) (0.525(-0.726))	4.54 (5.37)	0.09 (0.17)
	H6...O9	2.728 (2.721)	C1-H6...O9	103.7 (101.7)		0.190(-1.029) (0.189(-1.035))		
1EW4	H8...O4	1.952 (2.008)	O9-H8...O4	145.2 (152.3)	$q_H(q_O)$	0.524(-1.093) (0.516(-0.652))	4.60 (3.46)	0.24 (0.14)
	H6...O9	2.601 (2.546)	C1-H6...O9	103.9 (132.8)		0.186(-1.030) (0.189(-1.024))		
2ZW1	H10...X2	-	O8-H10...X2	-	$q_H(q_X)$	-	6.84	0.21

Contd......

Species	Hydrogen bond distances[a]	Hydrogen bond Angles[b]		Atomic charges		S.E.[c]	E_{Dis}[d]
	H7...O8	O4-H7...O8	(115.5)		(0.529(-0.316)		(7.92)	(0.21)
			174.2		0.563(-1.018)			
	(2.303)		(147.1)		(0.561(-1.024))			
	1.835			$q_H(q_O)$				
	(1.870)							
2ZW2	H5...O8	O3-H5...O8	173.4	$q_H(q_O)$	0.569(-1.021)		9.61	3.76
	1.763		(171.4)		(0.578(-1.037))		(7.32)	(1.82)
	(1.787)							
	H9...O4	O8-H9...O4	143.2	$q_H(q_O)$	0.544(-1.114)			
	1.870		(127.9)		(0.533(-0.718))			
	(2.087)							
2ZW3	H8...O3	O9-H8...O3	157.0	$q_H(q_O)$	0.517(-0.808)		3.86	0.01
	1.979		(133.1)		(0.518(-0.809))		(3.96)	(0.10)
	(2.106)							
	H6...O9	C1-H6...O9	-	$q_H(q_O)$	-			
	-		(113.3)					
	(2.517)				(0.238(-1.021))			
2EW1	H5...O9	O3-H5...O9	169.0	$q_H(q_O)$	0.561(-1.034)		7.71	0.28
	1.828		(159.6)		(0.554(-1.019))		(8.71)	(0.54)
	(1.815)							
	H8...X2	O9-H8...X2	103.1	$q_H(q_X)$	0.523(0.695)			
	2.982		(130.1)		(0.535(-0.368))			
	(2.124)							
2EW2	H10...X2	O8-H10...X2	-	$q_H(q_X)$	-		6.32	0.17
	-		(115.4)				(6.59)	(0.13)
	(2.343)				(0.524(-0.339))			
	H7...O8	O4-H7...O8	178.6	$q_H(q_O)$	0.552(-1.017)			
	1.842		(148.0)		(0.548(-1.021))			
	(1.902)							
2EW3	H6...O9	C1-H6...O9	-	$q_H(q_O)$	-		3.24	0.08
	-		(112.7)				(3.70)	(0.11)
	(2.560)				(0.229(-0.798))			
	H8...O3	O9-H8...O3	178.6	$q_H(q_O)$	0.516(-0.794)			
	1.975		(134.8)		(0.517(-0.798))			
	(2.098)							

a- in angstrom (Å)
b- in Degrees (°)
c, d- in kcal/mol

Table 6.37: Important hydrogen bond distances[a] and angles at bridging hydrogen[b] for intramolecular hydrogen bonds in adducts of FPA with water at MP2/6-31+G* [L3] theoretical level.

Species	Hydrogen bond distances[a]		Hydrogen bond angles[b]	
1ZW2	H5...O3	2.104	O4-H5...O3	121.2
1ZW3	H5...O3	2.170	O4-H5...O3	118.8
2ZW1	H5...O4	2.009	O3-H5...O4	123.4
2ZW3	H5...O4	2.005	O3-H5...O4	122.5

a- in angstrom (Å)
b- in Degrees (°)

Table 6.38: The second order stabilization energies $E^{(2)}$ (kcal mol^{-1}) and occupancies of acceptor orbitals associated with electron delocalizations responsible for hydrogen bond formation in water aggregates of FPA (FHA) at MP2/6-31+G* [L3] theoretical level.

System	Donor FPA(FHA)	Acceptor H$_2$O	$E^{(2)}$	Donor H$_2$O	Acceptor FPA(FHA)	$E^{(2)}$	Occupancies	
							Acceptor H$_2$O	Acceptor FPA(FHA)
1ZW1	n$_{O3(1)}$ → σ*$_{O8-H9}$		3.49 (4.52)	n$_{O8(1)}$ → σ*$_{O4-H5}$		0.23 (0.24)	σ*$_{O8-H9}$ 0.009 (0.015)	σ*$_{O4-H5}$ 0.036 (0.031)
	n$_{O3(2)}$ → σ*$_{O8-H9}$		2.22 (4.96)	n$_{O8(2)}$ → σ*$_{O4-H5}$		24.69 (23.08)		
1ZW2				n$_{O8(1)}$ → σ*$_{C1-H6}$		0.80		σ*$_{C1-H6}$ 0.040
				n$_{O8(2)}$ → σ*$_{C1-H6}$		0.52		
1ZW3	n$_{O3(1)}$ → σ*$_{O9-H8}$		1.87 (3.04)	n$_{O9(1)}$ → σ*$_{C1-H6}$		-	σ*$_{O9-H8}$ 0.010 (0.013)	σ*$_{C1-H6}$ 0.039 (0.038)
	n$_{O3(2)}$ → σ*$_{O9-H8}$		4.96 (6.48)	n$_{O9(2)}$ → σ*$_{C1-H6}$		0.68 (0.58)		
1EW1	n$_{O3(1)}$ → σ*$_{O8-H9}$		3.39 (3.04)	n$_{O8(1)}$ → σ*$_{X2-H7}$		0.07 (-)	σ*$_{O8-H9}$ 0.014 (0.014)	σ*$_{X2-H7}$ 0.020 (0.027)

Contd.....

System	Donor FPA(FHA)	Acceptor H₂O	$E^{(2)}$	Donor H₂O	Acceptor FPA(FHA)	$E^{(2)}$	Occupancies Acceptor H₂O	Acceptor FPA(FHA)
1EW2	$n_{O3(2)} \to \sigma^*_{O8-H9}$		6.40 (6.09)	$n_{O8(2)} \to \sigma^*_{X2-H7}$				
				$n_{O8(1)} \to \sigma^*_{O4-H5}$		0.33 (10.60)		σ^*_{O4-H5} 0.023 (0.024)
				$n_{O8(2)} \to \sigma^*_{O4-H5}$		0.10 (0.13) 19.76 (-)		
1EW3	$n_{O3(1)} \to \sigma^*_{O9-H8}$		2.20 (2.81)	$n_{O9(1)} \to \sigma^*_{C1-H6}$		0.32 (0.34)	σ^*_{O9-H8} 0.012 (0.014)	σ^*_{C1-H6} 0.043 (0.042)
	$n_{O3(2)} \to \sigma^*_{O9-H8}$		6.56 (7.47)					
1EW4	$n_{O4(1)} \to \sigma^*_{O9-H8}$		0.95 (2.35)	$n_{O9(1)} \to \sigma^*_{C1-H6}$		- (-)	σ^*_{O9-H8} 0.011 (0.011)	σ^*_{C1-H6} 0.045 (0.043)
	$n_{O4(2)} \to \sigma^*_{O9-H8}$		8.43 (6.91)	$n_{O9(2)} \to \sigma^*_{C1-H6}$		0.23 (1.40)		
2ZW1				$n_{O8(1)} \to \sigma^*_{O4-H7}$		0.11 (0.17)		σ^*_{O4-H7} 0.023 (0.023)
				$n_{O8(2)} \to \sigma^*_{O4-H7}$		19.33 (16.61)		
2ZW2	$n_{O4(1)} \to \sigma^*_{O8-H9}$		13.42 (4.36)	$n_{O8(1)} \to \sigma^*_{O3-H5}$		0.34 (0.29)	σ^*_{O8-H9} 0.017 (0.007)	σ^*_{O3-H5} 0.040 (0.038)
	$n_{O4(2)} \to \sigma^*_{O8-H9}$		- (0.41)	$n_{O8(2)} \to \sigma^*_{O3-H5}$		27.36 (24.81)		
2ZW3	$n_{O3(1)} \to \sigma^*_{O9-H8}$		6.36 (3.99)	$n_{O9(1)} \to \sigma^*_{C1-H6}$		0.10 (1.11)	σ^*_{O9-H8} 0.011 (0.005)	σ^*_{C1-H6} 0.012 (0.014)
2EW1	$n_{X2} \to \sigma^*_{O9-H8}$		0.81 (7.97)	$n_{O9(1)} \to \sigma^*_{O3-H5}$		0.42 (0.25)	σ^*_{O9-H8} 0.015	σ^*_{O3-H5} 0.035
				$n_{O9(2)} \to \sigma^*_{O3-H5}$		21.82 (22.36)		
2EW2				$n_{O8(1)} \to \sigma^*_{O4-H7}$		0.06 (0.14)		σ^*_{O4-H7} 0.023 (0.020)

Contd......

System	Donor FPA(FHA)	Acceptor H₂O	$E^{(2)}$	Donor H₂O	Acceptor FPA(FHA)	$E^{(2)}$	Occupancies Acceptor H₂O	Acceptor FPA(FHA)
2EW3	$n_{O3(1)} \to \sigma^*_{O9-H8}$ $n_{O3(2)} \to \sigma^*_{O9-H8}$		5.95 (4.22) 3.17 (0.11)	$n_{O8(2)} \to \sigma^*_{O4-H7}$		18.85 (14.86)	σ^*_{O9-H8}	0.012 (0.006)

Table 6.39: Second order delocalization energies ($E^{(2)}$ in kcal/mol) in adducts of FPA (FHA) with water at MP2/6-31+G* [L3] theoretical level.

Species	$n_{X2} \to \sigma^*_{C1-O3}$	$n_{X2} \to \pi^*_{C1-O3}$	$n_{O3} \to \sigma^*_{O4-H5}$	$n_{O3} \to \sigma^*_{C1-X2}$	$n_{O3} \to \pi^*_{C1-X2}$	$n_{O4} \to \sigma^*_{O3-H5}$	$n_{O4} \to \pi^*_{C1-X2}$	$n_{O4} \to \sigma^*_{C1-X2}$
1Z	2.93 (14.98)	5.57 (32.68)	1.09+3.31 (1.24+5.36)	20.73 (29.25)	- (-)	- (-)	- (-)	1.46+1.62 (4.36)
1ZW1	4.18 (1.13)	7.81 (86.45)	- (-)	2.08+19.19 (28.53+1.96)	- (-)	- (-)	- (-)	1.43+7.68 (1.50+6.45)
1ZW2	2.54 (13.09)	5.27 (36.31)	1.42+4.10 (1.41+6.23)	20.76 (1.38+28.39)	- (-)	- (-)	- (-)	1.50 (3.83)
1ZW3	3.04 (8.80)	5.84 (53.34)	1.23+2.71 (1.52+4.78)	20.00 (27.54)	- (-)	- (-)	- (-)	1.49+1.74 (4.65)
1E	- (12.11)	5.85 (20.90)	-	22.59 (32.25)	- (-)	- (-)	- (-)	1.29+6.70 (3.39)
1EW1	- (12.11)	6.15 (31.86)	- (-)	20.48 (1.48+28.08)	- (-)	- (-)	- (-)	6.07 (3.48)
1EW2	- (12.06)	6.29 (23.40)	- (-)	21.97 (31.38)	- (-)	- (-)	- (-)	8.26 (3.97)
1EW3	- (10.65)	6.39 (28.53)	- (-)	21.24 (29.91)	- (-)	- (-)	- (-)	7.02 (3.66)
1EW4	- (11.39)	5.60 (20.14)	- (-)	23.42 (32.96)	- (-)	- (-)	- (-)	6.84 (2.20)

Contd......

Species	$n_{X2} \rightarrow \sigma^*_{Cl-O3}$	$n_{X2} \rightarrow \pi^*_{Cl-O3}$	$n_{O3} \rightarrow \sigma^*_{O4-H5}$	$n_{O3} \rightarrow \sigma^*_{Cl-X2}$	$n_{O3} \rightarrow \pi^*_{Cl-X2}$	$n_{O4} \rightarrow \sigma^*_{O3-H5}$	$n_{O4} \rightarrow \pi^*_{Cl-X2}$	$n_{O4} \rightarrow \sigma^*_{Cl-X2}$
2Z	7.75 (13.50)	- (-)	- (-)	6.30+5.01 (8.63)	34.31 (54.42)	5.20+1.46 (5.06)	3.82 (13.22)	4.75 (-)
2ZW1	7.69 (13.33)	- (-)	- (-)	6.99 (8.54)	46.84 (54.35)	6.22+1.25 (-)	8.40 (14.53)	2.32 (-)
2ZW2	12.32 (15.18)	- (-)	- (-)	8.90 (7.63+17.21)	50.62 (21.97+6.78)	- (1.66)	5.96 (4.38)	2.86 (4.66)
2ZW3	7.77 (13.72)	- (-)	- (-)	5.67+7.43 (7.41)	24.01 (43.21)	5.99+1.28 (5.55)	2.28 (11.68)	5.59 (-)
2E	- (3.82)	- (-)	- (-)	11.03+4.41 (6.12)	18.03 (48.37)		9.49 (13.80)	
2EW1	- (2.82)	- (-)	- (-)	6.70+3.18 (5.98+11.30)	1.50+42.80 (3.71+27.45)	- (-)	1.26 (7.29)	6.91 (-)
2EW2	- (3.89)	- (-)	- (-)	5.80 (6.24)	35.13 (47.19)	- (-)	12.09 (15.28)	3.58 (-)
2EW3	- (4.15)	- (-)	- (-)	4.38+5.02 (-)	2.79+23.56 (44.07)	- (-)	10.92 (14.21)	3.37 (-)

6.2.10 HYDROGEN BONDING OF FPO WITH WATER:

Full geometry optimizations of water explicitly bound to FPO are also investigated. Figure 6.6 shows all the optimized aggregates. The data for characterizing intermolecular H-bonding is listed in Table 6.40. The aggregation of FPO with water has stabilization energies falling in the range 3.98-8.38 kcal/mol. The strongest bound aggregate **FPO I W1** has H8...O2 interactions with bond angle O9-H8...O2 of 147.5° along with C5-H7...O9 H-bond and stabilization energy of 8.38 kcal/mol which is 1.51 kcal/mol lower than for most stabilized aggregate of FPA with water (**1ZW1**). In **FPO I W1** oxygen of water acts as H-bond acceptor to C-H of formyl and hydrogen acts as H-bond donor to phosphoryl oxygen while in case of **1ZW1**, oxygen of water is H-bond acceptor to O-H of FPA and is H-bond donor to carbonyl oxygen. Both the H-bonds in FPA (**1ZW1**) are conventional H-bonds and strong stabilization of the aggregate is anticipated but the comparatively stronger electrostatic interaction in the O9-H8...O2 resulting from higher charge density at O2 favors the **FPO I W1** and the stabilization is only 1.51 kcal/mol lower than that in case of FPA (**1ZW1**).

The **FPO I W2** also suggests the presence of H8...O2 interactions with stabilization energy of 6.53 kcal/mol. In addition the non bonded electrostatic interaction between P of FPO and O of water is also indicated. The carbonyl oxygen acts as H-bond acceptor in **FPO I W3** but the H-bond angle of 134.3° certainly indicates the additional P...O interactions also. The distance between P and O is within the range of van der Waals radii. The **FPO II W1** with two H-bonds H8...O6 and H7...O9 has stabilization energy of 4.79 kcal/mol. **FPO II** with the two oxygens syn to one another does not allow non bonded P...O electrostatic interactions as observed in **FPO I** conformation. Attempts were made to optimize structure with water as H-bond donor toward carbonyl oxygen with near to ideally desired 180° H-bond angle in case of **FPO I** but it always resulted in considerable deviation from the angle. The highly polarized P-O bond of **FPO I** resulted in electrostatic interactions between P of **FPO I** and oxygen of water. The atomic charges on all atoms of **FPO I** and **FPO II** in aggregate form and monomeric form are listed in Table 6.41. The P-O bond is more polarized in FPO than FPA. Although the double bond character of phosphoryl group is assigned as sigma bond plus additional ionic interaction between strongly polarized P and O, but also assigned as P→O dative bond by several scientists. FPA acted as charge acceptor in its aggregates to water, the FPO acts as weak charge donor towards water as seen from $E^{(2)}$ values associated with intermolecular H-bonding (Table 6.42).

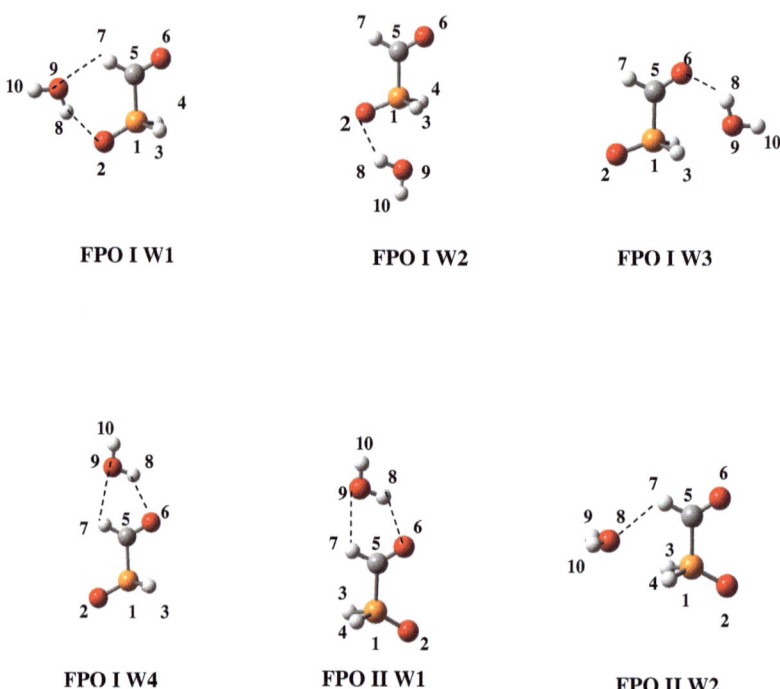

Figure 6.6: Hydrogen bonded complexes of FPO with water

Table 6.40: Important intermolecular hydrogen bonding parameters; hydrogen bond distances[a], hydrogen bond angles[b], charges on hydrogen bond acceptor and hydrogen and stabilization energies (S.E.[c]) of aggregates formylphosphine oxide (**FPO I** and **FPO II**) at MP2/6-31+G* [**L3**] theoretical level.

System	Hydrogen bond distances[a]		Hydrogen bond/Angles[b]		Atomic charges		S.E.[c]
FPO I W1	H8...O2	1.922	O9-H8...O2	147.5	q_O (q_H)	-1.184 (0.540)	8.38
	H7...O9	2.300	C5-H7...O9	127.3	q_H (q_O)	0.226 (-1.054)	
FPO I W2	H8...O2	2.060	O9-H8...O2	129.4	q_O (q_H)	-1.186 (0.534)	6.53
FPO I W3	H8...O6	2.060	O9-H8...O6	134.3	q_O (q_H)	-0.619 (0.528)	5.54
FPO I W4	H7...O9	2.469	C5-H7...O9	109.6	q_H (q_O)	0.214 (-1.026)	3.98
FPO II W1	H8...O6	2.140	O9-H8...O6	130.7	q_O (q_H)	-0.621 (0.518)	4.79
	H7...O9	2.132	C5-H7...O9	132.7	q_O (q_H)	-0.583 (0.520)	
	H8...O6	2.430	O9-H8...O6	110.9	q_H (q_O)	0.195 (-1.030)	
FPO II W2	H7...O8	2.490	C5-H7...O8	119.6	q_H (q_O)	0.184 (-1.025)	4.61

a- in Angstrom (Å).
b- in Degrees (°).
c- in kcal/mol.

Table 6.41: Atomic charges (NPA) on atoms of FPO in its aggregates with water at MP2/6-31+G* [**L3**] theoretical level.

System	P1	O2	H3	H4	C5	O6	H7
FPO I	1.558	-1.156	-0.074	-0.074	0.140	-0.580	0.187
FPO I W1	1.546	-1.184	-0.067	-0.067	0.146	-0.590	0.226
FPO I W2	1.554	-1.186	-0.059	-0.052	0.143	-0.581	0.185
FPO I W3	1.549	-1.169	-0.053	-0.053	0.164	-0.619	0.189
FPO I W4	1.560	-1.154	-0.075	-0.075	0.155	-0.621	0.214
FPO II	1.574	-1.135	-0.091	-0.091	0.120	-0.542	0.164
FPO II W1	1.574	-1.136	-0.088	-0.088	0.132	-0.583	0.195
FPO II W2	1.574	-1.152	-0.087	-0.087	0.122	-0.559	0.184

Table 6.42: Important orbital interactions, second order delocalization energies $E^{(2)}$ in kcal/mol and occupancies of acceptor antibonding orbitals for the hydrogen bonds present in the aggregates of **FPO I** and **FPO II** with water at MP2/6-31+G* [L3] theoretical level.

System	FPO Donor Acceptor	$E^{(2)}$	Water Donor Acceptor	$E^{(2)}$	Occupancies Acceptor water		Acceptor FPO	
FPO I W1	$n_{O2(1)} \rightarrow \sigma^*_{H8-O9}$	5.13	$n_{O9(2)} \rightarrow \sigma^*_{C5-H7}$	3.31	σ^*_{H8-O9}	0.016	σ^*_{C5-H7}	0.040
	$n_{O2(3)} \rightarrow \sigma^*_{H8-O9}$	7.88	-	-				
FPO I W2	$n_{O2(1)} \rightarrow \sigma^*_{H8-O9}$	1.98	-	-	σ^*_{H8-O9}	0.009		0.043
	$n_{O2(2)} \rightarrow \sigma^*_{H8-O9}$	0.07	-	-				
	$n_{O2(3)} \rightarrow \sigma^*_{H8-O9}$	4.82	-	-				
FPO I W3	$n_{O6(1)} \rightarrow \sigma^*_{H8-O9}$	2.62	-	-	σ^*_{H8-O9}	0.008		
	$n_{O6(2)} \rightarrow \sigma^*_{H8-O9}$	3.80	-	-				
FPO I W4	$n_{O6(1)} \rightarrow \sigma^*_{H8-O9}$	1.12	$n_{O9(1)} \rightarrow \sigma^*_{C5-H7}$	1.06	σ^*_{H8-O9}	0.006	σ^*_{C5-H7}	0.039
	$n_{O6(2)} \rightarrow \sigma^*_{H8-O9}$	3.37	-	-				
FPO II W1	$n_{O6(1)} \rightarrow \sigma^*_{H8-O9}$	1.16	$n_{O9(2)} \rightarrow \sigma^*_{C5-H7}$	1.14	σ^*_{H8-O9}	0.007	σ^*_{C5-H7}	0.044
	$n_{O6(2)} \rightarrow \sigma^*_{H8-O9}$	3.66	-	-				
FPO II W2	-	-	$n_{O8(1)} \rightarrow \sigma^*_{C5-H7}$	0.53	σ^*_{C5-H7}		σ^*_{C5-H7}	0.046
	-	-	$n_{O8(2)} \rightarrow \sigma^*_{C5-H7}$	1.26				

6.3 CONCLUSIONS:

The conformational stability of FPA, an isostere of FHA is analysed as regards to tautomeric and structural isomeric forms present for the molecule. The reasons for the stability differences between isomeric forms of FPA are explored. The probable interconversion barriers for the isomeric forms in the gas phase are also evaluated. Phosphinous acid and phosphine oxide exist in tautomeric equilibrium with each other. The theoretical calculations suggest phosphinous acids to be more stable than phosphine oxide and the difference is rationalized in terms of strengths of P-O single and double bonds and the relative strength of P-H and O-H in the two cases. The presence of -CHO group at P increases the relative energy difference between most stable phosphinous acid and phosphine oxide form to 4.48 kcal/mol (with respect to most stable conformer) at [L3] theoretical level, thereby suggesting that it stabilizes the phosphinous acid in gas phase. The relative energy differences in the various isomeric forms indicate that **1Z** is the most stable form of FPA. Intramolecular H-bonding is indicated by the structural parameters for **1Z** and **2Z** isomeric forms involving O-H...O hydrogen bond forming a five membered ring. The strength of intramolecular H-bond in FPA is reflected to be comparatively weaker by longer H-bond distances and smaller relative energies with respect to that in FHA. While CH_3 strengthens intramolecular H-bond in **1Z** and **2Z**, the presence of CF_3 tends to weaken it. The order of stability in the isomeric forms in case of both FPA and FHA is **1Z > 2Z > 1E > 2E > 3** at [L3] theoretical level. The stability order remains the same with the presence of CH_3 at carbonyl carbon. The presence of electron withdrawing CF_3 at the carbonyl carbon however changes the order to **2Z > 1Z > 1E > 2E > 3**. The rotational barrier **1Z → 1E** in FPA is half than similar barrier in FHA indicating that both the strength of intramolecular H-bond and the C-P π bond is comparatively weaker in **1Z**. The $E^{(2)}$ values for $n_{O3} \rightarrow \sigma^*_{C1-X2}$ and $n_{X2} \rightarrow \pi^*_{C1-O3}$ orbital interactions indicative of strong conjugation between lone pairs of electrons present on O3, X and π_{C-O} in FHA are comparatively weaker in FPA. The 1, 3 prototropic shift barriers for **1E → 2E** transition are larger in unsubstituted and CH_3, CF_3 substituted FPA over FHA since four membered ring FPA is more strained as ∠H-X-C is 12-15° lower than similar transition states in FHA. . Both CH_3 and CF_3 decrease this barrier in case of FPA.

The stability of anionic species of FPA decreases in the order **1ZP⁻ > 1ZO⁻ > 2ZO4⁻ > 1EO⁻ > 1EP⁻ = 2EO3⁻ > 2EO4⁻** which is different from that in case of FHA. Smaller deprotonation enthalpies of isomeric forms of FPA indicate that ionization is relatively easier in FPA in comparison to FHA. The **1ZP⁻** is the most stable anionic species in FPA while in FHA; **1ZN⁻** is most stable species. The anions of FPA do not show wide range of relative energies as the relative energies range between 4.81-14.85 kcal/mol. In comparison, relative energies for anions of FHA range between 8.75-27.97 kcal/mol. The smaller relative energy differences amongst all the anions of FPA reflect that presence of P in the anionic species stabilizes the anion because of its ability to hold additional charge density. The increase in intramolecular H-bond strength in **1ZX⁻** and **2ZO4⁻** is indicated by increase in $E^{(2)}$ value representing charge transfer from the hydrogen bond acceptor to hydrogen bond donor. Because of electronegativity, there is accumulation of charge density at N in neutral as well as in anionic species of FHA; phosphorus on the other hand in all the neutral and anionic species of FPA is positively charged.

The P is the most basic center in all the isomeric forms of FPA which is in contrast to the observations in FHA where O3 is the most basic site in **1Z** and **1E** conformers while in case of **2Z** and **2E** conformers it is N site which is most basic site. Proton affinity of P in keto forms (**1Z, 1E**) of FPA is larger than of N in similar forms of FHA. But for enolic forms (**2Z, 2E**) the proton affinity of P in FPA is lower than N in FHA. Much larger variation in charge (0.5 units) on P in both **1ZPH⁺** and **1ZP⁻** infers to greater charge holding capacity and larger polarizability of P in FPA than N in FHA, leading to higher protonation and deprotonation tendencies of FPA relative to FHA. NBO analysis explains the reason for P being most basic center in FPA while O3 being most basic site for **1Z** and **1E** of FHA. The formation of C=N π bond in place of C=O3 bond on O3 protonation stabilizes the protonated species of FHA. The protonation at P site in FPA stabilizes the protonated species more than O3 protonated species that requires change in hybridization at P center to form a C=P π bond.

1ZW1 is most stable aggregate of FPA with water. None of the aggregate with P-H as donor is observed but one aggregate **2EW1** with P as H-bond acceptor is optimized and the H-bond acceptor ability of P is lower than N. The aggregation of FPO with H_2O has stabilization energies in the range 3.98-8.38 kcal/mol.

6.4 REFERENCES:

1. W.A. Bridger, J.F. Henderson, Cell Adenosine Triphosphate Physiology, Wiley, Newyork, 1983.
2. A.L. Lehninger, D.L. Nelson, M.M. Cox, Principles of Biochemistry, Worth Publishers: New York, 1993.
3. Yu. V. Babin, A.V. Prisyazhnyuk, Yu. A. Ustynyuk, Russ. J. Phys. Chem. B 2 (2008) 684.
4. B. Hoge, P. Garcia, H. Willner, H. Oberhammer, Chem. Eur. J. 12 (2006) 3567.
5. D.E.C. Corbridge, Phosphorus: An Outline of its Chemistry, Biochemistry and Uses, Elsevier, Amsterdam, 5th ed. 1995.
6. J. Chatt, B.T. Heaton, J. Chem. Soc. A (1968) 2745.
7. J.E. Griffiths, A.B. Burg, J. Am. Chem. Soc. 82 (1960) 1507.
8. J.-L. Virlichie, P. Dagnac, Rev. Chim. Miner. 14 (1977) 355.
9. R.C. Dobbie, B.P. Straughan, Spectrochim. Acta 27 (1971)255.
10. N.V. Dubrovina, A. Borner, Angew. Chem. 116 (2004) 6007.
11. L.N. Heydorn, P.C. Burgers, P.J.A. Ruttink, J.K. Terlouw, Int. J. Mass Spectrom. 228 (2003) 759.
12. S.S. Wesolowski, N.R. Brinkmann, E.F. Valeev, H.F. Schaefer III, M.P. Repasky, W.L. Jorgensen, J. Chem. Phys. 116 (2002) 112.
13. B. Hoge, C. Thösen, T. Herrmann, P. Panne, I. Patenburg, J. Fluorine Chem. 125 (2004) 831.
14. B. Hoge, W. Wiebe, S. Hettl, S. Neufeind, C. Thösen, J. Organomet. Chem. 690 (2005) 2382.
15. D. Kaur, R. Kohli, Int. J. Quantum Chemistry 108 (2008)119.
16. D. Kaur, R. Kohli, R.P. Kaur, J. Mol. Structure (Theochem) 864 (2008) 72.

HYDROGEN BONDING INTERACTIONS OF HYDROXAMIC ACIDS WITH AMINO ACID SIDE CHAIN GROUPS

7.1 INTRODUCTION:

The building blocks of proteins are twenty naturally occurring L-amino acids, which are distinguished by their different side chain structures and compositions [1]. Most of the crystal structures on proteins report the geometries and thereby the close contacts suggesting interactions but the data is silent about the attractive and repulsive nature of these interactions [2]. Electrostatic interactions and hydrogen bonds (H-bonds) are considered as the main factors responsible for protein properties, including structural organization, stability of native state, as well as functional properties. Proper estimation of H-bonds between biopolymers and between macromolecules and a ligand is of crucial importance in drug design [3]. In many cases, ligand has multiple number of polar sites and forms more than one H-bond with the protein, then it becomes crucial to explore which combination of H-bonds come into existence, since it can dictate orientation of ligand at the binding site.

A large number of studies have analyzed the available 3D structural data in protein data bank (PDB) and have shown that side chains also have preferred interaction geometry; their packing is not entirely random [4-7]. The side chains of amino acid residues in proteins are of special importance to the formation and maintenance of the three dimensional structures. Amino acid side chains interact with their neighborhood predominantly with one or another type of non covalent interactions. The interactions play an important role in protein folding. There are a number of amino acid residues that can form H-bonds via side chains in addition to their peptide group. Side chain groups of some amino acids like phenylalanine, leucine etc. have small non polar side chains and hence are weakly hydrophobic. There are amino acids with polar side chains for instance histidine, glutamine etc. and there are other amino acids with polar charged side chains like aspartic acid and glutamic acids that are capable of H-bonding with other groups [11].

Stickle et al. suggested that on the average the CO...H-N H-bond formed by the protein backbone is most prevalent (68.1%). The C=O...side chain (10.9%), NH...side chain (10.4%) and side chain...side chain H-bonds (10.6%) account for the remainder [12].

Formohydroxamic acid (FHA) and thioformohydroxamic acids (TFHA) are representative of peptides and thiopeptides respectively. TFHA also models thiopurines. Hydroxamic acids (HAs) are known to react with both proteins and nucleic acids. They react with sulfahydryl group of side chain of cysteine residue at active site of various enzymes thereby inhibiting enzymes [13]. Active site of matrix metalloprotease (MMP) enzymes has histidine, glutamine, aspartic acid and water molecule bound to zinc ion [14]. Keeping in view, the biological importance of the FHA and TFHA molecules, the interactions between these molecules with the side chains of amino acids through H-bonding are modeled by selecting methanol (MeOH), methanethiol (MeSH), methaneselenol (MeSeH), methylamine (MeNH$_2$) and acetic acid (AcOH) respectively for the side chains of serine, cysteine, selenocysteine, lysine and aspartic/glutamic acid respectively. The study involves analysis of all possible H-bond interactions of FHA and TFHA molecules with model molecules as representative of side chains of amino acids.

The present study will help to gain knowledge about stabilization of proteins with H-bond involving side chains present on proteins and also the stabilization attained upon interactions of nucleobases with amino acids. Side chain water interactions are important for understanding the role of medium in the stability and behaviour of protein structure. The study has scope as it can also explain the effect of acidic and basic nature of side chains of amino acids on protein stabilization. Model molecule studies can lead to A) Understanding the substrate and inhibitor recognition. B) Designing artificial enzymes C) Determining mechanism of enzymes D) Developing new inhibitors.

7.2 RESULTS AND DISCUSSION:

1Z and **2Z** conformations represent the most stable keto and enol isomers of FHA and TFHA respectively. Both these forms of FHA and TFHA are selected for studying intermolecular interactions with model molecules of amino acid side chains. The amino acid side chain model molecule AcOH has two heteroatoms while MeNH$_2$, MeOH, MeSH and MeSeH have one heteroatom with lone pair of electrons to act as H-bond acceptor.

7.2.1 THE HYDROGEN BONDING OF HYDROXAMIC ACIDS WITH METHYLAMINE (MeNH$_2$):

Seven 1:1 aggregates each for FHA and TFHA (**1Z** and **2Z** conformations) with different spatial placement of MeNH$_2$ around the HA molecules have been optimized at MP2/6-31+G* **[L3]** theoretical level. All the optimized structures are shown in Figure 7.1. The intermolecular H-bond distances, angles and stabilization energies (S.E.) of the aggregates of HAs with MeNH$_2$ are listed in Table 7.1. Two H-bonds N8-H10...X3 (X=O, S in FHA and TFHA respectively) and O4-H5...N8, are present in the most stable adduct (**1Z MeNH$_2$ I**) of **1Z** of FHA and TFHA with MeNH$_2$. The N-H...S H-bonds are formed by cysteine with protein backbone in many enzymes whereas N-H...O are found in number of protein ligand complexes. The most stabilized adduct **1Z MeNH$_2$ I** has stabilization energy of 12.83 and 13.23 kcal/mol at **[L3]** theoretical level in case of FHA and TFHA respectively. It is interesting to note that the stabilization energies are higher than the values calculated for the most stabilized adduct of **1Z** of FHA and TFHA respectively with water at the same theoretical level (10.88 kcal/mol for **1ZW1** of FHA and 9.39 kcal/mol for **1ZW1** of TFHA). The **1Z MeNH$_2$ I** aggregate formation is accompanied by total distortion energy of 2.21 kcal/mol and 3.86 kcal/mol for both the constituting units of this aggregate in case of FHA and TFHA respectively which is also slightly higher than the distortion energy of **1ZW1** aggregate (1.72 kcal/mol and 3.14 kcal/mol for FHA and TFHA respectively).

1Z MeNH$_2$ II is the adduct with stabilization energy of 12.77 and 13.04 kcal/mol in FHA and TFHA respectively that are lower by 0.06 and 0.19 kcal/mol than the respective values for **1Z MeNH$_2$ I** at **[L3]** theoretical level. Both the adducts (**1Z MeNH$_2$ I** and **1Z MeNH$_2$ II**) have similar H-bond donor and acceptor groups but differ in relative orientation of MeNH$_2$ with respect to the acid. This is also reflected in the similar distortion energies associated with the formation of **1Z MeNH$_2$ I** and **1Z MeNH$_2$ II** adducts. The **1Z MeNH$_2$ III** adduct of TFHA has unconventional C1-H6...N9 bond with H-bond angle of 135.8°. On the other hand similar aggregate of FHA indicates the presence of two H-bonds N9-H8...O3 and C1-H6...N9 with H-bond distances of 2.344 Å and 2.466 Å respectively and angles 131.7° and 116.6°. The stabilization energies for the TFHA and FHA in this orientation are comparable.

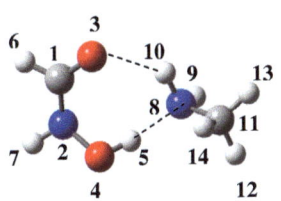 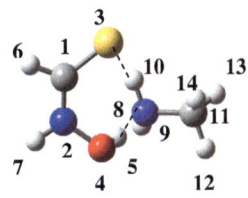

1Z MeNH₂ I

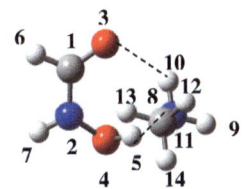 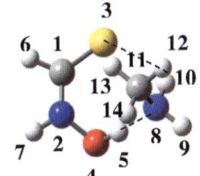

1Z MeNH₂ II

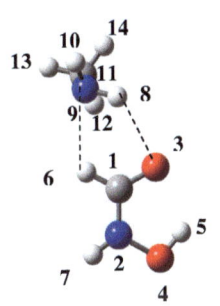 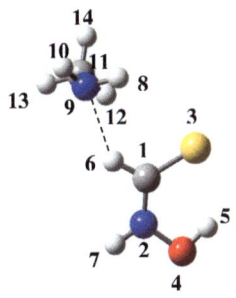

1Z MeNH₂ III

Contd…..

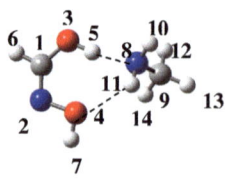 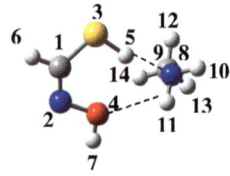

2Z MeNH$_2$ I

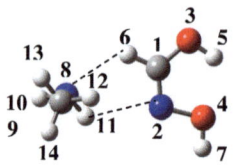 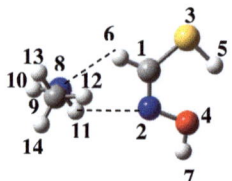

2Z MeNH$_2$ II

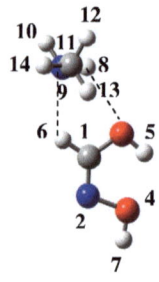 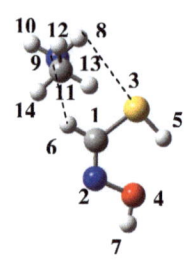

2Z MeNH$_2$ III

Contd…..

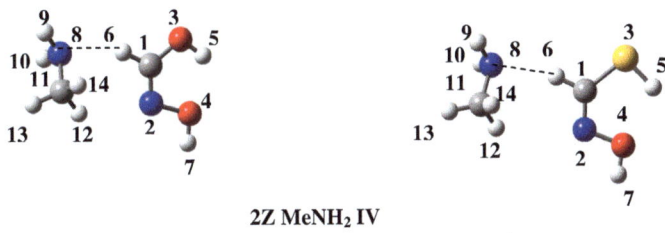

2Z MeNH₂ IV

Figure 7.1: Optimized aggregates of FHA and TFHA with methylamine at [L3] theoretical level.

Among the aggregates of 2Z with MeNH₂, the **2Z MeNH₂ I** is the most stable adduct of both FHA and TFHA with MeNH₂. This doubly H-bonded adduct has X3-H5...N8 and N8-H11...O4 H-bond interactions with H-bond angles closer to linearity (169.7°, 168.2° for the former bond and 102.0°, 108.0° for the latter in case of FHA and TFHA respectively). Hence X3-H5...N8 is chiefly responsible for the large stabilization energy with little contribution from second H-bond. A large difference is observed in the stabilization energies of **2Z MeNH₂ I** aggregate for the two HAs. For FHA, the stabilization energy of this aggregation is 11.18 kcal/mol while for TFHA; the stabilization is merely 4.61 kcal/mol. This large difference results due to different in H-bond donor ability of S-H of TFHA and O-H of FHA. Except for **2Z MeNH₂ I**, rest of the aggregates of 2Z have stabilization energies of the order of 3 to 4 kcal/mol. These aggregates involve at least one unconventional H-bond. In all the aggregates except **1Z MeNH₂ III** of TFHA and **2Z MeNH₂ IV** of both HAs, the N-H and N of MeNH₂ simultaneously act as H-bond donor and H-bond acceptor respectively.

The aggregates are examined for the presence of intramolecular H-bonding and the necessary parameters for characterization of intramolecular H-bonding are listed in Table 7.2. The **1Z MeNH₂ III** and all aggregates of 2Z conformation with MeNH₂ are stabilized by intramolecular H-bonding in addition to intermolecular H-bonding. Additionally Δr values are presented in Table 7.2, which refer to differences between sum of van der Waals radii of B (B; H-bond acceptor) and H atom of H-bond contact and H...B distances. Such differences are reflective of H-bond strength. Study of Δr values from Table 7.2 shows that intramolecular H-bond X3-H5...O4 in **1Z**

MeNH₂ III is stronger in aggregate of TFHA with MeNH₂ over FHA, but for all aggregates of **2Z** conformation, the intramolecular X3-H5...O4 H-bond interaction in FHA is stronger than in TFHA. The stronger intramolecular H-bond is anticipated as O-H is better H-bond donor.

Table 7.3 records the second order delocalization energies $E^{(2)}$ for the orbital interactions that are responsible for the covalent component of the H-bond. As can be seen from the table, the $E^{(2)}$ values are much higher in **1Z MeNH₂ I, 1Z MeNH₂ II** and **2Z MeNH₂ I**. Low $E^{(2)}$ values for the orbital interactions involving H-bonds in other aggregates explain the low stabilization energies evaluated for those aggregates. The charge transfer delocalization $n_{N8} \rightarrow \sigma^*_{O4-H5}$ in the most stable **1Z MeNH₂ I** aggregate has $E^{(2)}$ 43.22 and 51.00 kcal/mol in FHA and TFHA respectively while the most stable aggregate with water **1ZW1** has $E^{(2)}$ values for $n_{O8} \rightarrow \sigma^*_{O4-H5}$ orbital interactions to be 24.69 and 27.91 kcal/mol respectively for the two HAs, at **[L3]** theoretical level which further supports the larger stabilization offered by MeNH₂ than water upon aggregation. In other words H-bond donor acceptor ability of MeNH₂ is greater than water towards the HAs. The electron density transferability of N in MeNH₂ is larger over HAs as evident from larger $E^{(2)}$ in case of MeNH₂ acting as electron donor and HA as electron acceptor and much lower $E^{(2)}$ values for the reverse charge transfer delocalizations. Presence of greater electron density in acceptor antibond orbitals of HAs over antibond orbitals of MeNH₂ further supports the conclusion (Table 7.3). The orbital interactions involving other orbitals of the molecules are indicative of the electron delocalizations present in the molecule itself. The $E^{(2)}$ values for the important orbital interactions are recorded in Table 7.4. The table shows that $E^{(2)}$ value for delocalization $n_{N2} \rightarrow \pi^*_{C1-X3}$ is enhanced for all the aggregates of FHA with MeNH₂, that suggests that the conjugative interactions of **1Z** conformer of FHA are favored with the aggregate formation giving rise to cooperativity in the aggregate. The high $E^{(2)}$ value for $n_{N2} \rightarrow \pi^*_{C1-X3}$ in **1Z** of TFHA indicates the presence of strong conjugative interactions as already suggested in chapter 3. Insignificant variation in this interaction is observed on the aggregation of **1Z** conformer of TFHA with MeNH₂. The $E^{(2)}$ values for $n_{X3} \rightarrow \sigma^*_{O4-H5}$ indicates that electron delocalization resulting in imparting covalent character to the intramolecular H-bond are stronger in **1Z** and **1Z MeNH₂ III** of TFHA in comparison to FHA. The

$E^{(2)}$ values are in agreement to this conclusion derived earlier also on the basis of Δr values.

The atomic charges (NPA) for the HA unit of both FHA and TFHA in the various adducts from NBO analysis are recorded in Table 7.5. The atomic charge on N of $MeNH_2$ increases upon aggregation. Electrostatic component of H-bond is lower in case of aggregates of TFHA wherever there is S or S-H as H-bond acceptor and donor respectively because of small electronegativity difference of S from that of C and H, the S-H bond is only weakly polar and S also carries only a small negative charge character. As can be seen from the table, charge density on nitrogen of HA does not undergo large variation upon adduct formation in case of FHA. All other atomic charges show comparatively larger variation in adducts of FHA with $MeNH_2$. In case of TFHA, the charges on atoms other than C1 undergo variations as the result of adduct formation. Thus polarization of bonds is enhanced upon H-bonding with $MeNH_2$ and accounts for a part of cooperative stabilization.

The stronger stabilization energies in aggregates of HAs with $MeNH_2$ in comparison to the values for aggregates with water can be rationalized in terms of better H-bond acceptor ability of N of $MeNH_2$ and higher cooperativity in aggregates of HA with $MeNH_2$. Cooperativity in aggregates of HAs with $MeNH_2$ arises due to charge polarization, varied conjugations and in addition due to extension of conjugated system on aggregation.

Table 7.1: Important intermolecular hydrogen bonding parameters; hydrogen bond distances[a], hydrogen bond angles[b], charges on hydrogen bond acceptor and hydrogen, stabilization energies (S.E.[c]) and distortion energies (E_{Dis}[d]) of aggregates of FHA (TFHA) with methylamine at MP2/6-31+G* **[L3]** theoretical level.

Species	Hydrogen bond Distances[a]		Hydrogen bond Angles[b]		Atomic charges		S.E.[c]	E_{Dis}[d]
1Z MeNH$_2$ I	H10...X3	2.259 (2.928)	N8-H10...X3	133.2 (115.6)	$q_H(q_X)$	0.434(-0.749) (0.419(-0.258))	12.83 (13.23)	2.21 (3.86)
	H5...N8	1.733 (1.689)	O4-H5...N8	165.4 (171.6)	$q_H(q_N)$	0.557(-0.983) (0.558(-0.972))		
1Z MeNH$_2$ II	H10...X3	2.270 (2.868)	N8-H10...X3	130.0 (118.5)	$q_H(q_X)$	0.436(-0.746) (0.426(-0.253))	12.77 (13.04)	2.15 (3.75)
	H5...N8	1.739 (1.698)	O4-H5...N8	166.5 (172.7)	$q_H(q_N)$	0.557(-0.983) (0.559(-0.975))		
1Z MeNH$_2$ III	H8...X3	2.344 (-)	N9-H8...X3	131.7 (-)	$q_H(q_X)$	0.410(-0.762) (-)	4.86 (4.61)	0.14 (0.13)
	H6...N9	2.466 (2.273)	C1-H6...N9	116.6 (135.8)	$q_H(q_N)$	0.205(-0.974) (0.274(-0.972))		
2Z MeNH$_2$ I	H5...N8	1.716 (1.963)	X3-H5...N8	169.7 (168.2)	$q_H(q_N)$	0.570(-0.965) (0.215(-0.964))	11.18 (4.61)	2.76 (0.86)
	H11...O4	2.500 (2.543)	N8-H11...O4	102.0 (108.0)	$q_H(q_O)$	0.416(-0.704) (0.404(-0.683))		
2Z MeNH$_2$ II	H11...N2	2.503 (2.566)	N8-H11...N2	123.4 (120.0)	$q_H(q_N)$	0.401(-0.310) (0.399(-0.226))	3.39 (3.45)	0.07 (0.07)
	H6...N8	2.456 (2.383)	C1-H6...N8	123.6 (128.7)	$q_H(q_N)$	0.242(-0.966) (0.277(-0.965))		
2Z MeNH$_2$ III	H8...X3	2.513 (3.232)	N9-H8...X3	115.2 (90.9)	$q_H(q_X)$	0.398(-0.797) (0.392(0.010))	3.30 (3.40)	0.16 (0.13)
	H6...N9	2.441 (2.463)	C1-H6...N9	123.1 (126.5)	$q_H(q_N)$	0.242(-0.964) (0.272(-0.962))		

Contd.....

Species	Hydrogen bond Distances[a]	Hydrogen bond Angles[b]	Atomic charges		S.E.[c]	E_{Dis}[d]
	H6...N8	Cl-H6...N8	$q_H(q_N)$			
2Z MeNH2 IV	2.279 (2.241)	154.7 (162.1)	0.248(-0.960)	(0.281(-0.960))	3.27 (3.40)	0.10 (0.09)

a- in angstrom (Å)
b- in degrees (°)
c, d -in (kcal/mol)

Table 7.2: Intramolecular hydrogen bonding parameters in adducts of FHA (TFHA) with methylamine at MP2/6-31+G* [L3] theoretical level. Bond distances are in angstrom (Å) and angles are in degrees (°).

Species	Hydrogen bond distances (r) O4-H5...X3	Δr*	Hydrogen bond angles O4-H5...X3	Species	Hydrogen bond distances (r) X3-H5...O4	Δr*	Hydrogen bond angles X3-H5...O4
1Z	2.046 (2.307)	0.554 (0.743)	117.8 (124.4)	2Z	2.020 (2.399)	0.580 (0.201)	111.5 (108.6)
1Z MeNH2 I	- (-)	- (-)	- (-)	2Z MeNH2 I	2.309 (2.375)	0.291 (0.225)	99.6 (101.9)
1Z MeNH2 II	- (-)	- (-)	- (-)	2Z MeNH2 II	2.011 (2.190)	0.589 (0.410)	111.9 (108.5)
1Z MeNH2 III	2.035 (2.308)	0.565 (0.742)	118.1 (124.4)	2Z MeNH2 III	2.004 (2.200)	0.596 (0.400)	112.1 (108.3)
				2Z MeNH2 IV	2.004 (2.187)	0.596 (0.413)	112.3 (108.4)

Δr*= r_{vw}-r, r_{vw} = $r_O + r_H$ = 2.60 Å, $r_S + r_H$ = 3.05 Å

Table 7.3: The orbital interactions, second order delocalization energies $E^{(2)}$ in kcal/mol and occupancies of acceptor antibonding orbitals important for the hydrogen bonds present in the aggregates of FHA (TFHA) with methylamine at MP2/6-31+G* [L3] theoretical level.

Species	Donor HA	Acceptor MeNH$_2$	$E^{(2)}$	Donor MeNH$_2$	Acceptor HA	$E^{(2)}$	Occupancies	
							Acceptor MeNH$_2$	Acceptor HA
1Z MeNH$_2$ I	$n_{X3(1)} \rightarrow \sigma^*_{N8-H10}$		1.80 (0.30)	$n_{N8} \rightarrow \sigma^*_{O4-H5}$		43.22 (51.00)	σ^*_{N8-H10} 0.011 (0.009)	σ^*_{O4-H5} 0.066 (0.076)
	$n_{X3(2)} \rightarrow \sigma^*_{N8-H10}$		1.36 (0.40)					
1Z MeNH$_2$ II	$n_{X3(1)} \rightarrow \sigma^*_{N8-H10}$		1.80 (0.60)	$n_{N8} \rightarrow \sigma^*_{O4-H5}$		42.93 (50.19)	σ^*_{N8-H10} 0.010 (0.010)	σ^*_{O4-H5} 0.066 (0.075)
	$n_{X3(2)} \rightarrow \sigma^*_{N8-H10}$		1.52 (1.59)					
1Z MeNH$_2$ III	$n_{X3(1)} \rightarrow \sigma^*_{N9-H8}$		0.85 (0.19)	$n_{N9} \rightarrow \sigma^*_{C1-H6}$		2.64 (5.77)	σ^*_{N9-H8} 0.009 (0.008)	σ^*_{C1-H6} 0.042 (0.037)
	$n_{X3(2)} \rightarrow \sigma^*_{N9-H8}$		2.05 (1.12)					
2Z MeNH$_2$ I	$n_{O4(1)} \rightarrow \sigma^*_{N8-H11}$		0.63 (0.61)	$n_{N8} \rightarrow \sigma^*_{X3-H5}$		45.99 (20.71)	σ^*_{N8-H11} 0.007 (0.007)	σ^*_{X3-H5} 0.073 (0.044)
	$n_{O4(2)} \rightarrow \sigma^*_{N8-H11}$		0.10 (-)					
2Z MeNH$_2$ II	$n_{N2} \rightarrow \sigma^*_{N8-H11}$		1.95 (1.47)	$n_{N8} \rightarrow \sigma^*_{C1-H6}$		3.03 (3.80)	σ^*_{N8-H11} 0.009 (0.008)	σ^*_{C1-H6} 0.020 (0.019)
2Z MeNH$_2$ III	$n_{X3(1)} \rightarrow \sigma^*_{N9-H8}$		0.92 (0.12)	$n_{N9} \rightarrow \sigma^*_{C1-H6}$		3.20 (3.10)	σ^*_{N9-H8} 0.007 (0.006)	σ^*_{C1-H6} 0.020 (0.017)
2Z MeNH$_2$ IV				$n_{N8} \rightarrow \sigma^*_{C1-H6}$		7.60 (8.39)		σ^*_{C1-H6} 0.025 (0.024)

Table 7.4: Important orbital interactions and second order delocalization energies $E^{(2)}$ in kcal/mol in aggregates of FHA (TFHA) with methylamine at MP2/6-31+G* [L3] theoretical level.

Species	$n_{N2} \to \sigma^*_{C1-X3}$	$n_{N2} \to \pi^*_{C1-X3}$	$n_{X3} \to \sigma^*_{C1-N2}$	$n_{X3} \to \pi^*_{C1-N2}$	$n_{O4} \to \sigma^*_{C1-N2}$	$n_{O4} \to \pi^*_{C1-N2}$	$n_{X3} \to \sigma^*_{O4-H5}$	$n_{O4} \to \sigma^*_{X3-H5}$
1Z	14.98 (-)	32.68 (145.34)	29.25 (15.92)	- (-)	4.36 (6.56)	- (-)	1.24+5.36 (13.71)	- (-)
1Z MeNH$_2$ I	6.36 (-)	64.78 (140.50)	29.96 (19.38)	- (-)	6.57 (6.11)	- (-)	-	- (-)
1Z MeNH$_2$ II	10.23 (19.19)	52.08 (138.3)	30.05 (19.36)	- (-)	6.62 (6.28)	- (-)	- (-)	- (-)
1Z MeNH$_2$ III	12.00 (-)	39.95 (147.62)	28.19 (15.34)	- (-)	4.39 (6.46)	-	1.44+5.50 (13.52)	- (-)
2Z	13.50 (14.57)	- (-)	8.63 (6.45)	54.42 (39.47)	- (-)	13.22 (18.29)	- (-)	5.06 (2.61)
2Z MeNH$_2$ I	14.84 (14.49)	- (-)	12.27 (7.94)	63.93 (41.11)	-	12.33 (16.03)	- (-)	2.19 (1.28)
2Z MeNH$_2$ II	12.91 (14.25)	- (-)	8.59 (6.47)	55.52 (39.78)	- (-)	12.44 (16.98)	- (-)	5.22 (2.57)
2Z MeNH$_2$ III	13.60 (15.03)	- (-)	5.20+15.80 (6.27)	4.85+18.11 (37.90)	4.47 (-)	4.91 (17.27)	- (-)	5.44 (2.44)
2Z MeNH$_2$ IV	13.05 (14.52)	- (-)	8.23 (6.24)	51.66 (38.54)	- (-)	11.68 (16.67)	- (-)	5.35 (2.56)

Table 7.5: Atomic charges (NPA) on atoms of FHA (TFHA) in their aggregates with methylamine at MP2/6-31+G* [L3] theoretical level.

Species	C1	N2	X3	O4	H5	H6	H7
1Z	0.636	-0.426	-0.732	-0.616	0.535	0.173	0.430
	(-0.019)	(-0.341)	(-0.251)	(-0.603)	(0.531)	(0.242)	(0.447)
1Z MeNH$_2$ I	0.650	-0.436	-0.749	-0.655	0.557	0.152	0.423
	(-0.007)	(-0.369)	(-0.258)	(-0.647)	(0.558)	(0.227)	(0.429)
1Z MeNH$_2$ II	0.649	-0.439	-0.746	-0.654	0.557	0.152	0.423
	(-0.009)	(-0.372)	(-0.253)	(-0.647)	(0.559)	(0.227)	(0.429)
1Z MeNH$_2$ III	0.628	-0.421	-0.762	-0.617	0.536	0.205	0.431
	(-0.011)	(-0.344)	(-0.297)	(-0.606)	(0.531)	(0.274)	(0.446)
2Z	0.478	-0.282	-0.783	-0.703	0.547	0.214	0.528
	(-0.099)	(-0.197)	(0.025)	(-0.677)	(0.176)	(0.248)	(0.524)
2Z MeNH$_2$ I	0.497	-0.313	-0.829	-0.704	0.570	0.202	0.516
	(-0.089)	(-0.217)	(-0.016)	(-0.683)	(0.215)	(0.244)	(0.517)
2Z MeNH$_2$ II	0.482	-0.310	-0.781	-0.704	0.545	0.242	0.525
	(-0.094)	(-0.226)	(0.024)	(-0.678)	(0.173)	(0.277)	(0.521)
2Z MeNH$_2$ III	0.469	-0.285	-0.797	-0.708	0.547	0.242	0.527
	(-0.090)	(-0.209)	(0.010)	(-0.681)	(0.171)	(0.272)	(0.521)
2Z MeNH$_2$ IV	0.471	-0.303	-0.785	-0.709	0.544	0.248	0.524
	(-0.098)	(-0.218)	(0.014)	(-0.682)	(0.172)	(0.281)	(0.520)

7.2.2 THE HYDROGEN BONDING OF HYDROXAMIC ACIDS WITH ACETIC ACID (AcOH):

The optimization of AcOH at **[L3]** theoretical level resulted in two energy minimum structures that differ in relative orientation of O-H with respect to C=O and more stable of the two conformers of AcOH (syn arrangement of C=O and O-H) is selected for present study. Five aggregates each of **1Z** and **2Z** conformers of both the HAs with AcOH have been optimized at **[L3]** theoretical level. All the optimized molecular aggregates of HAs with AcOH are shown in Figure 7.2. Of all the twenty aggregate structures of HAs with AcOH, the salient features of the important H-bonded aggregates are discussed. In Table 7.6, intramolecular H-bonding parameters like H-bond distances, angles and differences of H-bond contact distance from sum of van der Waals radii (Δr) are recorded for aggregates of HAs with AcOH. Except for the aggregates **1Z AcOH I** and **1Z AcOH II** in both FHA and TFHA, the intramolecular H-bonding remains intact in all the optimized aggregates of **1Z** and **2Z** conformations with AcOH.

The geometrical parameters important for characterizing the intermolecular H-bonding in these aggregates along with the stabilization energies (S.E.) associated with the formation of aggregates are listed in the Table 7.7. Upon aggregation both the units undergo variation in structure which causes their distortion. The change in conjugative interactions involving lone pair of electrons of nitrogen with C=X group along with rupturing of intramolecular H-bond is mainly responsible for distortion energies (E_{Dis}). The energy changes accompanying the distortion are also listed in the Table. The energy change associated with distortion of acetic acid is relatively smaller than observed in **1Z** and **2Z** monomer units.

As can be seen from Table 7.7, the **1Z AcOH I** is the most stable aggregate out of all the five aggregates between **1Z** and AcOH for both FHA and TFHA with stabilization energies 16.08 and 13.40 kcal/mol respectively at **[L3]** theoretical level. Distortion energy is also maximum for **1Z AcOH I** aggregate and amounts to ~ 34 % and ~19 % of the total stabilization energy in **1Z AcOH I** for TFHA and FHA respectively at **[L3]** theoretical level. One of the H-bond in this aggregate, involves O-H of HAs as H-bond donor while carbonyl oxygen of AcOH as H-bond acceptor. The second H-bond has O-H of AcOH as H-bond donor with carbonyl/thiocarbonyl group of FHA and TFHA respectively as H-bond acceptor. Obtention of larger stabilization energies for **1Z AcOH I** in FHA indicate stronger strength of O10-H11…O3 H-bond than O10-H11…S3 H-bonds in similar aggregate of TFHA. The stabilization energy of **1Z AcOH I** is only 0.04 and 0.48 kcal/mol i.e. invariably different from the stabilization energies of homodimer of **1Z** (**1Z-1Z (1)**) dimer in HAs) with two similar H-bond interactions among H-bond donors and acceptors at [**L3**] theoretical level.

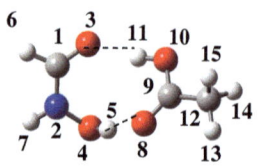 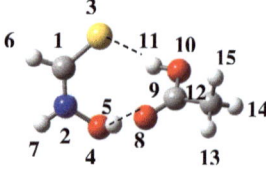

1Z AcOH I

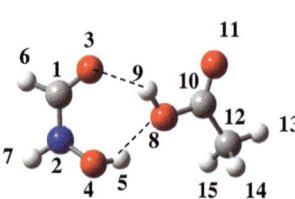 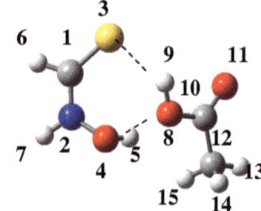

1Z AcOH II

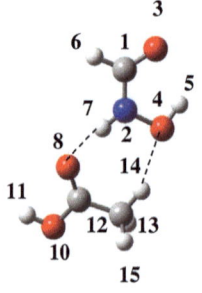 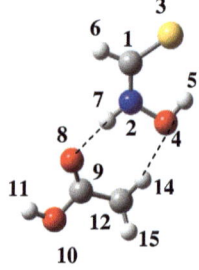

1Z AcOH III

Contd…..

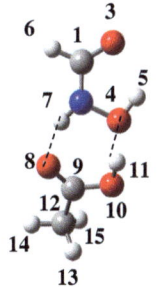
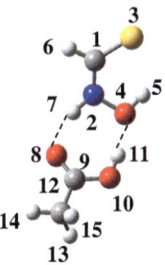

1Z AcOH IV

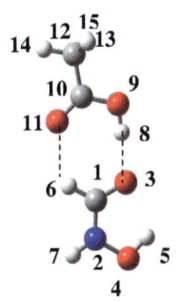
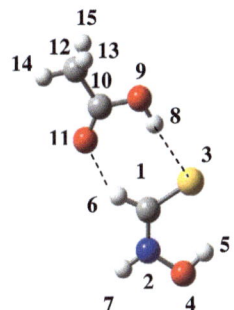

1Z AcOH V

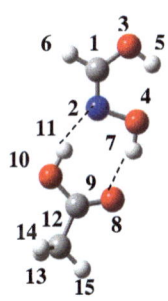
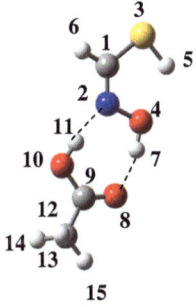

2Z AcOH I

Contd…..

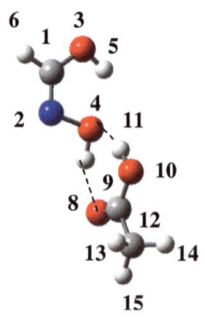
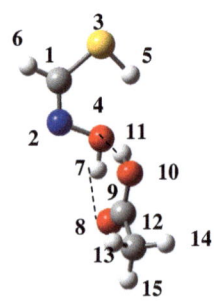

2Z AcOH II

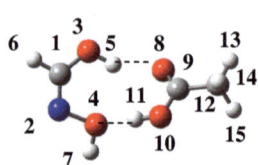
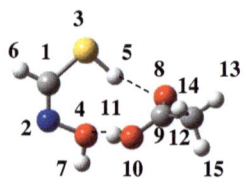

2Z AcOH III

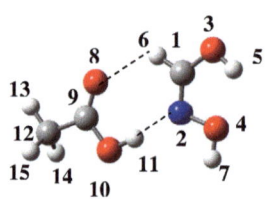
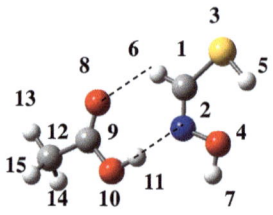

2Z AcOH IV

Contd…..

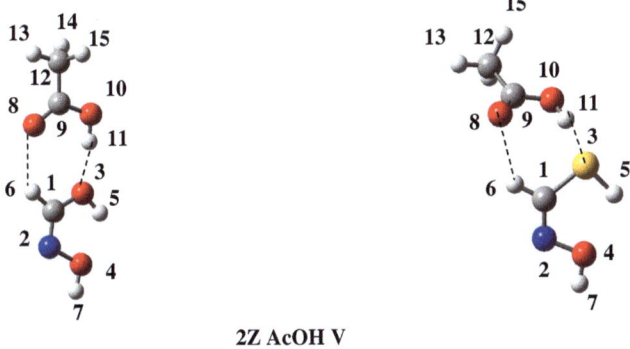

2Z AcOH V

Figure 7.2: Optimized aggregates of FHA and TFHA with acetic acid at **[L3]** theoretical level.

In both **1Z AcOH I** and **1Z AcOH II** adducts, the H-bond donor and acceptor sites of HAs remain the same but difference arises in H-bond acceptor from AcOH. The carbonyl oxygen of both AcOH and HA acts as H-bond acceptor in the former while it is hydroxylic oxygen of AcOH and carbonyl oxygen of HA that are H-bond acceptors in the latter aggregate. As a result, the H-bond distances are longer in **1Z AcOH II** due to constraints imposed by the H-bond donor and acceptor of AcOH. The stabilization energy values are decreased by 7.56 and 6.73 kcal/mol in the adducts of FHA and TFHA relative to those in **1Z AcOH I**. Interaction of **1Z** of FHA and TFHA with water in the same orientation (**1ZW1**) and same donor acceptors as in **1Z AcOH II** evolved stabilization of 10.88 and 9.39 kcal/mol, which is 2.36 and 2.72 kcal/mol higher than their respective AcOH aggregate at **[L3]** theoretical level.

With the presence of C12-H14...O4 H-bond interaction along with N2-H7...O8 H-bond in **1Z AcOH III** aggregate, the stabilization energy in case of FHA is 7.53 kcal/mol while for TFHA it is 8.62 kcal/mol that are 8.55 and 4.78 kcal/mol lower than respective values in **1Z AcOH I**. The high stabilization energy values in **1Z AcOH IV** and **1Z AcOH V** of FHA are understandable as at least one strong conventional H-bond is present in each of the aggregate.

Out of the five aggregates of **2Z** with AcOH, **2Z AcOH I** aggregate is the most stable one involving hydroxyl of HA as donor to carbonyl oxygen of AcOH and N of HAs as acceptor toward OH of carboxylic group of AcOH. The stabilization energies are 14.22 and 13.88 kcal/mol for this aggregate of FHA and TFHA respectively. The values are lower by 1.86 kcal/mol and higher by 0.48 kcal/mol than the value associated with the respective most stable aggregate of AcOH with **1Z** conformation (**1Z AcOH I**) in FHA and TFHA respectively. The intramolecular H-bond in **2Z AcOH I** adduct remains preserved upon adduct formation which is reflected by the geometrical parameters and the low distortion energy (E_{Dis}) values associated with the adduct formation.

Mostly the structures are stabilized by conventional H-bonds except for **1Z AcOH III, 1Z AcOH V, 2Z AcOH IV** and **2Z AcOH V**, in which primary stabilization is due to one conventional H-bond interaction and in addition the stabilization is also contributed by the unconventional C-H...O H-bond. There are two types of C-H H-bond donors, one being part of CH_3 group of AcOH and second from H-C=O of HA. Only **1Z AcOH III** aggregate involves C-H of AcOH as H-bond donor. The C-H...O bond distance range between 2.186 to 2.465 Å for TFHA and 2.315 to 2.422 Å for FHA which agrees with well established range of C-H...O H-bond lengths [15].

In the **2Z AcOH II** adduct, the O-H of HAs is H-bond donor to C=O of AcOH and oxygen of HAs is H-bond acceptor to hydroxyl of AcOH. The stabilization energy values are considerably lower than the respective values in **2Z AcOH I** both for FHA and TFHA as in this aggregate the angle O4-H7...O8 deviates significantly from 180°. The X-H group as H-bond donor to carbonyl oxygen of AcOH in **2Z AcOH III** along with O10-H11...O4 H-bond interactions results in stabilization energy of 10.06 kcal/mol in case of FHA and 6.09 kcal/mol in case of TFHA. The difference in stabilization energy of these two reflects the relatively weaker H-bond donor ability of S-H group relative to that of O-H and is also in agreement with the known fact that homonuclear H-bonds are stronger than heteronuclear ones. **2Z AcOH III** in FHA has two homonuclear O-H...O H-bonds while in TFHA, it has one heteronuclear S-H...O H-bond. The C1-H6...O8 H-bond being common in both **2Z AcOH IV** and **2Z AcOH V** aggregates with similar H-bond distances and angles, it can be suggested that the difference in stabilization energies of the two aggregates reflects the difference in H-bond acceptor ability of N and X toward OH of AcOH.

Charges on all the atoms of HAs in their aggregate form and isolated monomeric form are recorded in Table 7.8. The AcOH molecule with two oxygen atoms is highly polar and thus electrostatic component of the H-bond is anticipated to be significant. The atomic charges on hydrogen and H-bond acceptor atoms (Table 7.7) suggest that electrostatic component of the H-bond is significant with the exception of aggregates with sulfur as H-bond acceptor as discussed in previous section. The charge transfer interactions that are responsible for imparting covalent character to the H-bond are represented by the orbital interactions associated with the relevant H-bond acceptor orbital and antibond of H-bond donor are evaluated using NBO analysis. The $E^{(2)}$ values for such orbital interactions are recorded in Table 7.9. The highest sum of the $E^{(2)}$ values for the two lone pairs present at X indicate that the $n_{X3} \rightarrow \sigma^*_{H11-O10}$ orbital interactions representing covalent component of the O10-H11...X3 H-bond is the strongest interaction in **1Z AcOH I** in comparison to the orbital interactions involving H-bonds in the aggregates of **1Z**. The second H-bond O4-H5...O8 in **1Z AcOH I** also has strong covalent component as suggested by $E^{(2)}$ values for $n_{O8} \rightarrow \sigma^*_{O4-H5}$ interaction in both FHA and TFHA. The high occupancies of acceptor orbitals in case of **1Z AcOH I** in comparison to other aggregates of **1Z** also support the strongest H-bonds in the aggregate. The $E^{(2)}$ values for $n_{N2} \rightarrow \sigma^*_{O11-H10}$ orbital interactions in **2Z AcOH I** (30.66 and 29.94 kcal/mol in case of FHA and TFHA respectively) along with those for $n_{O8} \rightarrow \sigma^*_{O4-H7}$ orbital interactions for the second H-bond explain the higher stability of the aggregate in comparison to other aggregates of **2Z**.

Various orbital interactions present in the HA unit of the aggregates of HAs with AcOH are recorded in Table 7.10. As can be seen from the table, in aggregates of **1Z** of FHA, the $n_{N2} \rightarrow \pi^*_{C1-X3}$ orbital interaction has enhanced $E^{(2)}$ values in the aggregates suggesting that conjugative interactions are strengthened as a result of aggregate formation. The $n_{N2} \rightarrow \pi^*_{C1-X3}$ orbital interactions in **1Z** of TFHA however indicate increase in two aggregates and decrease in three aggregates. There is decrease in $E^{(2)}$ value for the most stabilized **1Z AcOH I** of TFHA. The decrease in conjugative interactions can be held responsible for large distortion energies.

With S of TFHA as H-bond acceptor, the comparable or sometimes slightly larger stabilization energies in aggregates with MeNH$_2$ were seen relative to those in FHA, while stabilization energies for such aggregates of TFHA with AcOH are smaller than the aggregates of FHA which shows that the strength of H-bonds to S as acceptor depends on nature of H-bond donor. Upon interaction with HAs, the AcOH molecule in **1Z AcOH I** evolve larger stabilization than any of the MeNH$_2$ aggregate. The difference in stabilization energies of 3.25 and 0.17 kcal/mol are observed for most stable **1Z AcOH I** in FHA and TFHA evaluated with respect to the most stable MeNH$_2$ aggregate **1Z MeNH$_2$ I.**

The stabilization energies as well as the distortion energies on aggregation of model molecules with HAs in their most stable H-bonded structures follow the sequence AcOH > MeNH$_2$ > H$_2$O. $E^{(2)}$ values on aggregation with MeNH$_2$, H$_2$O and AcOH suggest that in all the aggregates of HAs with MeNH$_2$ and H$_2$O, these simple molecules clearly act as strong charge donors towards HAs while charge donor and charge acceptor ability to a good extent of AcOH is also reflected in its aggregates with HAs. The conjugative interactions in AcOH part of the aggregates are also contributing towards stability of aggregate.

Table 7.6: Intramolecular hydrogen bonding parameters in adducts of FHA (TFHA) with acetic acid at MP2/6-31+G* [L3] theoretical level. Bond distances are in angstrom (Å) and angles are in degrees (°).

Species	Hydrogen bond distances (r) O4-H5...X3	Δr*	Hydrogen bond angles O4-H5...X3	Species	Hydrogen bond distances (r) X3-H5...O4	Δr*	Hydrogen bond angles X3-H5...O4
1Z	2.046	0.554	117.8	2Z	2.020	0.580	111.5
	(2.307)	(0.743)	124.4		(2.399)	(0.201)	108.6
1Z AcOH I	-	-	-	2Z AcOH I	2.006	0.594	112.5
	(-)	(-)	(-)		(2.177)	(0.873)	(109.5)
1Z AcOH II	-	-	-	2Z AcOH II	2.010	0.590	112.1
	(-)	(-)	(-)		(2.193)	(0.857)	(108.0)
1Z AcOH III	2.017	0.583	118.7	2Z AcOH III	2.258	0.342	102.1
	(2.263)	(0.787)	(125.7)		(2.300)	(0.750)	(105.0)
1Z AcOH IV	1.973	0.627	119.3	2Z AcOH IV	2.027	0.573	111.4
	(2.228)	(0.822)	(126.1)		(2.199)	(0.851)	(108.5)
1Z AcOH V	2.053	0.547	116.9	2Z AcOH V	1.989	0.611	111.9
	(2.341)	(0.709)	(122.9)		(2.188)	(0.862)	(108.2)

Δr* = r_{VW}-r,

r_{VW} = r_O+r_H=2.60 Å, r_S+r_H=3.05 Å

Table 7.7: Important intermolecular hydrogen bonding parameters; hydrogen bond distances[a], hydrogen bond angles[b], charges, stabilization energies[c] (S.E.) and distortion energies[d] (E_{Dis}) of aggregates of FHA (TFHA) with acetic acid at MP2/6-31+G* [L3] theoretical level.

Species	Hydrogen bond Distances[a]		Hydrogen bond Angles[b]	Atomic charges		S.E.[c]	E_{Dis}[d]
1Z AcOH I	H11...X3	1.760 (2.257)	O10-H11...X3	167.7 (162.7)	$q_H(q_X)$	0.570(-0.772) (0.537(-0.230))	16.08 (13.40)	3.07 (4.50)
	H5...O8	1.755 (1.734)	O4-H5...O8	168.3 (169.2)	$q_H(q_O)$	0.560(-0.779) (0.559(-0.773))		
1Z AcOH II	H9...X3	1.904 (2.415)	O8-H9...X3	143.2 (138.3)	$q_H(q_X)$	0.572(-0.746) (0.553(-0.227))	8.52 (6.67)	1.74 (3.19)
	H5...O8	1.908 (1.828)	O4-H5...O8	147.3 (158.8)	$q_H(q_O)$	0.548(-0.871) (0.552(-0.868))		
1Z AcOH III	H7...O8	1.916 (1.864)	N2-H7...O8	169.0 (168.6)	$q_H(q_O)$	0.469(-0.754) (0.485(-0.758))	7.53 (8.62)	0.32 (0.41)
	H14...O4	2.423 (2.445)	C12-H14...O4	157.8 (156.2)	$q_H(q_O)$	0.277(-0.637) (0.274(-0.626))		
1Z AcOH IV	H7...O8	1.919 (1.885)	N2-H7...O8	152.3 (150.5)	$q_H(q_O)$	0.471(-0.764) (0.487(-0.766))	11.38 (11.62)	0.91 (1.06)
	H11...O4	1.826 (1.848)	O10-H11...O4	173.0 (170.5)	$q_H(q_O)$	0.556(-0.662) (0.555(-0.652))		
1Z AcOH V	H8...X3	1.775 (2.297)	O9-H8...X3	177.9 (177.2)	$q_H(q_X)$	0.564(-0.802) (0.535(-0.304))	13.40 (7.99)	1.00 (0.62)
	H6...O11	2.315 (2.186)	C1-H6...O11	127.0 (144.4)	$q_H(q_O)$	0.221(-0.760) (0.282(-0.756))		
2Z AcOH I	H11...N2	1.844 (1.843)	O10-H11...N2	168.9 (169.3)	$q_H(q_N)$	0.557(-0.345) (0.555(-0.258))	14.22 (13.88)	1.29 (1.27)
	H7...O8	1.794 (1.798)	O4-H7...O8	163.5 (163.8)	$q_H(q_O)$	0.567(-0.766) (0.563(-0.765))		
2Z AcOH II	H11...O4	1.925	O10-H11...O4	153.9	$q_H(q_O)$	0.551(-0.753)	8.14	0.68

Contd.....

Species	Hydrogen bond Distances[a]		Hydrogen bond Angles[b]		Atomic charges		S.E.[c]	E_{Dis}[d]
	H7...O8	(1.906) 1.876 (1.904)	O4-H7...O8	(154.6) 110.1 (110.3)	$q_H(q_O)$	(0.550(-0.726)) 0.568(-0.749) (0.566(-0.749))	(7.82)	(0.54)
2Z AcOH III	H5...O8	1.833 (2.160)	X3-H5...O8	155.8 (147.9)	$q_H(q_O)$	0.571(-0.754) (0.203(-0.739))	10.06 (6.09)	2.11 (0.78)
	H11...O4	1.805 (1.849)	O10-H11...O4	165.4 (171.1)	$q_H(q_O)$	0.555(-0.723) (0.551(-0.701))		
2Z AcOH IV	H6...O8	2.422 (2.361)	C1-H6...O8	129.0 (134.8)	$q_H(q_O)$	0.248(-0.744) (0.283(-0.743))	8.03 (7.96)	0.45 (0.44)
	H11...N2	1.894 (1.893)	O10-H11...N2	176.3 (177.1)	$q_H(q_N)$	0.551(-0.349) (0.549(-0.269))		
2Z AcOH V	H6...O8	2.416 (2.465)	C1-H6...O8	129.9 (123.6)	$q_H(q_O)$	0.248(-0.739) (0.274(-0.735))	6.35 (3.94)	0.35 (0.18)
	H11...X3	1.883 (2.440)	O10-H11...X3	176.0 (167.0)	$q_H(q_X)$	0.553(-0.820) (0.532(0.002))		

a- in angstrom (Å)
b- in degrees (°)
c, d- in kcal/mol

Table 7.8: Atomic charges (NPA) on atoms of FHA (TFHA) in their aggregates with acetic acid at MP2/6-31+G* [L3] theoretical level.

Species	C1	N2	X3	O4	H5	H6	H7
1Z	0.636 (-0.019)	-0.426 (-0.347)	-0.732 (-0.251)	-0.616 (-0.603)	0.535 (0.531)	0.173 (0.242)	0.430 (0.447)
1Z AcOH I	0.672 (0.012)	-0.423 (-0.365)	-0.772 (-0.230)	-0.632 (-0.628)	0.560 (0.559)	0.166 (0.235)	0.430 (0.434)
1Z AcOH II	0.664 (0.002)	-0.435 (-0.373)	-0.746 (-0.227)	-0.618 (-0.612)	0.548 (0.552)	0.165 (0.234)	0.429 (0.435)
1Z AcOH III	0.635 (-0.010)	-0.439 (-0.358)	-0.756 (-0.285)	-0.637 (-0.626)	0.539 (0.530)	0.173 (0.242)	0.469 (0.485)
1Z AcOH IV	0.640 (-0.010)	-0.436 (-0.367)	-0.739 (-0.243)	-0.662 (-0.652)	0.550 (0.536)	0.180 (0.247)	0.471 (0.487)
1Z AcOH V	0.641 (0.004)	-0.405 (-0.334)	-0.802 (-0.304)	-0.607 (-0.598)	0.539 (0.533)	0.221 (0.282)	0.442 (0.451)
2Z	0.478 (-0.099)	-0.282 (-0.197)	-0.783 (0.025)	-0.703 (-0.677)	0.547 (0.176)	0.214 (0.248)	0.528 (0.524)
2Z AcOH I	0.510 (-0.071)	-0.345 (-0.258)	-0.780 (0.031)	-0.708 (-0.683)	0.552 (0.184)	0.214 (0.246)	0.567 (0.563)
2Z AcOH II	0.483 (-0.098)	-0.280 (-0.188)	-0.783 (0.024)	-0.753 (-0.726)	0.547 (0.175)	0.215 (0.249)	0.568 (0.566)
2Z AcOH III	0.507 (-0.082)	-0.301 (-0.212)	-0.796 (0.028)	-0.723 (-0.701)	0.571 (0.203)	0.212 (0.250)	0.532 (0.531)
2Z AcOH IV	0.517 (-0.060)	-0.349 (-0.269)	-0.773 (0.044)	-0.686 (-0.662)	0.547 (0.174)	0.248 (0.283)	0.528 (0.523)
2Z AcOH V	0.468 (-0.093)	-0.268 (-0.190)	-0.820 (0.002)	-0.705 (-0.678)	0.559 (0.180)	0.248 (0.274)	0.530 (0.524)

Table 7.9: The orbital interactions and second order delocalization energies $E^{(2)}$ in kcal/mol and occupancies of acceptor antibonding orbitals important for the hydrogen bonds present in the aggregates of FHA (TFHA) with acetic acid at **[L3]** theoretical level.

Species	Donor HA	Acceptor AcOH	$E^{(2)}$	Donor AcOH	Acceptor HA	$E^{(2)}$	Occupancies Acceptor AcOH		Acceptor HA	
1Z AcOH I	$n_{X3(1)} \to \sigma^*_{O10-H11}$		12.83 (4.53)	$n_{O8(1)} \to \sigma^*_{O4-H5}$		11.54 (11.33)	$\sigma^*_{O10-H11}$	0.044 (0.062)	σ^*_{O4-H5}	0.039 (0.043)
	$n_{X3(2)} \to \sigma^*_{O10-H11}$		15.33 (25.08)	$n_{O8(2)} \to \sigma^*_{O4-H5}$		16.13 (18.77)				
1Z AcOH II	$n_{X3(1)} \to \sigma^*_{O8-H9}$		6.87 (2.87)	$n_{O8(1)} \to \sigma^*_{O4-H5}$		11.65 (16.52)	σ^*_{O8-H9}	0.029 (0.036)	σ^*_{O4-H5}	0.018 (0.025)
	$n_{X3(2)} \to \sigma^*_{O8-H9}$		6.36 (7.59)	$n_{O8(2)} \to \sigma^*_{O4-H5}$		(0.34)				
1Z AcOH III	$n_{O4(1)} \to \sigma^*_{C12-H14}$		1.87 (2.05)	$n_{O8(1)} \to \sigma^*_{N2-H7}$		8.49 (9.82)	$\sigma^*_{C12-H14}$	0.007 (0.007)	σ^*_{N2-H7}	0.033 (0.039)
	$n_{O4(2)} \to \sigma^*_{C12-H14}$		1.19 (0.54)	$n_{O8(2)} \to \sigma^*_{N2-H7}$		10.16 (12.89)				
1Z AcOH IV	$n_{O4(1)} \to \sigma^*_{O10-H11}$		0.94 (3.14)	$n_{O8(1)} \to \sigma^*_{N2-H7}$		8.61 (9.64)	$\sigma^*_{O10-H11}$	0.036 (0.034)	σ^*_{N2-H7}	0.034 (0.043)
	$n_{O4(2)} \to \sigma^*_{O10-H11}$		21.48 (17.86)	$n_{O8(2)} \to \sigma^*_{N2-H7}$		9.59 (10.7)				
1Z AcOH V	$n_{X3(1)} \to \sigma^*_{O9-H8}$		9.71 (3.71)	$n_{O11(1)} \to \sigma^*_{C1-H6}$		2.08 (3.35)	σ^*_{O9-H8}	0.042 (0.051)	σ^*_{C1-H6}	0.038 (0.033)
	$n_{X3(2)} \to \sigma^*_{O9-H8}$		18.10 (21.0)	$n_{O11(2)} \to \sigma^*_{C1-H6}$		1.50 (2.87)				
2Z AcOH I	$n_{N2} \to \sigma^*_{O10-H11}$		30.66 (29.94)	$n_{O8(1)} \to \sigma^*_{O4-H7}$		9.82 (9.67)	$\sigma^*_{O10-H11}$	0.052 (0.053)	σ^*_{O4-H7}	0.035 (0.034)
				$n_{O8(2)} \to \sigma^*_{O4-H7}$		15.89 (15.92)				

Contd.....

Species	Donor HA	Acceptor AcOH	$E^{(2)}$	Donor AcOH	Acceptor HA	$E^{(2)}$	Occupancies Acceptor AcOH		Acceptor HA	
2Z AcOH II	$n_{O4(1)} \to$	$\sigma^*_{O10-H11}$	1.13 (2.47)	$n_{O8(1)} \to$	σ^*_{O4-H7}	6.20 (5.71)	$\sigma^*_{O10-H11}$	0.029 (0.031)	σ^*_{O4-H7}	0.028 (0.025)
	$n_{O4(2)} \to$	$\sigma^*_{O10-H11}$	13.17 (12.95)	$n_{O8(2)} \to$	σ^*_{O4-H7}	11.86 (10.44)				
2Z AcOH III	$n_{O4(1)} \to$	$\sigma^*_{O10-H11}$	3.74 (4.46)	$n_{O8(1)} \to$	σ^*_{X3-H5}	9.19 (3.17)	$\sigma^*_{O10-H11}$	0.038 (0.035)	σ^*_{X3-H5}	0.037 (0.019)
	$n_{O4(2)} \to$	$\sigma^*_{O10-H11}$	20.61 (16.70)	$n_{O8(2)} \to$	σ^*_{X3-H5}	12.23 (3.90)				
2Z AcOH IV	$n_{N2} \to$	$\sigma^*_{O10-H11}$	24.76 (24.08)	$n_{O8(1)} \to$	σ^*_{C1-H6}	1.56 (1.81)	$\sigma^*_{O10-H11}$	0.045 (0.046)	σ^*_{C1-H6}	0.018 (0.016)
				$n_{O8(2)} \to$	σ^*_{C1-H6}	1.08 (1.54)				
2Z AcOH V	$n_{X3(1)} \to$	$\sigma^*_{O10-H11}$	14.79 (3.52)	$n_{O8(1)} \to$	σ^*_{C1-H6}	1.64 (1.17)	$\sigma^*_{O10-H11}$	0.028 (0.032)	σ^*_{C1-H6}	0.017 (0.014)
	$n_{X3(2)} \to$	$\sigma^*_{O10-H11}$	0.64 (9.96)	$n_{O8(2)} \to$	σ^*_{C1-H6}	1.06 (0.36)				

Table 7.10: Important orbital interactions and second order delocalization energies $E^{(2)}$ in kcal/mol in the aggregates of FHA (TFHA) with acetic acid at [**L3**] theoretical level.

Species	$n_{N2} \to \pi^*_{C1-X3}$	$n_{N2} \to \sigma^*_{C1-X3}$	$n_{X3} \to \sigma^*_{C1-N2}$	$n_{X3} \to \pi^*_{C1-N2}$	$n_{O4} \to \pi^*_{C1-N2}$	$n_{X3} \to \sigma^*_{O4-H5}$	$n_{O4} \to \sigma^*_{X3-H5}$
1Z	32.68 (145.34)	14.98 (-)	29.25 (15.92)	- (-)	- (-)	- (-)	- (-)
1Z AcOH I	83.07 (119.21)	3.18 (-)	24.31 (5.65+13.74)	- (-)	- (-)	1.24+5.36 (13.71)	- (-)
1Z AcOH II	84.12 (60.30)	-	28.36 (3.85+17.12)	- (-)	-	-	-
1Z AcOH III	50.50 (160.63)	15.45 (-)	28.02 (14.86)	- (-)	- (-)	1.37+6.28 (1.94+16.50)	- (-)
1Z AcOH IV	31.57 (124.80)	8.82 (-)	28.24 (14.85)	-	-	1.74+7.50 (2.22+19.02)	-
1Z AcOH V	75.91 (156.96)	14.47 (-)	26.60 (16.54)	- (-)	- (-)	2.14+3.80 (2.88+9.62)	- (-)
2Z	- (-)	4.76 (-)	8.63 (6.45)	54.42 (39.47)	13.22 (18.29)	- (-)	5.06 (2.61)
2Z AcOH I	- (-)	13.50 (14.57)	8.45 (-)	57.51 (41.60)	15.66 (21.89)	-	5.92 (3.06)
2Z AcOH II	- (-)	12.80 (13.80)	6.48+10.72 (6.13)	4.05+26.60 (34.10)	4.27 (13.73)	- (-)	5.84 (2.71)
2Z AcOH III	- (-)	13.51 (14.55)	7.71+17.39 (6.42)	6.67+23.20 (26.83)	2.13 (6.03)	- (-)	2.18 (1.71)
2Z AcOH IV	- (-)	14.24 (13.94)	8.55 (6.40)	59.06 (43.53)	12.51 (17.01)	-	4.86 (2.48)
2Z AcOH V	- (-)	12.42 (13.52)	7.18 (5.75)	46.86 (34.74)	13.50 (18.35)	- (-)	5.86 (2.58)

7.2.3 THE HYDROGEN BONDING OF HYDROXAMIC ACIDS WITH METHANOL (MeOH):

To determine the ability of HAs to interact with another model molecule methanol (MeOH) as side chain representative of serine and threonine amino acids, the aggregation of both FHA and TFHA with MeOH has been analysed. Three aggregates of **1Z** conformer each of FHA and TFHA with MeOH have been optimized at **[L3]** theoretical level. The interaction of MeOH with **2Z** conformer of FHA and TFHA also resulted in three energy minima. The geometries of optimized aggregates are shown in Figure 7.3.

Table 7.11 records the parameters important for the description of intermolecular H-bonds present in the aggregates of HAs with MeOH. The charges on the atoms participating in H-bonding and resulting stabilization energies are also included in the table. The highest stabilization results in **1Z MeOH I** with two H-bonds, both in case of FHA and TFHA. In the aggregate, one H-bond involve OH of HAs as H-bond donor to oxygen of MeOH as H-bond acceptor while the second H-bond results from OH of MeOH as H-bond donor toward chalcogen of C=X group (X=O, S) of HAs. The stabilization energies of aggregation both in case of FHA and TFHA are comparable (11.44 and 10.78 kcal/mol respectively). It is interesting to note that the H-bonding of **1Z** conformer of FHA and TFHA with MeOH generates 0.56 and 1.39 kcal/mol higher stabilization energy respectively in comparison to that in case of aggregation with water in similar orientation (**1ZW1**) at the same **[L3]** theoretical level.

Structural changes that occur in geometry of both HA and MeOH are manifested in the form of distortion energies. The distortion of monomeric units upon aggregation is also high in case of **1Z MeOH I** as apparent from E_{Dis} values present in Table 7.11. The distortion energy in **1Z MeOH I** of TFHA is larger over that in case of FHA. The stabilization energy associated with the H-bonded adduct formation in **1Z MeOH II** is nearly half of that in case of **1Z MeOH I** for both TFHA and FHA as in this adduct one of the H-bond involves unconventional H-bond donor C-H. With N-H of HAs acting as H-bond donor to O of MeOH and hydroxyl of MeOH as H-bond donor to O4, the adduct **1Z MeOH III** indicates stabilization of 7.60 and 8.77 kcal/mol in FHA and TFHA respectively.

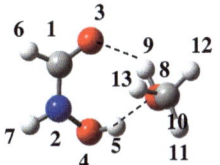
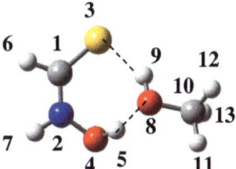

1Z MeOH I

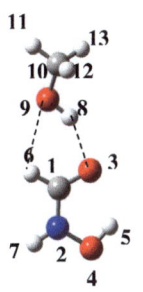
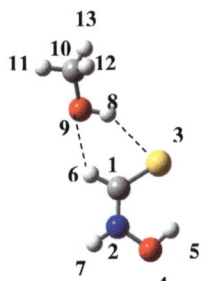

1Z MeOH II

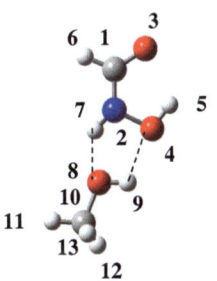
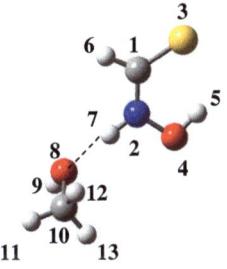

1Z MeOH III

Contd…..

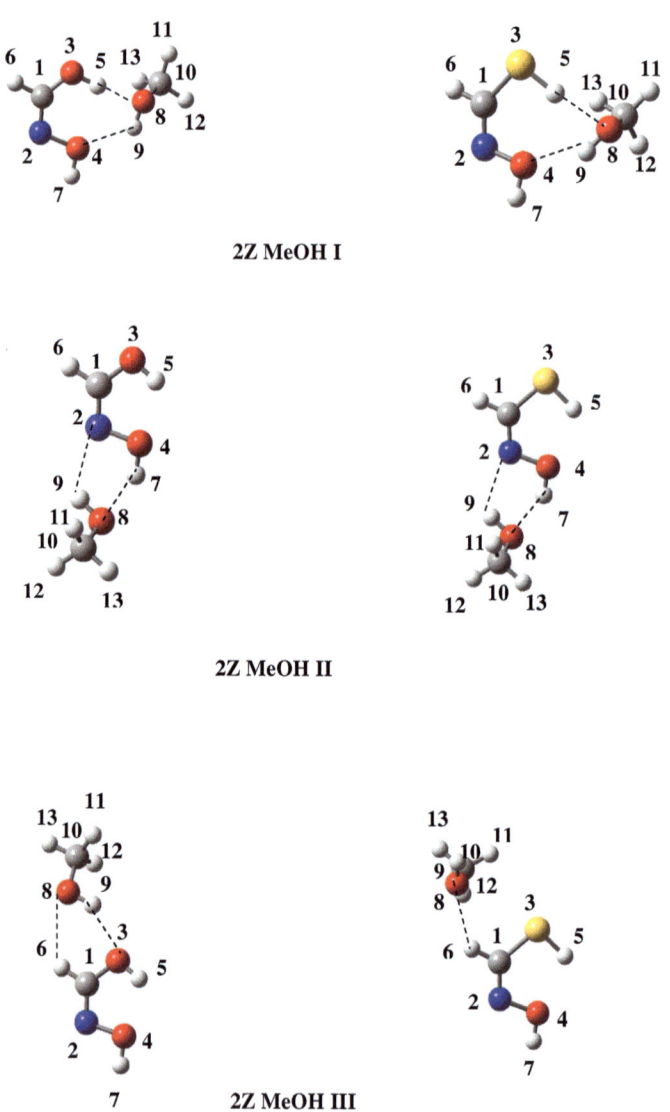

Figure 7.3: Optimized aggregates of FHA and TFHA with methanol at [L3] theoretical level.

The **2Z MeOH I** aggregate indicates the presence of two X3-H5...O8 and O8-H9...O4 H-bond interactions. The O8-H9...O4 interactions are comparatively weaker in comparison to H-bond interactions (X3-H5...O8) as reflected by the H-bond distances and angles in the two (Table 7.11). The stabilization energy of **2Z MeOH I** aggregate is 2.92 kcal/mol lower than that in **1Z MeOH I** in case of FHA while the difference increases to 6.81 kcal/mol in case of TFHA because of S-H being H-bond donor in one of the two H-bonds. Another aggregate of **2Z** conformer, **2Z MeOH II** with N as H-bond acceptor and OH of MeOH as H-bond donor has stabilization of 8.44 and 8.21 kcal/mol for the FHA and TFHA respectively. The aggregation of HAs with the MeOH is more favored than the aggregation of MeOH with water or with other methanol molecule. The stabilization energy of MeOH with water is 5.44 kcal/mol and for methanol dimer it is 5.27 kcal/mol at **[L3]** theoretical level. The higher stabilization energy in aggregates of HAs with MeOH thus indicates the cooperativity arising as the result of H-bonding.

The atomic charges (Table 7.11) suggest that electrostatic component for O4-H5...O8 H-bond in **1Z MeOH I** is nearly same in case of FHA and TFHA but differ in case of the second H-bond (O8-H9...X3). The electrostatic component of H-bond is lower where ever X=S acts as H-bond acceptor or unconventional H-bond donors are involved as can be expected. The covalent component of H-bond results from electron density transfer from the lone pair of electrons present at H-bond acceptor toward the σ^* of H-bond donor. The orbital interactions signifying these delocalizations are listed in Table 7.12. The second order delocalization energies $E^{(2)}$ values and the occupancies of acceptor orbitals both indicate that covalent component is stronger in the H-bonds with HAs acting as H-bond donor. The comparison of occupancies of acceptor orbitals also reflects that the covalent components in both the H-bonds are relatively higher in **1Z MeOH I** aggregate of both FHA and TFHA. The $E^{(2)}$ value for $n_{O8} \rightarrow \sigma^*_{O4-H5}$ in most stable adduct **1Z MeOH I** (Table 7.12) being 25.97 and 32.61 kcal/mol in case of FHA and TFHA respectively show MeOH to possess better charge donor ability relative to FHA.

The geometrical parameters are also analyzed for the intramolecular H-bonds. The parameters indicating presence of intramolecular H-bonding are listed in Table 7.13. The analysis of these parameters indicates that intramolecular H-bonding remains intact in all the aggregates of FHA and TFHA with MeOH except for **1Z MeOH I**.

Atomic charges on all atoms of **1Z** and **2Z** components of aggregates are listed in Table 7.14. The analysis of atomic charges (Table 7.14) obtained at MP2/6-31+G* using NBO analysis reflects the variation in charge polarities of several bonds upon aggregation. It has been inferred in the previous chapters that cooperativity in H-bonded adducts arises as the result of enhanced polarization of bonds and enhanced conjugative interactions. The interactions arising from the enhanced polarity of bonds add to the stabilization energy. The $E^{(2)}$ values for the important second order orbital interactions in aggregates of HA with MeOH are reported in Table 7.15. The $E^{(2)}$ values indicate enhancement of $n_{N2} \rightarrow \pi^*_{C1-X3}$ orbital interaction relative to that in **1Z** in all the adducts of **1Z** of FHA and TFHA with MeOH except for **1Z MeOH I** of TFHA. The presence of π^*_{C-N} molecular orbital in the NBO analysis of **1Z MeOH I** of TFHA suggests that the lone pair of N is completely delocalized to form the bond and hence result in increasing the stabilization energy. Thus H-bonding has synergic effect on conjugative interactions that affects the stabilization. The stabilization energy of most stable H-bonded aggregates follow the order AcOH > MeNH$_2$ > MeOH > H$_2$O for both FHA and TFHA. Larger ring system with extended conjugation is formed when AcOH H-bonds to HAs than when MeOH, MeNH$_2$ and H$_2$O interact with HAs. Most stable aggregate with water and methanol does not differ much in stabilization energy.

Table 7.11: Important intermolecular hydrogen bonding parameters; hydrogen bond distances[a], hydrogen bonding angles[b], charges on hydrogen bond acceptor and hydrogen along with stabilization energies S.E[c] and distortion energies E_{Dis}[d] of various hydrogen bonded aggregates of FHA (TFHA) with methanol at MP2/6-31+G* [L3] theoretical level.

Species	Hydrogen bond Distances[a]	Hydrogen bond Angles[b]	Atomic charges		S.E.[c]	E_{Dis}[d]
1Z MeOH I	H9…X3 1.962 (2.387)	O8-H9…X3 143.3 (143.8)	$q_H(q_X)$	0.546(-0.759) (0.527(-0.266))	11.44 (10.78)	1.91 (3.52)
	H5…O8 1.746 (1.695)	O4-H5…O8 158.3 (166.5)	$q_H(q_O)$	0.559(-0.868) (0.560(-0.864))		
1Z MeOH II	H8…X3 1.965 (2.506)	O9-H8…X3 149.1 (141.5)	$q_H(q_X)$	0.526(-0.775) (0.512(-0.296))	5.71 (4.80)	0.24 (1.44)
	H6…O9 2.619 (2.365)	C1-H6…O9 103.8 (122.0)	$q_H(q_O)$	0.199(-0.860) (0.268(-0.853))		
1Z MeOH III	H7…O8 1.962 (1.810)	N2-H7…O8 135.7 (177.2)	$q_H(q_O)$	0.465(-0.854) (0.485(-0.846))	7.60 (8.77)	1.18 (0.31)
2Z MeOH I	H9…O4 2.294	O8-H9…O4 117.0	$q_H(q_O)$	0.523(-0.645)	8.52 (3.97)	2.11 (0.89)
	H5…O8 1.737 (2.094)	X3-H5…O8 172.4 (152.5)	$q_H(q_O)$	0.568(-0.849) (0.197(-0.847))		
	H9…O4 2.067 (2.000)	O8-H9…O4 129.4 (145.3)	$q_H(q_O)$	0.532(-0.716) (0.521(-0.695))		
2Z MeOH II	H9…N2 2.331 (2.351)	O8-H9…N2 114.5 (113.5)	$q_H(q_N)$	0.525(-0.315) (0.524(-0.225))	8.44 (8.21)	0.33 (1.66)
	H7…O4 1.825 (1.826)	O8-H7…O4 149.7 (150.3)	$q_H(q_O)$	0.559(-0.847) (0.557(-0.846))		
2Z MeOH III	H9…X3 2.073 (-)	O8-H9…X3 138.9 (-)	$q_H(q_X)$	0.516(-0.808) (-)	3.66 (3.03)	0.13 (0.09)
	H6…O8 2.590 (2.320)	C1-H6…O8 112.2 (126.2)	$q_H(q_O)$	0.235(-0.844) (0.271(-0.836))		

a- in angstrom (Å)
b- in degrees (°)
c, d- in (kcal/mol)

Table 7.12: The orbital interactions, second order electron delocalization energies $E^{(2)}$ in kcal/mol and occupancies of acceptor antibonding orbitals important for the hydrogen bonds present in the aggregates of FHA (TFHA) with MeOH at MP2/6-31+G* [L3] theoretical level.

Species	Donor HA	Acceptor MeOH	$E^{(2)}$	Donor MeOH	Acceptor HA	$E^{(2)}$	Occupancies Acceptor MeOH		Acceptor HA	
1Z MeOH I	$n_{X3(1)} \to$	σ^*_{O8-H9}	5.17 (3.02)	$n_{O8(1)} \to$	σ^*_{O4-H5}	1.61 (1.40)	σ^*_{O8-H9}	0.020 (0.002)	σ^*_{O4-H5}	0.039 (0.048)
	$n_{X3(2)} \to$	σ^*_{O8-H9}	5.43 (8.24)	$n_{O8(2)} \to$	σ^*_{O4-H5}	25.97 (32.61)				
1Z MeOH II	$n_{X3(1)} \to$	σ^*_{O8-H9}	3.76 (1.50)	$n_{O9(1)} \to$	σ^*_{C1-H6}	0.31 (0.87)	σ^*_{O8-H9}	0.019 (0.019)	σ^*_{C1-H6}	0.038 (0.032)
	$n_{X3(2)} \to$	σ^*_{O8-H9}	7.48 (6.96)	$n_{O9(2)} \to$	σ^*_{C1-H6}	0.30 (1.28)				
1Z MeOH III	$n_{O4(1)} \to$	σ^*_{O8-H9}	1.12	$n_{O8(1)} \to$	σ^*_{N2-H7}	2.35 (1.62)	σ^*_{O8-H9}	0.008 (-)	σ^*_{N2-H7}	0.028 (0.042)
	$n_{O4(2)} \to$	σ^*_{O8-H9}	1.64	$n_{O8(2)} \to$	σ^*_{N2-H7}	10.54 (24.31)				
2Z MeOH I	$n_{O4(1)} \to$	σ^*_{O8-H9}	5.88 (4.98)	$n_{O8(1)} \to$	σ^*_{X3-H5}	1.46 (0.89)	σ^*_{O8-H9}	0.013 (0.017)	σ^*_{X3-H5}	0.045 (0.022)
	$n_{O4(2)} \to$	σ^*_{O8-H9}	0.09 (5.03)	$n_{O8(2)} \to$	σ^*_{X3-H5}	27.58 (8.14)				
2Z MeOH II	$n_{N2} \to$	σ^*_{O8-H9}	3.49 (3.18)	$n_{O8(1)} \to$	σ^*_{O4-H7}	1.47 (1.49)	σ^*_{O8-H9}	0.010 (0.010)	σ^*_{O4-H7}	0.028 (0.028)
				$n_{O8(2)} \to$	σ^*_{O4-H7}	18.32 (18.47)				
2Z MeOH III	$n_{X3(1)} \to$	σ^*_{O8-H9}	5.43 (-)	$n_{O8(1)} \to$	σ^*_{C1-H6}	0.45 (2.48)	σ^*_{O8-H9}	0.011 (-)	σ^*_{C1-H6}	0.016 (0.016)
	$n_{X3(2)} \to$	σ^*_{O8-H9}	0.12 (-)	$n_{O8(2)} \to$	σ^*_{C1-H6}	0.55 (0.39)				

Table 7.13: Intramolecular hydrogen bonding parameters in adducts of FHA (TFHA) with methanol at MP2/6-31+G* [L3] theoretical level. Bond distances are in angstrom (Å) and angles are in degrees (°).

Species	Hydrogen bond distances (r) O4-H5...X3	Δr*	Hydrogen bond angles O4-H5...X3	Species	Hydrogen bond distances (r) X3-H5...O4	Δr*	Hydrogen bond angles X3-H5...O4
1Z	2.046 (2.307)	0.554 (0.743)	117.8 124.4	2Z	2.020 (2.399)	0.580 (0.201)	111.5 108.6
1Z MeOH I	- (-)	- (-)	- (-)	2Z MeOH I	2.329 (2.383)	0.271 (0.217)	100.9 (103.1)
1Z MeOH II	2.051 (2.321)	0.549 (0.729)	117.3 (123.7)	2Z MeOH II	2.000 (2.175)	0.600 (0.425)	112.7 (109.4)
1Z MeOH III	2.017 (2.277)	0.583 (0.773)	118.5 (125.3)	2Z MeOH III	2.001 (2.193)	0.599 (0.407)	111.8 (108.7)

Δr* = r_{VW}-r,
r_{VW} = r_O+r_H=2.60 Å, r_S+r_H=3.05 Å

Table 7.14: Atomic charges (NPA) in aggregates of **1Z** and **2Z** conformers of FHA (TFHA) with MeOH at MP2/6-31+G* [L3] theoretical level.

Species	C1	N2	X3	O4	H5	H6	H7
1Z	0.636	-0.426	-0.732	-0.616	0.535	0.173	0.430
	(-0.019)	(-0.347)	(-0.251)	(-0.603)	(0.531)	(0.242)	(0.447)
1Z MeOH I	0.656	-0.433	-0.758	-0.632	0.559	0.161	0.428
	(0.008)	(-0.357)	(-0.266)	(-0.628)	(0.560)	(0.233)	(0.435)
1Z MeOH II	0.637	-0.414	-0.775	-0.610	0.537	0.199	0.437
	(-0.003)	(-0.339)	(-0.296)	(-0.600)	(0.532)	(0.268)	(0.449)
1Z MeOH III	0.638	-0.438	-0.746	-0.645	0.539	0.173	0.465
	(-0.013)	(-0.358)	(-0.293)	(-0.617)	(0.528)	(0.239)	(0.485)
2Z	0.478	-0.282	-0.783	-0.703	0.547	0.214	0.528
	(-0.099)	(-0.197)	(0.025)	(-0.677)	(0.176)	(0.248)	(0.524)
2Z MeOH I	0.497	-0.301	-0.808	-0.716	0.568	0.208	0.524
	(-0.089)	(-0.208)	(0.018)	(-0.695)	(0.197)	(0.249)	(0.527)
2Z MeOH II	0.483	-0.315	-0.786	-0.722	0.549	0.210	0.559
	(-0.096)	(-0.225)	(0.017)	(-0.697)	(0.181)	(0.243)	(0.557)
2Z MeOH III	0.470	-0.273	-0.808	-0.705	0.554	0.236	0.529
	(-0.092)	(-0.208)	(0.016)	(-0.681)	(0.168)	(0.271)	(0.521)

Table 7.15: Important orbital interactions and second order electron delocalization energies $E^{(2)}$ in kcal/mol for orbital interactions in aggregates of FHA (TFHA) with MeOH at MP2/6-31+G* [L3] theoretical level.

Species	$n_{N2} \to \sigma^*_{C1-X3}$	$n_{N2} \to \pi^*_{C1-X3}$	$n_{X3} \to \sigma^*_{C1-N2}$	$n_{X3} \to \pi^*_{C1-N2}$	$n_{O4} \to \sigma^*_{C1-N2}$	$n_{O4} \to \pi^*_{C1-N2}$	$n_{X3} \to \sigma^*_{O4-H5}$	$n_{O4} \to \sigma^*_{X3-H5}$
1Z	14.98	32.68	29.25	-	4.36	-	1.24+5.36	-
	(-)	(145.34)	(15.92)	(-)	(6.56)	(-)	(13.71)	(-)
1Z MeOH I	-	91.13	28.35	-	6.31	-	-	-
	(-)	(-)	(16.25)	(-)	(6.73)	(2.23)	(-)	(-)
1Z MeOH II	9.47	51.67	27.53	-	4.65	-	1.56+4.73	-
	(-)	(152.51)	(15.41)	(-)	(6.58)	(-)	(2.13+12.08)	(-)
1Z MeOH III	12.77	37.70	28.29	-	3.89	-	6.17	-
	(-)	(164.00)	(14.89)	(-)	(6.20)	(-)	(1.85+15.43)	(-)
2Z	13.50	-	8.63	54.42	-	13.22	-	5.06
	(14.57)	(-)	(6.45)	(39.47)	(-)	(18.29)	(-)	(2.61)
2Z MeOH I	15.24	-	7.63+18.61	7.12+20.80	-	5.26	-	1.67
	(14.60)	(-)	(6.79+4.03)	(28.55)	(-)	(10.31)	(-)	(0.99)
2Z MeOH II	13.35	-	8.53	54.81	-	14.88	-	5.98
	(14.58)	(-)	(6.36)	(39.04)	(-)	(21.06)	(-)	(3.03)
2Z MeOH III	13.73	-	7.35	42.75	-	11.70	-	5.51
	(14.91)	(-)	(6.16)	(38.00)	(-)	(17.17)	(-)	(2.52)

7.2.4 THE HYDROGEN BONDING OF HYDROXAMIC ACIDS WITH METHANETHIOL (MeSH):

The S-H functionality of the side chains of cysteine and methionine is represented by methanethiol model molecule. Fazary et al. suggested the reactivity of HAs towards SH group of proteins to be reason for their inhibitory effect on various enzymes [13]. In order to analyze the interactions between the HAs and the MeSH, the aggregation behavior of **1Z** and **2Z** conformations of FHA and TFHA with MeSH has been studied at **[L3]** theoretical level. Five aggregates in total of **1Z** and **2Z** conformations for each FHA and TFHA with MeSH have been optimized at **[L3]** theoretical level and are shown in Figure 7.4. The Table 7.16 displays the geometrical parameters important for H-bonding, stabilization energies of aggregate formation and distortion energies associated with the distortion of the monomer units in the aggregate geometry.

The **1Z MeSH I** aggregate of FHA and TFHA is the most stabilized aggregate formed between **1Z** and MeSH. In case of FHA, **1Z MeSH I** has two H-bonds S8-H9...O3 and O4-H5...S8 with H-bond distances and angles of 2.428, 2.345 Å and 127.9°, 158.9° respectively. The stabilization energy of 5.48 kcal/mol associated with this aggregate is 5.96 kcal/mol lower than that in **1Z MeOH I** of FHA. The O4-H5...S8 H-bond with a distance of 2.345Å and H-bond angle O4-H5...S8 of 158.9° is stronger in comparison to the second H-bond S8-H9...O3 (2.428 Å and angle S8-H9...O3 127.9°). Δr in the former is 0.705 Å while in case of latter the value is 0.172 Å. Only single H-bond is indicated in case of **1Z MeSH I** of TFHA. The H-bond O4-H5...S8 with distance of 2.270 Å and angle of 168.3° is stronger than the similar H-bond in case of FHA. In spite of the single bond in **1Z MeSH I** of TFHA, the stabilization energy is only 0.84 kcal/mol lower than that in case of FHA in the similar orientation. It is important to note that E_{Dis} in case of **1Z MeSH I** of TFHA is 1.41 kcal/mol higher than that in case of similar aggregate of FHA.

1Z MeSH II has nearly half of the stabilization energy than similar aggregate **1Z MeOH II** in both the HAs that depicts relatively weaker strength of S9-H8...X3 and C1-H6...S9 bonds in this aggregate with MeSH over O9-H8...X3 and C1-H6...O9 bonds in aggregate with MeOH. The MeSH tends to form only single H-bond of type N2-H7...S8 in **1Z MeSH III** aggregate. This aggregate has 3.54

kcal/mol lower strength than for the corresponding **1Z MeOH III** aggregate of TFHA which shows that S of MeSH is weaker acceptor than O of MeOH.

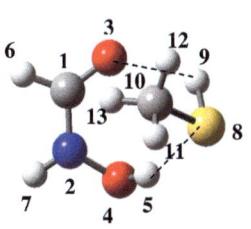
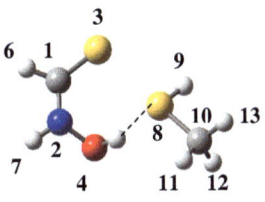

1Z MeSH I

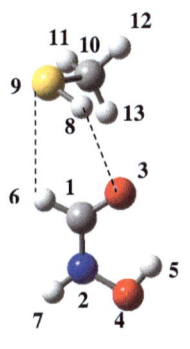
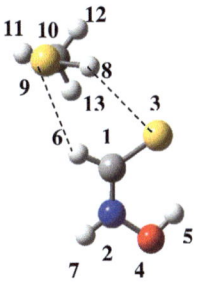

1Z MeSH II

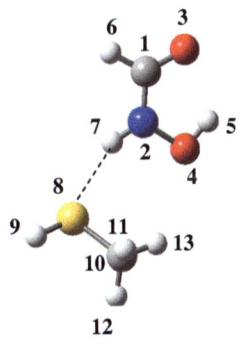
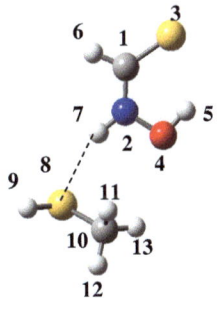

1Z MeSH III

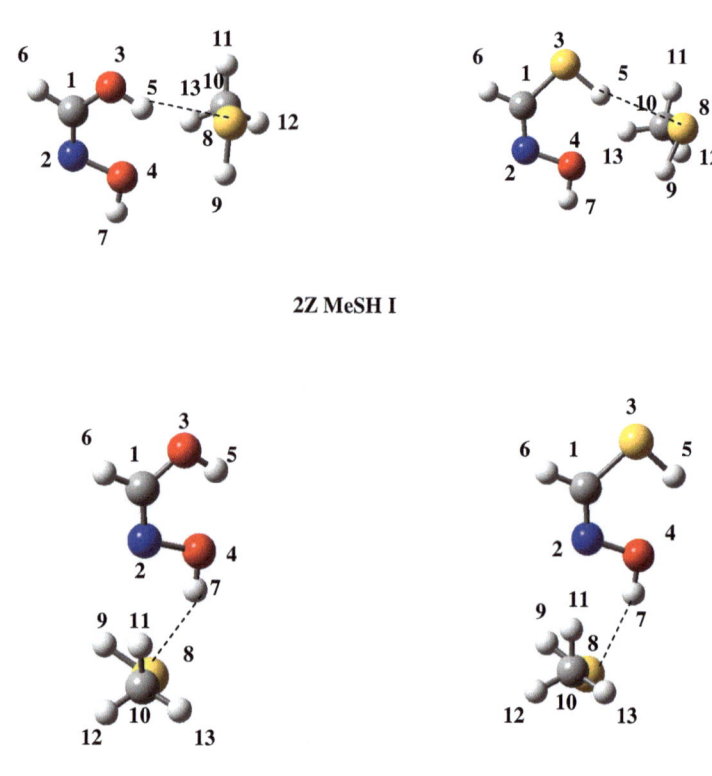

2Z MeSH I

2Z MeSH II

Figure 7.4: Optimized aggregates of FHA and TFHA with methanethiol at [L3] theoretical level.

Since **2Z MeSH I** and **2Z MeSH II** of both the HAs are stabilized by single intermolecular H-bond only, the stabilization energy of aggregation can be taken as the reflective of the strength of H-bond present in the aggregates. The O3-H5...S8 H-bond present in **2Z MeSH I** of FHA has stabilization energy of 3.34 kcal/mol which is higher than the stabilization associated with aggregation of MeSH with water (2.48 kcal/mol). The **2Z MeSH II** of FHA also indicates the presence of O4-H7...S8 H-bond interaction but with stabilization energy of 5.08 kcal/mol. In case of TFHA, the **2Z MeSH I** with single H-bond O3-H5...S8 shows much lower stabilization of 1.75

kcal/mol. The **2Z MeSH II** of TFHA with single O4-H7...S8 H-bond however leads to 4.93 kcal/mol stabilized aggregate. Thus cooperative interactions play important role in overall stability of the aggregate.

Atomic charges (Table 7.17) indicate that bond polarizations are increased upon aggregation that contributes toward the strengthening of the bonds. This variation in bond polarization is significantly large in **1Z MeSH I** particularly that in case of TFHA. Comparatively smaller variations in bond polarizations are reflected in aggregates of **2Z**. The SH bond of MeSH also undergoes polarization upon aggregation. Insignificant to small changes in atomic charges for rest of atoms in the MeSH are observed as the result of aggregation. The S of MeSH carries only small negative charge and H attached to S is also insignificantly polarized. Hence comparatively weaker electrostatic interactions are anticipated in aggregates of MeSH with HAs relative to those of MeOH with HAs.

NBO analysis on all optimized structures is performed at **[L3]** theoretical level to investigate and quantify the possible charge transfer during interactions with MeSH and the $E^{(2)}$ values for charge transfer interactions are presented in Table 7.18. NBO analysis shows that MeSH acts primarily as a charge donor toward HAs. The $n_{S8} \rightarrow \sigma^*_{O4-H5}$ in **1Z MeSH I** is associated with highest $E^{(2)}$ of 24.30 and 18.97 kcal/mol in TFHA and FHA respectively. The $E^{(2)}$ values are much higher than the $n_S \rightarrow \sigma^*_{O-H}$ orbital interactions in MeSH-H_2O aggregate ($E^{(2)}$=8.44 kcal/mol at the same theoretical level). Value of charge transfer is small for C1-H6...S9 interaction in **1Z MeSH II**, thus indicating that charge transfer plays minor role in C1-H6...S9 H-bond stabilization. S-H and C-H unconventional bonds acting as H-bond donors in MeSH have very low charge transfer from HAs to MeSH in comparison to charge transfer from HAs to MeOH in aggregates with MeOH, thereby with C-H and S-H as H-bond donors, H-bond strengths are not favored by covalent component also.

Though the orbital interaction $n_{N2} \rightarrow \pi^*_{C1-X3}$ (Table 7.19) is enhanced in most stabilized aggregate of **1Z** of FHA with MeSH (**1Z MeSH I**), but decrease in $E^{(2)}$ value for the similar interactions in case of (**1Z MeSH I**) of TFHA accounts for the lesser stabilization in this case. Thus the cooperativity in **1Z MeSH I** of FHA results from change in conjugative interactions while in case of TFHA, it is the variation in bond polarizations which is responsible for it.

In earlier analysis on aggregation with water, we found comparable H-bond acceptor abilities of O and S of FHA and TFHA units but the H-bond donor and acceptor ability of MeSH to form H-bonding is definitely lower than MeOH as seen from the aggregates with lower stabilization energies.

Stabilization energy of most stable aggregate of HAs with MeSH is significantly lower than most stable aggregate with MeOH. Here charge transfer occurs in one direction i.e. from MeSH to HA. In spite of low stabilization energies, the H-bond interactions are significant enough to favor specific conformational behavior.

Table 7.16: Important intermolecular hydrogen bonding parameters; hydrogen bond distances[a], hydrogen bond angles[b], charges on hydrogen bond acceptor and hydrogen, stabilization energies (S.E.[c]) and distortion energies (E_{Dis}[d]) of aggregates of FHA (TFHA) with methanethiol at MP2/6-31+G* [L3] theoretical level.

Species	Hydrogen bond Distances[a]		Hydrogen bond Angles[b]	Atomic charges		S.E.[c]	E_{Dis}[d]
1Z MeSH I	H9...X3	2.428 (-)	S8-H9...X3	127.9 (-)	$q_H(q_X)$	0.171(-0.719) (-)	5.48 (4.64)	1.17 (2.58)
	H5...S8	2.345 (2.270)	O4-H5...S8	158.9 (168.3)	$q_H(q_S)$	0.526(-0.069) (0.526(-0.041))		
1Z MeSH II	H8...X3	2.603 (2.935)	S9-H8...X3	118.4 (133.1)	$q_H(q_X)$	0.146(-0.755) (0.142(-0.271))	2.62 (2.72)	0.15 (0.05)
	H6...S9	3.061 (2.863)	C1-H6...S9	112.0 (131.2)	$q_H(q_S)$	0.186 (-0.074) (0.249(-0.069))		
1Z MeSH III	H7...S8	2.448 (2.378)	N2-H7...S8	150.7 (154.9)	$q_H(q_S)$	0.440(-0.056) (0.455(-0.054))	4.48 (5.23)	0.12 (0.16)
2Z MeSH I	H5...S8	2.382 (2.678)	O3-H5...S8	156.0 (156.8)	$q_H(q_S)$	0.545(-0.044) (0.179(-0.048))	3.34 (1.75)	0.66 (0.22)
2Z MeSH II	H7...S8	2.346 (2.349)	O4-H7...S8	155.1 (155.5)	$q_H(q_S)$	0.532(-0.055) (0.529(-0.054))	5.08 (4.93)	0.19 (0.17)

a- in angstrom (Å)
b- in degrees (°)
c, d- in (kcal/mol)

Table 7.17: Atomic charges (NPA) on atoms of FHA (TFHA) in their aggregates with methanethiol at MP2/6-31+G* [L3] theoretical level.

Species	C1	N2	X3	O4	H5	H6	H7
1Z	0.636	-0.426	-0.732	-0.616	0.535	0.173	0.430
	(-0.019)	(-0.347)	(-0.251)	(-0.603)	(0.531)	(0.242)	(0.447)
1Z MeSH I	0.650	-0.446	-0.719	-0.625	0.526	0.155	0.423
	(-0.016)	(-0.386)	(-0.203)	(-0.621)	(0.526)	(0.229)	(0.430)
1Z MeSH II	0.631	-0.420	-0.755	-0.614	0.536	0.186	0.433
	(-0.015)	(-0.343)	(-0.271)	(-0.601)	(0.531)	(0.249)	(0.448)
1Z MeSH III	0.637	-0.439	-0.744	-0.632	0.538	0.173	0.440
	(-0.013)	(-0.358)	(-0.273)	(-0.619)	(0.531)	(0.241)	(0.455)
2Z	0.478	-0.282	-0.783	-0.703	0.547	0.214	0.528
	(-0.099)	(-0.197)	(0.025)	(-0.677)	(0.176)	(0.248)	(0.524)
2Z MeSH I	0.486	-0.294	-0.797	-0.702	0.545	0.211	0.522
	(-0.094)	(-0.205)	(0.021)	(-0.683)	(0.179)	(0.248)	(0.521)
2Z MeSH II	0.479	-0.300	-0.785	-0.719	0.549	0.211	0.532
	(-0.099)	(-0.212)	(0.019)	(-0.693)	(0.179)	(0.244)	(0.529)

Table 7.18: The orbital interactions and second order delocalization energies $E^{(2)}$ in kcal/mol and occupancies of acceptor antibonding orbitals important for the hydrogen bonds present in the aggregates of FHA (TFHA) with methanethiol at MP2/6-31+G* [L3] theoretical level.

Species	Donor HA	Acceptor MeSH	$E^{(2)}$	Donor MeSH	Acceptor HA	$E^{(2)}$	Occupancies Acceptor MeSH		Acceptor HA	
1Z MeSH I	$n_{X3(1)} \to \sigma^*_{S8-H9}$		1.22 (0.06)	$n_{S8(1)} \to$	σ^*_{O4-H5}	0.62 (1.10)	σ^*_{S8-H9}	0.008 (0.006)	σ^*_{O4-H5}	0.038 (0.047)
	$n_{X3(2)} \to \sigma^*_{S8-H9}$		0.78 (0.44)	$n_{S8(2)} \to$	σ^*_{O4-H5}	18.97 (24.30)				
1Z MeSH II	$n_{X3(1)} \to \sigma^*_{S9-H8}$		0.37 (0.49)	$n_{S9(1)} \to$	σ^*_{C1-H6}	0.10 (0.08)	σ^*_{S9-H8}	0.007 (0.011)	σ^*_{C1-H6}	0.043 (0.036)
	$n_{X3(2)} \to \sigma^*_{S9-H8}$		0.60 (2.06)	$n_{S9(2)} \to$	σ^*_{C1-H6}	1.28 (2.37)				
1Z MeSH III	-		-	$n_{S8(1)} \to$	σ^*_{N2-H7}	0.97 (1.07)	-	-	σ^*_{N2-H7}	0.038 (0.048)
				$n_{S8(2)} \to$	σ^*_{N2-H7}	14.57 (19.37)				
2Z MeSH I	-		-	$n_{S8(1)} \to$	σ^*_{X3-H5}	0.54 (0.27)	-	-	σ^*_{X3-H5}	0.040 (0.022)
				$n_{S8(2)} \to$	σ^*_{X3-H5}	15.89 (6.09)				
2Z MeSH II	$n_{N2} \to \sigma^*_{S8-H9}$		0.18 (0.16)	$n_{S8(1)} \to$	σ^*_{O4-H7}	0.50 (0.50)	σ^*_{S8-H9}	0.006 (0.006)	σ^*_{O4-H7}	0.035 (0.034)
				$n_{S8(2)} \to$	σ^*_{O4-H7}	17.58 (17.54)				

Table 7.19: Important orbital interactions and second order electron delocalization energies $E^{(2)}$ in kcal/mol in aggregates of FHA (TFHA) with methanethiol at MP2/6-31+G* **[L3]** theoretical level.

Species	$n_{N2} \to \sigma^*_{C1-X3}$	$n_{N2} \to \pi^*_{C1-X3}$	$n_{X3} \to \sigma^*_{C1-N2}$	$n_{X3} \to \pi^*_{C1-N2}$	$n_{O4} \to \sigma^*_{C1-N2}$	$n_{O4} \to \pi^*_{C1-N2}$	$n_{X3} \to \sigma^*_{O4-H5}$	$n_{O4} \to \sigma^*_{X3-H5}$
1Z	14.98	32.68	29.25	-	4.36	-	1.24+5.36	-
	(-)	(145.34)	(15.92)	(-)	(6.56)	(-)	(13.71)	(-)
1Z MeSH I	5.93	52.33	31.34	-	6.58	-	-	-
	(-)	(100.39)	(19.79)	(-)	(2.38+5.61)	(-)	(-)	(-)
1Z MeSH II	24.46	25.28	28.43	-	4.45	-	1.33+5.26	-
	(-)	(146.23)	(15.72)	(-)	(6.58)	(-)	(1.90+12.78)	(-)
1Z MeSH III	11.63	40.75	28.56	-	4.15	-	1.31+5.89	-
	(-)	(156.05)	(15.25)	(-)	(6.25)	(-)	(1.87+15.47)	(-)
2Z	13.50	-	8.63	54.42	-	13.22	-	5.06
	(14.57)	(-)	(6.47)	(39.47)	(-)	(18.29)	(-)	(2.61)
2Z MeSH I	14.29	-	6.74	6.09	5.14	4.23	-	3.34
	(14.36)	(-)	(6.42)	(30.36)	(-)	(12.08)	(-)	(1.98)
2Z MeSH II	13.52	-	8.09	46.11	-	11.52	-	5.78
	(14.70)	(-)	(6.30)	(37.70)	(-)	(19.12)	(-)	(2.92)

7.2.5 HYDROGEN BONDING OF HYDROXAMIC ACIDS WITH METHANESELENOL (MeSeH):

Association of **1Z** and **2Z** conformations of HAs (FHA and TFHA) with methaneselenol in 1:1 ratio leads to formation of three H-bonded aggregates for each conformation denoted by **1Z MeSeH I**, **1Z MeSeH II**, **1Z MeSeH III**, **2Z MeSeH I**, **2Z MeSeH II** and **2Z MeSeH III** and are shown in Figure 7.5. Table 7.20 displays the geometrical parameters important for characterizing intermolecular H-bonding. The stabilization energies and the distortion energies associated with the aggregate formation are also reported in the table. The **1Z MeSeH I** aggregate has highest stabilization energy both in case of FHA as well as TFHA. The aggregate in case of TFHA is 0.99 kcal/mol more stabilized than that in case of FHA. The **1Z MeSeH I** of FHA has Se8-H9...O3 and O4-H5...Se8 H-bond interactions with H-bond distance of 2.432 and 2.427 Å and the H-bond angles of 127.0° and 162.6° respectively. The parameters however suggest the presence of single H-bond in **1Z MeSeH I** of TFHA. The O4-H5...Se8 H-bond in the aggregate with distance 2.366 Å and angle 163.4° stabilizes the aggregate more than the similar interaction in case of FHA. The E_{Dis} is also high in this case (2.99 kcal/mol) that can be understood in terms of more highly conjugated **1Z** conformer and intramolecular H-bond breakage upon aggregate formation.

The stabilization energies for **1Z MeSeH III** of FHA and TFHA are only 0.60 and 0.72 kcal/mol lower than the respective value for **1Z MeSeH I**. It is interesting to note that although NH of HAs acts as H-bond donor to Se of MeSeH in **1Z MeSeH III** aggregate for both the acids, yet the stabilization energy in case of FHA is 4.57 kcal/mol while in case of TFHA it is 5.44 kcal/mol. Similarly the **2Z MeSeH II** with Se of MeSeH as H-bond acceptor to O4-H7 of HAs also indicates high stabilization energy of 5.04 and 4.88 kcal/mol in case of FHA and TFHA respectively. Inspite of non conventional H-bond donor C-H group to Se of MeSeH in **1Z MeSeH II** and **2Z MeSeH III** the stabilization energies of 2.19 (2.43) and 0.94 (1.43) kcal/mol are evaluated in case of FHA (TFHA) which suggest larger stabilization energy for **1Z** than **2Z** conformer in both cases. Thus according to traditional facts, though Se is considered to be among weak H-bond acceptors yet the obtention of magnificent stabilization energy in aggregates with Se as acceptor towards conventional H-bond donors indicate its importance in stabilization of protein structure.

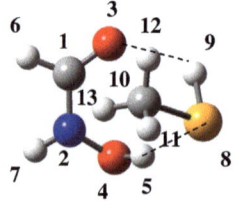 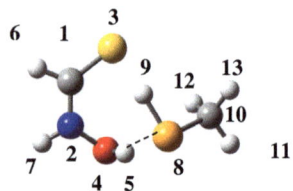

1Z MeSeH I

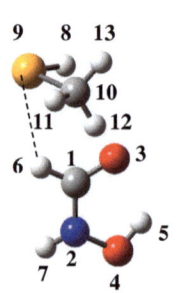 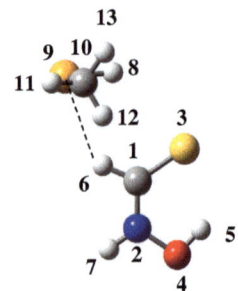

1Z MeSeH II

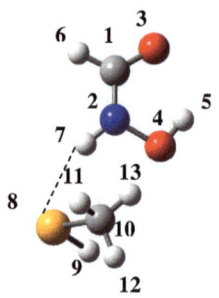 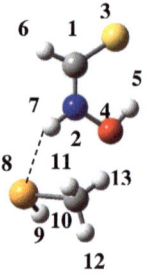

1Z MeSeH III

Contd…..

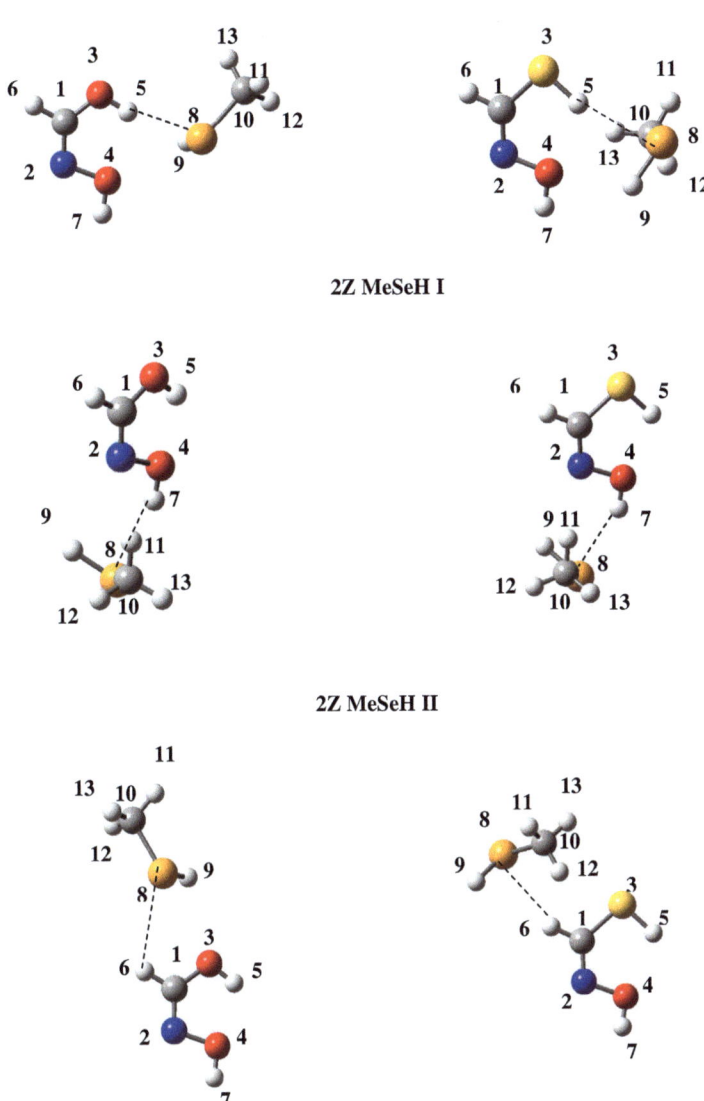

2Z MeSeH I

2Z MeSeH II

2Z MeSeH III

Figure 7.5: Optimized aggregates of FHA and TFHA with methaneselenol at MP2/6-31+G* **[L3]** theoretical level

The orbital interactions reflective of charge transfer between H-bond acceptor and donor suggesting the extent of covalency present in the H-bonds are recorded in Table 7.21. The $E^{(2)}$ values for the adducts **1Z MeSeH I, 1Z MeSeH III** and **2Z MeSeH II** are magnificently high for charge transfer from Se towards HAs in comparison to the values for other adducts suggesting presence of high covalent character in the H-bond of these three adducts. Indeed these are highly stabilized adducts.

The atomic charges on H-bond acceptor and the H of H-bond donor are also listed in the Table 7.20 and charges on all atoms of HAs upon aggregation are listed in Table 7.22. From atomic charge analysis of MeSeH, MeOH and MeSH, the charge on Se is smaller in comparison to O in MeOH and S of MeSH thus comparatively weakest electrostatic interactions are expected in H-bonds of MeSeH. The variation in atomic charges arising upon aggregate formation suggests that most of bonds of HAs are more polarized in the aggregates. Thus the ionic component of all such bonds favors the stability of the adduct.

As discussed in previous chapters, there are magnificent electron delocalizations involving lone pair of electrons present at N2, X3 and O4 of HAs that are clearly reflected by the $E^{(2)}$ values associated with the orbital interactions. Lone pair of electrons at the nitrogen is the most delocalized pair of electrons. The increase in $E^{(2)}$ values for most of the orbital interactions on aggregation (Table 7.23) suggests that these electron delocalizations are favored by aggregation and tend to stregthen the adduct thereby resulting in cooperativity. Thus the extra stability arising from the strengthening of bonds as a result of polarization of bonds and enhanced conjugative interactions accounts for the H-bond cooperativity observed in the adducts.

At the end of this section, it can be concluded that the stabilization produced by all the amino acid side chain model molecules on interaction with HAs follows the order AcOH > $MeNH_2$ > MeOH > MeSH > MeSeH for FHA while the order is AcOH > $MeNH_2$ > MeOH > MeSeH > MeSH for TFHA.

Table 7.20: Important intermolecular hydrogen bonding parameters; hydrogen bond distances[a], hydrogen bond angles[b], charges on hydrogen bond acceptor and hydrogen, stabilization energies (S.E.[c]) and distortion energies (E_{Dis}[d]) of aggregates of FHA (TFHA) with MeSeH at MP2/6-31+G* [L3] theoretical level.

Species	Hydrogen bond distances[a]		Hydrogen bond angles[b]	Atomic charges		S. E.[c]	E_{Dis}[d]
1Z MeSeH I	H9...X3	2.432	Se8-H9...X3 127.0	$q_H(q_X)$	0.115(-0.716)	5.17	1.26
		(-)	(-)		(-)	(6.16)	(2.99)
	H5...Se8	2.427	O4-H5...Se8 162.6	$q_H(q_O)$	0.523(0.030)		
		(2.366)	(163.4)		(0.519(0.043))		
1Z MeSeH II	H6...Se9	2.838	C1-H6...Se9 123.4	$q_H(q_O)$	0.188(0.031)	2.19	0.09
		(2.734)	(136.7)		(0.249(0.031))	(2.43)	(0.09)
1Z MeSeH III	H7...Se8	2.500	N2-H7...Se8 144.4	$q_H(q_O)$	0.441(0.045)	4.57	0.28
		(2.436)	(152.8)		(0.450(0.052))	(5.44)	(0.18)
2Z MeSeH I	H5...Se8	2.434	X3-H5...Se8 169.2	$q_H(q_O)$	0.540(0.081)	3.04	0.63
		(2.649)	(161.3)		(0.172(0.056))	(1.46)	(2.12)
2Z MeSeH II	H7...Se8	2.411	O4-H7...Se8 155.4	$q_H(q_O)$	0.529(0.050)	5.04	0.19
		(2.419)	(155.0)		(0.526(0.050))	(4.88)	(0.17)
2Z MeSeH III	H6...Se8	3.051	C1-H6...Se8 109.0	$q_H(q_O)$	0.223(0.064)	0.94	0.02
		(2.781)	(147.8)		(0.255(0.059))	(1.43)	(0.05)

a- in angstrom (Å)
b- in degrees (°)
c, d- in (kcal/mol)

Table 7.21: The orbital interactions and second order delocalization energies $E^{(2)}$ in kcal/mol and occupancies of acceptor antibonding orbitals important for the hydrogen bonds present in the aggregates of FHA (TFHA) with MeSeH at MP2/6-31+G* [L3] theoretical level.

Species	Donor HA	Acceptor MeSeH	$E^{(2)}$	Donor MeSeH	Acceptor HA	$E^{(2)}$	Occupancies Acceptor MeSeH	Acceptor HA
1Z MeSeH I	$n_{X3(1)} \rightarrow$	σ^*_{Se8-H9}	1.17 (0.15)	$n_{Se8(1)} \rightarrow$	σ^*_{O4-H5}	0.54 (0.60)	σ^*_{Se8-H9} 0.009	σ^*_{O4-H5} 0.051
	$n_{X3(2)} \rightarrow$	σ^*_{Se8-H9}	0.85 (0.09)	$n_{Se8(2)} \rightarrow$	σ^*_{O4-H5}	20.08 (24.66)		
1Z MeSeH II	$n_{X3(1)} \rightarrow$	σ^*_{Se9-H8}	0.08 (0.29)	$n_{Se9(1)} \rightarrow$	σ^*_{C1-H6}	0.30 (0.23)	σ^*_{Se9-H8} 0.007 (0.010)	σ^*_{C1-H6} 0.048 (0.042)
	$n_{X3(2)} \rightarrow$	σ^*_{Se9-H8}	0.22 (1.43)	$n_{Se9(2)} \rightarrow$	σ^*_{C1-H6}	4.19 (5.46)		
1Z MeSeH III	-	-	-	$n_{Se8(1)} \rightarrow$	σ^*_{N2-H7}	0.61 (0.73)	-	σ^*_{N2-H7} 0.044 (0.056)
				$n_{Se8(2)} \rightarrow$	σ^*_{N2-H7}	17.32 (22.73)		
2Z MeSeH I	-	-	-	$n_{Se8(1)} \rightarrow$	σ^*_{X3-H5}	0.71 (0.37)	-	σ^*_{X3-H5} 0.045
				$n_{Se8(2)} \rightarrow$	σ^*_{X3-H5}	19.65 (9.11)		
2Z MeSeH II	$n_{N2} \rightarrow$	σ^*_{Se8-H9}	0.11 (0.12)	$n_{Se8(1)} \rightarrow$	σ^*_{O4-H7}	0.52 (0.51)	σ^*_{Se8-H9} 0.007 (0.007)	σ^*_{O4-H7} 0.040 (0.007)
				$n_{Se8(2)} \rightarrow$	σ^*_{O4-H7}	20.35 (19.86)		
2Z MeSeH III	-	-	-	$n_{Se8(1)} \rightarrow$	σ^*_{C1-H6}	0.54 (0.32)	-	σ^*_{C1-H6} 0.019 (0.023)
				$n_{Se8(2)} \rightarrow$	σ^*_{C1-H6}	1.25 (6.30)		

Table 7.22: Atomic charges (NPA) on atoms of FHA (TFHA) in their aggregates with MeSeH at MP2/6-31+G* [L3] theoretical level.

Species	C1	N2	X3	O4	H5	H6	H7
1Z	0.636	-0.426	-0.732	-0.616	0.535	0.173	0.430
	(-0.019)	(-0.347)	(-0.251)	(-0.603)	(0.531)	(0.242)	(0.447)
1Z MeSeH I	0.650	-0.448	-0.716	-0.624	0.523	0.155	0.423
	(-0.015)	(-0.390)	(-0.198)	(-0.619)	(0.519)	(0.230)	(0.429)
1Z MeSeH II	0.627	-0.420	-0.756	-0.615	0.535	0.188	0.432
	(-0.015)	(-0.343)	(-0.277)	(-0.602)	(0.532)	(0.249)	(0.448)
1Z MeSeH III	0.634	-0.440	-0.754	-0.628	0.540	0.174	0.441
	(-0.013)	(-0.361)	(-0.272)	(-0.620)	(0.531)	(0.241)	(0.450)
2Z	0.478	-0.282	-0.783	-0.703	0.547	0.214	0.528
	(-0.099)	(-0.197)	(0.025)	(-0.677)	(0.176)	(0.248)	(0.524)
2Z MeSeH I	0.048	-0.292	-0.800	-0.700	0.541	0.209	0.523
	(-0.094)	(-0.205)	(0.021)	(-0.684)	(0.172)	(0.248)	(0.521)
2Z MeSeH II	0.479	-0.299	-0.785	-0.720	0.549	0.210	0.529
	(-0.099)	(-0.212)	(0.018)	(-0.694)	(0.180)	(0.244)	(0.526)
2Z MeSeH III	0.475	-0.281	0.792	-0.705	0.550	0.223	0.528
	(-0.101)	(-0.201)	(0.012)	(-0.678)	(0.176)	(0.255)	(0.522)

Table 7.23: Important orbital interactions and second order electron delocalization energies $E^{(2)}$ in kcal/mol in aggregates of FHA (TFHA) with methaneselenol at MP2/6-31+G* **[L.3]** theoretical level.

Species	$n_{N2} \to \sigma^*_{C1-X3}$	$n_{N2} \to \pi^*_{C1-X3}$	$n_{X3} \to \sigma^*_{C1-N2}$	$n_{X3} \to \pi^*_{C1-N2}$	$n_{O4} \to \sigma^*_{C1-N2}$	$n_{O4} \to \pi^*_{C1-N2}$	$n_{X3} \to \sigma^*_{O4-H5}$	$n_{O4} \to \sigma^*_{X3-H5}$
1Z	14.98	32.68	29.25	-	4.36	-	1.24+5.36	-
	(-)	(145.34)	(15.92)	(-)	(6.56)	(-)	(13.71)	(-)
1Z MeSeH I	6.09	50.96	31.35	-	6.61	-	-	-
	(-)	(96.39)	(19.72)	(-)	(6.19)	(-)	(-)	(-)
1Z MeSeH II	22.29	27.66	28.55	-	4.42	-	1.30+5.27	-
	(-)	(147.11)	(15.74)	(-)	(6.58)	(-)	(1.83+12.62)	(-)
1Z MeSeH III	8.16	56.56	28.55	-	4.18	-	1.25+5.77	-
	(-)	(155.61)	(15.27)	(-)	(6.16)	(-)	(1.89+15.57)	(-)
2Z	13.50	-	8.63	54.42	-	13.22	-	5.06
	(14.57)	(-)	(6.47)	(39.47)	(-)	(18.29)	(-)	(2.61)
2Z MeSeH I	14.60	-	7.02+15.50	5.87+21.66	4.67	4.91	-	2.91
	(14.51)	(-)	(6.59)	(29.47)	(2.03)	(11.82)	(-)	(1.55)
2Z MeSeH II	13.52	-	8.23	48.57	-	12.34	-	5.87
	(14.73)	(-)	(6.29)	(37.81)	(-)	(19.32)	(-)	(2.97)
2Z MeSeH III	13.62	-	6.00	4.09	3.65	6.06	-	5.30
	(14.94)	(-)	(6.41)	(37.67)	(-)	(17.77)	(-)	(2.45)

7.3 CONCLUSIONS:

Study of H-bonding of amino acid side chain groups with HAs (FHA and TFHA) gives interesting results. Out of **1Z** and **2Z** conformations, **1Z** conformer results in greater stabilization on interaction with all the amino acid side chain groups studied in this chapter. The greater stabilization in case of **1Z** in comparison to **2Z** with the model molecules also indicates the preference for the existence of **1Z** conformer in the presence of specific interactions with solvent and other molecules. The chosen side chain analogues produce stabilization energy in the order AcOH > $MeNH_2$ > MeOH > MeSH > MeSeH for FHA while for TFHA the order is AcOH > $MeNH_2$ > MeOH > MeSeH > MeSH at MP2/6-31+G*. The most stable aggregate in all model molecules in the **1Z** conformer of FHA involves the carbonyl oxygen as H-bond acceptor and hydroxyl group as H-bond donor towards the model molecule. For the most stable aggregates of **1Z** conformation of TFHA, the thiocarbonyl sulphur is H-bond acceptor and hydroxyl group is H-bond donor towards $MeNH_2$, AcOH, MeOH. Only single H-bond is reflected in the most stable aggregate of **1Z** of TFHA with MeSH and MeSeH. In addition to electrostatic interactions, charge transfer interactions from H-bond acceptor to H-bond donor play very important role in the stabilization energies. Nearly in all the aggregates the dominant charge transfer interactions are those involving drift of electron density from H-bond acceptor of model molecules toward HAs.

While $MeNH_2$, MeOH, MeSH and MeSeH mainly act as charge donor towards HAs, the AcOH acts as good charge acceptor as well as donor in a number of aggregates. Inherent conjugation present in AcOH itself also augments the stabilization. The amino acid side chain groups can be broadly categorized into two groups 1) Highly polar type (AcOH, $MeNH_2$ and MeOH) 2) Relatively less polar type (MeSH, MeSeH). Stabilization energy for group I range from 16.08 to 6.35 kcal/mol, 12.83 to 3.27 kcal/mol and 11.44 to 3.66 kcal/mol for the interaction of FHA with AcOH, $MeNH_2$, MeOH respectively. The range is 13.40 to 3.94 kcal/mol, 13.23 to 3.40 kcal/mol and 10.78 to 3.03 kcal/mol for aggregates of TFHA with these molecules in respective order as mentioned earlier. The group II amino acid side chain analogues produce stabilization in the range 5.48 to 2.62 kcal/mol and 5.08 to 1.75 kcal/mol on H-bond interactions with MeSH and 5.17 to 0.94 kcal/mol and 6.16 to 1.43 kcal/mol for interaction of MeSeH with FHA and TFHA respectively.

H-bonding with Group I amino acid side chain model molecules has electrostatics as the chief component while it is charge transfer, cooperativity and bond polarizations which have larger share towards stabilization energy for group II side chain model molecules. S-H and Se-H though generally accepted as weak H-bonding groups yet evolve magnificent stabilization energy on aggregation. Thus it highlights the role of these groups in stabilizing the protein structure.

The difference in stabilization energies associated with the aggregation of **1Z** conformer of FHA and TFHA with four model molecules (MeNH$_2$, MeOH, MeSH, MeSeH) is less than 1.00 kcal/mol in the most stable orientation. The only exception being aggregate with AcOH (**1Z AcOH I**) where difference in stabilization energy is 2.68 kcal/mol. The results have been rationalized in terms of electrostatics, conjugative interactions and bond polarization arising with the H-bond formation giving rise to cooperativity in the aggregates.

7.4 REFERENCES:

1. K. Berka, R. Laskowski, K.E. Riley, P. Hobza, J. Vondràšek, J. Chem. Theory Comput. 5 (2009) 982.
2. S. Scheiner, T. Kar, J. Pattanayak, J. Am. Chem. Soc. 124 (2002) 13257.
3. P.I. Nagy, P.W. Erhardt, J. Phys. Chem. A 112 (2008) 4342.
4. R. Banerjee, M. Sen, D. Bhattacharya, P. Saha, J. Mol. Biol. 333 (2003) 211.
5. P. Chakrabarti, R. Bhattacharya, Prog. Biophys. Mol. Biol. 95 (2007) 83.
6. K.M.S. Misura, A.V. Morozov, D. Baker, J. Mol. Biol. 342 (2004) 651.
7. J.B.O. Mitchell, R.A. Laskowski, J.M. Thornton, Proteins: Struct., Funct., Bioinf. 29 (1998) 370.
8. M.S. Braiman, D.M. Briercheck, K.M. Kriger, J. Phys. Chem. B 103 (1999) 4744.
9. J.M. Yang, C.H. Tsai, M.J. Hwang, H.K. Tsai, J.K. Hwang, C.Y. Kao, Protein Sci. 11 (2002) 1897.
10. S.T. Phillips, G. Piersanti, P.A. Bartlett, Proc. Natl. Acad. Sci. USA 102 (2005) 13727.
11. J.K. Myers, C.N. Pace, Biophys. J. 71 (1996) 2033.
12. D.F. Stickle, L.G. Presta, K.A. Dill, G.D. Rose, J. Mol. Biol. 226 (1992) 1143.
13. A.E. Fazary, M.M. Khalil, A. Fahmy, T.A. Tantawy, Medical Journal of Islamic Academy of Sciences 14 (2001) 107.
14. Y.-H. Chiu, G.J. Gabriel, J.W. Canary, Inorg. Chem. 44 (2005) 40.
15. G.R. Desiraju, T. Steiner, The Weak Hydrogen Bond In Structural Chemistry and Biology, Oxford University Press, 1999.

CONCLUSIONS:

With the aim to understand intra- and intermolecular hydrogen bonding (H-bonding) ability of simplest hydroxamic acid (FHA) and its isosteres, namely thiformohydroxamic acid (TFHA) and formylphosphinous acid (FPA), the optimization of molecules and H-bonded aggregates of the selected molecules with water in 1:1 ratio is done. The dimerization studies of FHA and TFHA has also been carried out and resulting stabilization energies are evaluated. The aggregation with model molecules representing side chains of a few amino acids has also been analyzed across various isomeric forms of FHA and TFHA with a view to understand the underlying role of H-bonding in protein-protein and protein-drug interactions. The molecular orbital (MO) and density functional theory (DFT) methods are used for optimization of molecules. The electrostatic interactions, charge transfer interactions (covalency) and van der Waals are the contributors towards H-bond strengths in intermolecular H-bonded complexes. However the stabilization energies associated with the aggregate formation indicated the presence of additional factors as contributors.

Various conformers of keto and enol form of FHA and its isosteres have been explored for existence of intramolecular H-bonding. The study inferred the presence of intramolecular H-bond in keto tautomer (**1Z**) between C=X (X=O, S, O respectively in case of FHA, TFHA and FPA) and N-OH groups placed in syn conformation and for enol tautomer (**2Z**) between X-H and O-H groups of FHA, TFHA and FPA. Presence of intra molecular H-bonding makes **1Z** and **2Z** to be relatively more stable keto and enol isomer respectively among all other possible conformers of these forms. To analyze the intermolecular H-bonding ability of the HAs and isosteres, fifteen, eighteen and thirteen 1:1 aggregates of FHA, TFHA and FPA respectively with water are optimized. The highest stabilization energy amongst all the aggregates of FHA and its isosteres with water is for the aggregate labeled **1ZW1** in each case. In this aggregate the N-OH group acts as H-bond donor towards water and X of C=X group acts as H-bond acceptor. As both N-OH and C=X are intramolecularly H-bonded in **1Z** conformer, the stabilization energy of **1ZW1** indicates that stabilization resulting from intermolecular H-bonding is stronger than that associated with the intramolecular H-bond. The higher stability of **1Z** conformer in gas phase as well as in aggregate state resulting from strong intramolecular H-bonding in the former case and intermolecular H-bonding in the latter case is responsible for dominance of this form in the tautomeric mixture.

Though the **2Z** form of TFHA is the most stable in the gas phase but amongst the aggregates of TFHA with water, **1ZW1** is the most stable aggregate suggesting the prevalence of the **1Z** conformer of TFHA in aqueous medium. The **1ZW1** aggregate of FHA and TFHA has stabilization energy of 11.30 kcal/mol and 10.51 kcal/mol respectively that differ only by 0.79 kcal/mol at MP2/Aug-cc-pVDZ level. The electrostatic and covalent components of the H-bond strengths have been studied through the atomic charges on the H-bond donors and acceptors and second order delocalization energies ($E^{(2)}$ values) associated with charge transfer from the H-bond acceptors to H-bond donor orbital respectively. The comparison indicates that the electrostatic component of H-bond is weaker while the covalent component is stronger in aggregates of TFHA relative to those for FHA. The study also concluded that the H-bond donor ability of S-H in TFHA is less than that of O-H in FHA but the H-bond acceptor abilities of sulfur and oxygen of thiocarbonyl and carbonyl group of TFHA and FHA respectively are comparable. The observed results are explained in terms of larger size of S, its higher charge holding capacity, greater charge transfer, hence stronger covalent character of the H-bond and larger conjugation effects of thiocarbonyl of TFHA than carbonyl of FHA. The H-bond ability of the TFHA makes **1Z** conformer to dominate over others in the solution phase.

The stabilization energies associated with the dimer formation between the tautomeric forms of FHA and TFHA have also been analyzed to know more about the intermolecular H-bonding ability of these molecules. The **1Z-1Z (1)** dimer of both FHA and TFHA is the most stabilized dimer with two symmetrical C=X...H-O (N) H-bonds. The stabilization energy for the most stable dimer in FHA and TFHA is higher in comparison to most stabilized aggregate of both FHA and TFHA with water respectively. Gas phase results on dimerization are extended to solution phase employing polarized continuum model (PCM) using water as solvent. The enthalpy changes accompanying dimerization are reduced considerably in aqueous phase as the monomeric units interact with solvent more strongly, thus disfavoring the dimerization in some of the isomeric forms in solution although the dimer labeled **1Z-1Z (1)** remained the most stable dimer for both FHA and TFHA. The C=X is the strongest H-bond acceptor site and the hydroxyl is the strongest H-bond donor site of FHA and TFHA as is suggested by the results on dimerization and aggregate formation with water in both gas and aqueous phase. From the second order delocalization energies ($E^{(2)}$ values) evaluated using natural bond orbital analysis (NBO), FHA and TFHA are concluded to be better charge acceptor than water in the dimers and aggregates with water.

An interesting result of dimerization and aggregation with water is that the calculated stabilization energies in both FHA and TFHA are non additive in nature i.e. the energies are higher than anticipated for H-bonded similar functional groups, which hints that the additional stabilization is outcome of H-bond cooperativity which has been explained in terms of variation in bond polarizations, electron delocalizations and extension of the conjugated system.

The results of protonation and deprotonation at the probable sites indicate that X3 of C=X group is observed to be the most basic site while the N-H is the most acidic site in both FHA and TFHA. The inferences obtained from protonation and deprotonation data are correlated to H-bonding dimerization results. The comparison of the stabilization energies of dimerization with the proton affinities and deprotonation enthalpies for different sites in TFHA and FHA respectively indicates that the H-bonds of the dimer that involve the most basic site of TFHA and FHA are comparatively stronger than those of the other basic sites. However the dimer with strongest basic and strongest acidic site is not the most stable dimer.

The effect of medium on the deprotonation behavior of FHA and TFHA has been studied by evaluating free energies of deprotonation (ΔG) values by applying PCM on isolated conformers and by placing aggregates of HAs and their anions with water in the continuum of medium. The studies suggest that N-H deprotonation requires lower energy in comparison to O-H deprotonation in both gas phase as well as that for deprotonation of the aggregates in medium. The anionic species have stronger H-bonding interactions with water in comparison to those of isolated molecules. The acidity of TFHA is higher than that of FHA. The free energy required for various deprotonation processes is approximately halved in aqueous phase than in the gas phase and through aggregate formation it is further reduced, thus evidently the medium facilitates the deprotonation processes in HAs both explicitly and implicitly.

FPA is another isostere of FHA selected for study. Its tautomer formylphosphinous oxide (FPO) is also studied along with FPA beecause of reported tautomerism between parent phosphinous acids and phosphine oxide. The presence of formyl group increases the stability of FPA to 4.48 kcal/mol relative to the most stable conformation of FPO from the initially smaller relative energy difference of just 1.04 kcal/mol between phosphinous acid and phosphine oxide at MP2/6-31+G*. The order of stability of isomeric forms of FPA is same as that observed in case of FHA. The rotational transition state interconnecting **1Z** with **1E** has rotational barrier of 7.41 kcal/mol which is less than half the value for **1Z** → **1E** interconversion in case of FHA pointing that intramolecular H-bond in FPA is weak. The proton affinity values for all potential sites of FPA indicate P to be the most basic site which is in contrast to the observation in FHA and TFHA where the carbonyl chalcogen was most basic in nature. The large size and polarizability of P is responsible for the same. Deprotonation enthalpies also suggest that the deprotonation in FPA is relatively easier in comparison to FHA because of the presence of P with ability to hold additional charge density.

Results of aggregation of FPA and FPO with water also show FPA to be more stabilized and the stabilization energy associated with most stable aggregate of FPA with similar H-bond donor and acceptor groups as in case of aggregate of FHA with water in similar orientation (**1ZW1**) is only 0.99 kcal/mol lower at MP2/6-31+G*. In none of the aggregates of FPA and FPO with water, P-H act as H-bond donor but in one aggregate **2EW1** of FPA, P acts as H-bond acceptor that has been rationalized as P is partially positively charged and hence is electrostatically disfavored to H-bond formation. The Δr (difference of H-bond distances from sum of van der Waals radii), $E^{(2)}$ values from NBO analysis and charge analysis reflect H-bonding to be weak.

Intermolecular H-bonding of FHA and TFHA with amino acid side chain model molecules ($MeNH_2$, AcOH, MeOH, MeSH and MeSeH) has been explored. The chosen side chain analogues result in stabilization energies in the order AcOH > $MeNH_2$ > MeOH > MeSH > MeSeH for FHA while for TFHA the order is AcOH > $MeNH_2$ > MeOH > MeSeH > MeSH at MP2/6-31+G*. The results have been rationalized in terms of electrostatics, conjugative interactions and bond polarization arising with the H-bond formation giving rise to cooperativity in the aggregates. The strong cooperativity reflected in aggregates of FHA and TFHA with AcOH is rationalized in terms of enhanced conjugative interactions in both hydroxamic acid and AcOH units. In the aggregates involving H-bonding of model molecules AcOH, $MeNH_2$ and MeOH, the stabilization energies range in 16.08 to 6.35, 12.83 to 3.27

and 11.44 to 3.66 kcal/mol in case of FHA respectively and 13.40 to 3.94, 13.23 to 3.40 and 10.78 to 3.03 kcal/mol in case of TFHA respectively. The stabilization energies for aggregation of FHA with MeSH range in 5.48 to 2.62, with MeSeH the range is 5.17 to 0.94 kcal/mol and that of TFHA with MeSH the range is 5.08 to 1.75 kcal/mol, with MeSeH the range is 6.16 to 1.43 kcal/mol. Both electrostatic and covalent component of H-bond strength are magnificent in aggregates of FHA with AcOH, MeNH$_2$ and MeOH. The charge transfer component plays dominant role in strengthening the H-bonded aggregate through S-H or Se-H as H-bond donors and S as H-bond acceptor. The orbital interactions in the aggregates of FHA and TFHA with model molecules of amino acid side chains suggest that nearly in all the aggregates the most significant charge transfer interactions are those involving electron density shift from H-bond acceptor of model molecules toward HAs. In the end it can be concluded that following factors contribute toward the stabilization energy of aggregate formation (including dimer formation)

1. Electrostatic interactions.
2. Charge transfer and hence covalent interactions.
3. Variation in bond polarizations.
4. Variation in electron delocalizations particularly conjugative interactions.
5. The cooperativity arising from the H-bonding in the FHA and its isosteres makes stabilization much more significant than expected.